Micro-Macro-Interactions

Albrecht Bertram
Jürgen Tomas

Micro-Macro-Interactions

In Structured Media and Particle Systems

Springer

Authors

Prof. Dr.-Ing. Albrecht Bertram
Otto-von-Guericke-Universität Magdeburg
Inst. für Mechanik
Universitätsplatz 2
39106 Magdeburg
Germany
Email: albrecht.bertram@ovgu.de

Prof. Dr.-Ing. habil. Jürgen Tomas
Otto-von-Guericke-Universität Magdeburg
Inst. für Verfahrenstechnik
Universitätsplatz 2
39106 Magdeburg
Email: juergen.tomas@ovgu.de

ISBN 978-3-540-85714-3 e-ISBN 978-3-540-85715-0

Library of Congress Control Number: 2008933387

Typesetting: Scientific Publishing Services Pvt. Ltd., Chennai, India.
Cover Design: Mönnich, Max

Printed in acid-free paper

9 8 7 6 5 4 3 2 1

springer.com

Preface

It is a common feature of many materials and media that their behaviour is much better understood on a micro level than on the macro level, on which engineers normally work for designing technical parts, machines, apparatuses, or engineering systems. In such cases it is advantageous to study and compare the behaviour on both scales, the micro and the macro scale. Such micro-macro transitions have become rather successful and customary during the last decade, perhaps stimulated by the increasing computational power nowadays available. We are now capable to model the micro behaviour in detail, and to accomplish the micro-macro transition or homogenisation numerically. One of the most successful methods in many technological branches is the Representative Volume Element technique (RVE), where only a small representative part on the micro scale is modelled and used for the determination of the macro values of the physical variables like, e.g., forces, strains, heat fluxes, solid or liquid phase distributions, etc.

In the present volume we have collected results of such micro-macro investigations from more than half a decade of joint research work in different branches of engineering, in particular on

- the inelastic material behaviour of polycrystals;
- fibre and particle reinforced composites;
- solids under thermal loads;
- particle contacts and dynamics of particle systems.

The distributions of this book have been collected in four parts, according to these four fields of application, although many of them combine methods from different approaches and, thus, belong to more than just one of the above topics.

In all contributions, the behaviour of micro structures determines the macro behaviour of the system or medium. This micro structure may consist of

- different grains or phases of solids in polycrystalline materials,
- assemblies of matrix material and reinforcement in composites,
- solid particles moving in fluids,
- mixtures of interacting particles or liquids and gases in porous solid media,

etc. However, despite of the large manifold of different media and fields, the reader will detect common methods, algorithms, and models, which have been applied in rather different areas. The range of the applications spans from microphysics, material science, to mechanical engineering and process engineering, including numerical and mathematical methods. Such results can only be expected, if experts from physics, mathematics, and engineering cooperate and exchange knowledge, methods, and models within one joint research project.

All the presented results stem from investigations which have been performed at the Graduate School within the Otto von Guericke University Magdeburg during the period of 2001 - 2008 supported by the German Science Foundation (DFG) and the Land Sachsen-Anhalt under grant GK 828.

Magdeburg, June 2008

A. Bertram
J. Tomas

Contents

Part I

Inelastic Material Behaviour of Polycrystals

Normal Grain Growth: Monte Carlo Potts Model Simulation and Mean-Field Theory

D. Zöllner and P. Streitenberger

Institut für Experimentelle Physik, Otto-von-Guericke-Universität Magdeburg

Abstract. Grain growth in polycrystals is modelled using an improved Monte Carlo Potts model algorithm. By extensive simulation of three-dimensional normal grain growth it is shown that the simulated microstructure reaches a quasi-stationary self-similar coarsening state, where especially the growth of grains can be described by an average self-similar growth law, which depends only on the number of faces described by a square-root law. Together with topological considerations a non-linear effective growth law results. A generalized analytic mean-field theory based on the growth law yields a scaled grain size distribution function that is in excellent agreement with the simulation results. Additionally, a comparison of simulation and theory with experimental results is performed.

1 Introduction

Many technical properties of polycrystalline materials depend strongly on the grain size of the microstructure. The control of the microstructure is a key to improve material's properties like, e.g., strength, toughness, diffusivity and electrical conductivity in processing. On the other hand, many technical processes lead, e.g., by thermal influence, to grain growth, which is the migration of a grain boundary driven by the boundary energy. The associated thermodynamic driving force is the decrease in the Gibbs free interface energy. In this process the mean grain size increases with a simultaneous decrease of the total inner interface leading to a minimization of the total interface free energy. In order to study the phenomenon of grain growth more closely, first physically motivated grain growth models have been developed in the early 1950s (Smith 1952; Burke and Turnbull 1952). However, there remained clear discrepancies between the theories and experiments.

A new approach has been provided in the 1980s by computer simulations as new possibilities to model the grain microstructure and its temporal evolution under realistic conditions allowing for the observation of features that are difficult to observe experimentally, like, e.g., the surface or the rate of volume change of individual grains. Due to the broad field of applicability a number of different simulation methods have been developed throughout the years like, e.g., the Monte Carlo Potts model, the phase-field method, the Surface Evolver, and the vertex method (compare, e.g., (Miodownik 2002; Atkinson 1988; Thompson 2001)).

Among the above methods, the Monte Carlo Potts model is the most widely used one. The model is in its basics simple but in its specifics rather complex and therewith flexible. Hence, it can be applied efficiently to complex microstructures. Within the

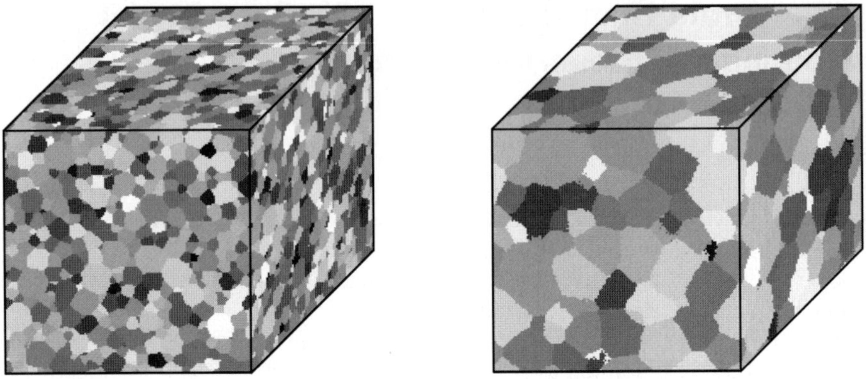

Fig. 1. Simulated 3D grain structures at different grain growth stages

scope of the authors present work the Monte Carlo Potts model method has been implemented in two and three dimensions based on the original works of (Anderson et al. 1984; Srolovitz et al. 1984; Anderson et al. 1989) including improvements in the algorithm, which have been suggested recently (e.g., (Yu and Esche 2003; Kim et al. 2005; Zöllner 2006) and the references within).

2 Monte Carlo Potts Model Simulation

Before a simulation can be started, the continuously given microstructure has to be mapped onto a discrete lattice. In three dimensions a cubic lattice is usually used with 26 nearest neighbours (first, second, and third nearest neighbours). Although the cubic 3D resp. quadratic 2D lattice is the simplest one for implementation, there has been a discussion throughout the years, whether this underlying lattice constrains the simulation results. In Figure 2 one can see that the grain boundaries cling to the underlying lattice. The reason for this effect is substantiated in the Potts model itself as the driving force places the boundaries along the lattice facets yielding a growth kinetics that differs from the expected normal grain growth (Holm et al. 1991 and 2001).

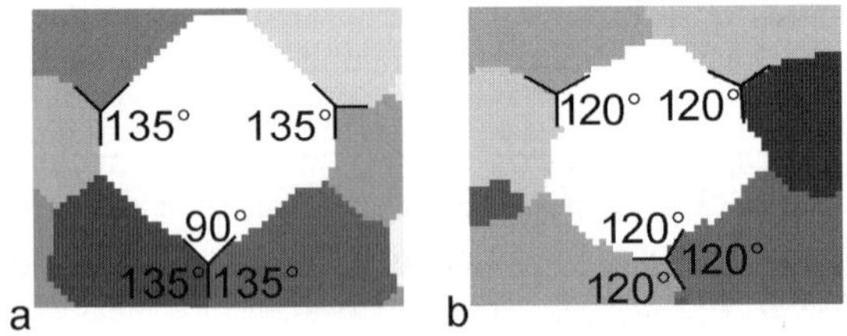

Fig. 2. a – clinging of the grain boundaries to the lattice at zero simulation temperature; b – grain boundaries with 120° angles at high simulation temperatures

Since these lattice effects depend strictly on the simulation algorithm and are highly non-physical, one has to eliminate them, e.g., following (Holm et al. 1991) by increasing the defined number of neighbouring lattice points or the simulation temperature T activating thermal fluctuations. In Figure 2b one can see that independent of the underlying lattice the angles adjust at 120° in the triple points for a non-zero temperature.

Each lattice point represents in the simulation a Monte Carlo unit (MCU), to which a crystallographic orientation is assigned specified by the rotation angles in the three-dimensional Euler space. The rotation angles specify the relative orientation to the given fixed coordinate system of the lattice (Ivasishin et al. 2003). In the simulation the orientation is usually represented by natural numbers.

The smallest time unit of the Monte Carlo Potts model simulation is called a Monte Carlo step (MCS) and defined as N reorientation attempts, where N is equal to the total number of MCUs, i.e., lattice points of the lattice.

The basic Potts algorithm (Anderson et al. 1984 and 1989; Srolovitz et al. 1984) shows some disadvantages that are inherent to the technique, e.g., unrealistic nucleation events and a growth exponent in Eq. (1) smaller than the expected value of n = 0.5. Furthermore, the basic algorithm is very time consuming. Especially in recent times, changes in the algorithm have been suggested improving the accuracy of the simulation results and reducing the runtime of the simulations (Yu and Esche 2003; Zöllner and Streitenberger 2004; Kim et al. 2005; Zöllner 2006).

Each of the N reorientation attempts consists of the following steps.

<u>In the first step</u> a MCU is chosen in a probabilistic way. This MCU has an orientation Q_μ (old state). But mostly the chosen MCU will be inside a grain and not on a boundary. Hence, unrealistic nucleation events may occur due to fluctuations induced by the simulation temperature. Seeing that grain growth always means grain boundary migration, a change of orientation can only occur if the chosen MCU is on the boundary. In this case the simulation algorithm proceeds with Step 2, otherwise the algorithm terminates this loop (Step 1).

<u>In the second step</u> a new orientation Q_ν different from the old orientation Q_μ is assigned on probation to the chosen MCU. This new orientation is chosen from all other (Q-1) orientations, where Q is the total number of orientations. However, most reorientation attempts will fail or again unrealistic nucleation events happen. Therefore, only orientations of the neighbouring MCUs are considered, because of the above grain boundary migration argument.

<u>In the third step</u> the energy of both states is given by the Hamiltonian

$$H = J \cdot \sum_{i=1}^{N} \sum_{j=1}^{nn} \left(1 - \delta_{Q_i Q_j}\right).$$

The inner sum sums up over the nearest neighbours of the i-th MCU and the outer sum over all N MCUs of the lattice. Due to the Kronecker delta each pair of nearest neighbours contributes J to the system energy, if they do not have the same orientation, and 0 otherwise, where J measures the interaction of the i-th MCU with all neighbouring MCUs as a function of the misorientation angle θ between two grains calculated by the Read-Shockley equation (Read and Shockley 1950)

$$J = \begin{cases} \frac{\theta}{\theta^*} \cdot \left(1 - \ln \frac{\theta}{\theta^*}\right), & \text{if } \theta \le \theta^* \\ 1, & \text{if } \theta \le \theta^* \end{cases},$$

where θ^* is the maximal value of a low angle grain boundary. From experiments it is known to be between 10° and 30° depending on the material (Sutton and Balluffi 1995) (compare also (Read and Shockley 1950; Hui et al. 2003) and the literature therein). For simulations of normal grain growth, where only high angle grain boundaries occur, it holds $J = 1$ (Fig. 3b).

In the fourth step the difference in energy ΔE between new and old state is calculated. However, due to the fact that from the whole lattice only one MCU is reoriented at a time, the difference in energy can simply be calculated as

$$\Delta E = J \cdot \sum_{j=1}^{nn} \left(\delta_{Q_j Q_\mu} - \delta_{Q_j Q_\nu}\right).$$

Finally – in the fifth step – the final state with the final orientation Q_μ^* of the selected MCU is chosen with the probability p given by

$$p = \begin{cases} m, & \Delta E \le 0 \\ m \cdot \exp \frac{-\Delta E}{k_B T}, & \Delta E > 0 \end{cases},$$

where k_B is Boltzmann's constant and T the simulation temperature (Fig. 3a).

It should be mentioned here again that the factor T does not measure a real temperature but rather represents a parameter of the probability-jump-function introduced to avoid lattice pinning.

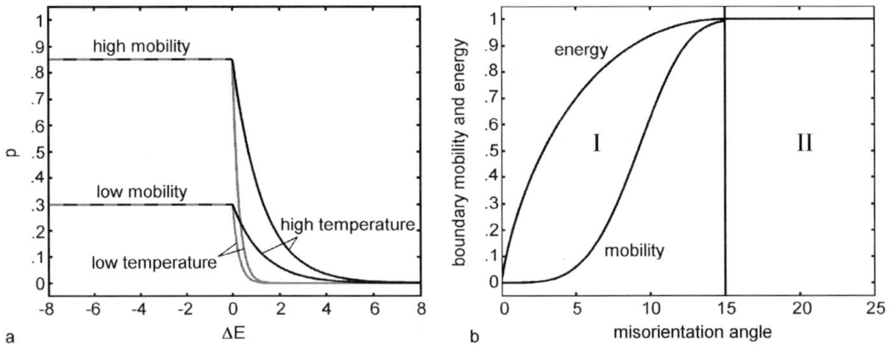

Fig. 3. a – Probability p for acceptance of orientation; b – boundary mobility and energy depending on misorientation angle

The boundary mobility is also a function of the misorientation angle (Fig. 3b)

$$m = \begin{cases} 1 - \exp\left(-B \cdot \left(\dfrac{\theta}{\theta^*} \right)^n \right), & \text{if } \theta \leq \theta^* \\ 1, & \text{if } \theta > \theta^* \end{cases}$$

given by (Huang and Humphreys 2000). The constants are $B = 5$ and $n = 4$.

3 Monte Carlo Simulation Results

The coarsening process has been investigated by following the temporal development of 3D grain microstructures simulated by the Monte Carlo Potts model simulation. These microstructures are initially either Rayleigh distributed structures or Voronoi Tessellations (Zöllner 2006).

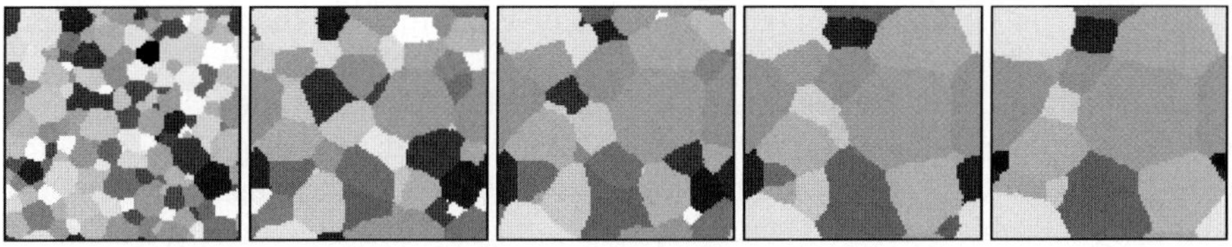

Fig. 4. 3D coarsening process shown through temporal development of a 2D section

The size of the lattice has been chosen as 200×200×200 with periodic boundary conditions and the simulation temperature is $k_BT = 2.6$. For analyses the simulation results are averaged for each simulation step over ten simulation runs.

3.1 Coarsening Process: Growth Law, Scaling Regime and Grain Size Distribution

Curvature-driven normal grain growth – as simulated by the Monte Carlo Potts model – can essentially be characterized by a parabolic growth law and statistical self-similarity (cf. the review articles (Atkinson 1988; Thompson 2001)).

The average grain size $\langle R \rangle$ of an ensemble of grains of a polycrystalline solid increases with time t according to the parabolic growth law

$$\langle R \rangle^n - \langle R \rangle_0^n = b \cdot t\,, \tag{1}$$

where R is the radius of a grain volume equivalent sphere, b is the growth factor and n is the growth exponent, which is theoretically supposed to be 0.5. The volume of each grain is equal to the number of MCUs representing the grain.

Both 3D grain structures as they have been simulated (Zöllner and Streitenberger 2008) by the Monte Carlo Potts model follow after an initial period of time (Figure 5a and b, part I) the well-known growth law, Eq. (1), (Figure 5a and b, part II). For both structures the numerically given growth exponent n is in very good agreement with

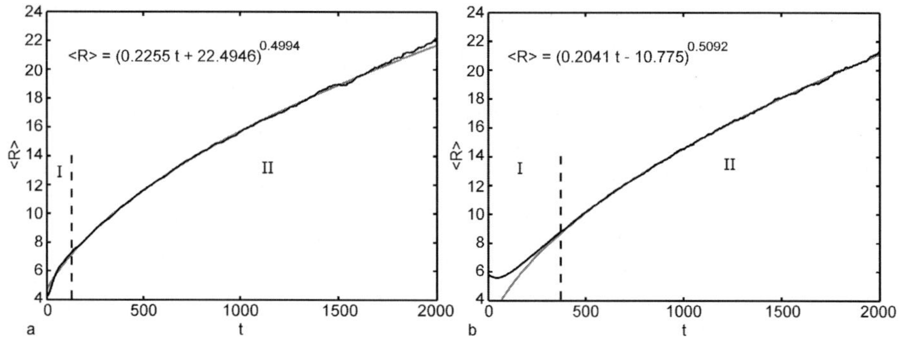

Fig. 5. Temporal development of the mean grain size (black) with initial period (I) and self-similar coarsening regime (II) together with fit (grey) of growth law, Eq. (1): a – for the Rayleigh distributed grain ensemble; b – for the Voronoi tessellated grain ensemble

the expected value of $n = 0.5$ in Eq. (1), which can be found in all three fields of investigation of normal grain growth, namely experiments, theory and computer simulations (compare (Yu and Esche 2003)).

The initially Rayleigh distributed grain ensemble reaches faster than the initial Voronoi tessellation the state characterised by the growth law, Eq. (1). The authors have shown (cf. Fig. 3b in (Zöllner and Streitenberger 2008)) that the number of grains reaching this state is significant larger for the initially Rayleigh distributed structure (approx. 34% grains left) than for the initial Voronoi structure (with approx. 20% grains left). Therefore, the initially Rayleigh distributed grain structure is used for further statistical analyses.

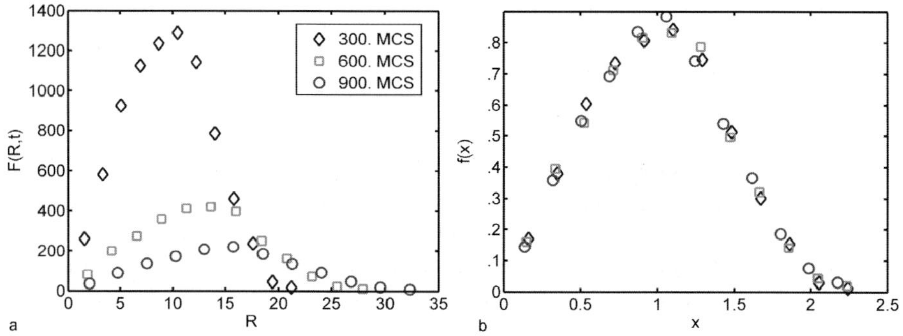

Fig. 6. a – Absolute grain size distribution showing number of grains vs. grain size; b - relative size distribution with relative number of grains vs. relative grain size $x = R/\langle R\rangle$

Contemporary, the coarsening process of the grain structure develops towards a quasi-stationary state that exhibits statistical self-similarity (Burke 1949; Mullins 1986). The grain size distribution function $F(R,t)$ in the quasi-stationary state is characterized by the scaling form

$$F(R,t) = g(t) \cdot f(x), \; x = \frac{R}{\langle R \rangle}. \tag{2}$$

All scaled grain size distribution functions *f*(*x*) within this quasi-stationary self-similar coarsening regime collapse to a single universal, time-independent size distribution as shown in Figure 6b, where three time steps out of the quasi-stationary self-similar state indeed coincide. The temporal development of the size distribution can be seen best by looking at the absolute size distribution in Figure 6a (compare (Zöllner 2006; Zöllner and Streitenberger 2006, 2007b and 2008)).

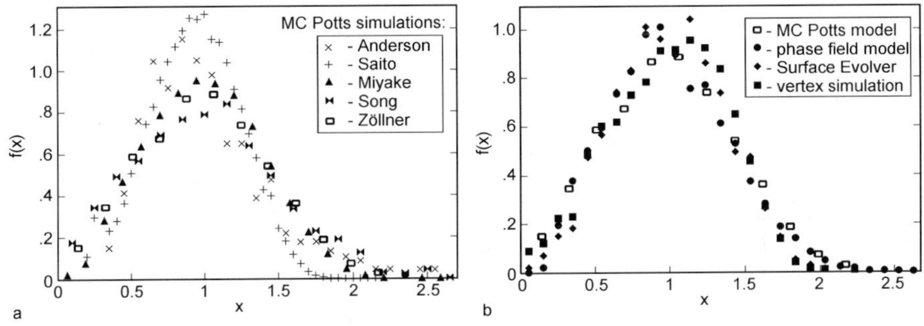

Fig. 7. Comparison of the simulated grain size distribution with the results of: a – other Monte Carlo Potts model simulations; b – other simulation methods both taken from literature (Krill and Chen 2002)

Figure 7a shows the simulated self-similar grain size distribution within the quasi-stationary coarsening regime in comparison to Monte Carlo Potts model simulations of (Anderson et al. 1989; Saito 1998; Miyake 1998; Song and Liu 1998), which have been taken from (Krill and Chen 2002). The grain size distribution of the simulation of the authors (blank squares) is very similar to that of other simulations. Deviations can be explained by the use of different simulation parameters like, e.g., simulation temperature or underlying lattice.

The comparison between the grain size distribution obtained in this work with results of other simulation methods, namely the phase-field simulation of (Krill and Chen 2002), the Surface Evolver approach of (Wakai et al. 2000) and the vertex method of (Weygand et al. 1999), shows an even better agreement (Figure 7b).

3.2 Topology: Number of Faces vs. Grain Size

The correlation between the number of faces *s* per grain and the relative grain size *x* is an important topological feature of the microstructure. Due to the relaxation process connected with grain growth this correlation changes with time. However, it is known that the average number of faces of a grain of given size can be described within the quasi-stationary self-similar state by a time-invariant function of the relative grain size *x* (cf. Fig. 8b).

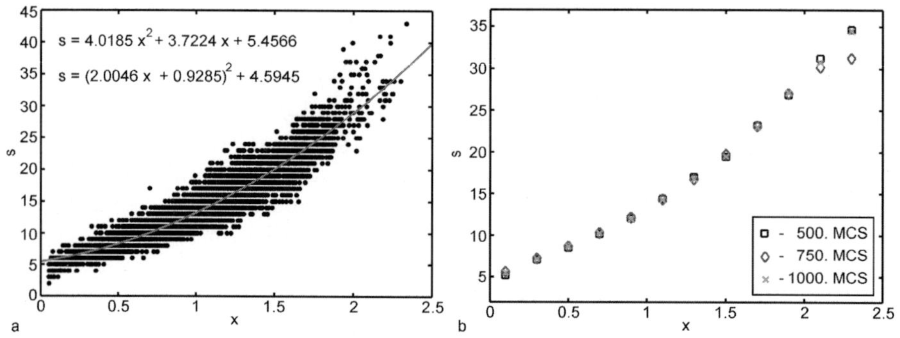

Fig. 8. a – Number of faces vs. relative grain size for all grains of an ensemble for the 500th MCS together with quadratic least-squares fit; b – average number of faces vs. relative grain size divided into size classes for three different time steps

Within the quasi-stationary state the number of faces of the individual grains depends non-linearly on the relative grain size (compare Fig. 8a). The relation can be approximated in the average by a non-binomial parabolic function

$$s(x) = s_2 x^2 + s_1 x + s_0 = (px + q)^2 + \delta \,. \tag{3}$$

This is consistent with experimental observations (Zhang et al. 2004), geometrical considerations following (Abbruzzese and Lücke 1996; Streitenberger and Zöllner 2006; Zöllner 2006) and 3D computer simulations (Wakai et al. 2000).

3.3 Volumetric Rate of Change

In the quasi-stationary state the growth of each grain can be described by the average self-similar growth law (Streitenberger 1998; Streitenberger and Zöllner 2006 and 2007)

$$\dot{R} = \frac{dR}{dt} = \frac{k}{R} \cdot H(x). \tag{4}$$

$H(x)$ is a time-invariant dimensionless function of the relative grain size, and k is the kinetic constant of curvature driven grain boundary motion. According to this equation $R\dot{R}$ is directly linked to the volume change rate with

$$V^{-1/3}\dot{V} = \zeta \cdot R\dot{R} = k \cdot \zeta \cdot H(x); \quad \zeta = \left(48\pi^2\right)^{1/3} \tag{5}$$

In recent times, (Hilgenfeldt et. al. 2001) and (Glicksman 2005) have shown by considering three-dimensional space filling polyhedral networks that the average volumetric rate of change is solely a function of its average number of faces or neighbours $s = s(x)$ and can be approximated by the expression

$$R\dot{R} = C_0 + C_1\sqrt{s} \,, \tag{6}$$

which can be considered as the 3D analogue to the von Neumann–Mullins law.

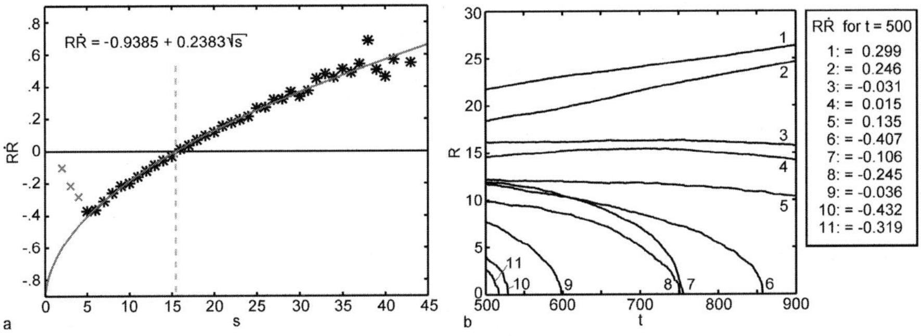

Fig. 9. a – Volumetric rate of change vs. number of faces for 500th MCS with fit of Eq. (6); b – development of grain size for some selected grains together with volume change rate

Figure 9a shows $R\dot{R}$ vs. s as it follows from the Monte Carlo simulation, where $\dot{R} \approx \Delta R / \Delta t$ is approximated (Zöllner and Streitenberger 2004 and 2006). The least-square fit of Eq. (6) to the simulation data yields a very good representation. Deviations for small grains are inherent to the simulation technique. It can be seen that the fitted values C_0 = -0.9385 and C_1 = 0.23829 are close to Glicksman's theoretical values $C_{0\ \mathrm{Glicksman}}$ = -1.0583 and $C_{1\ \mathrm{Glicksman}}$ = 0.2886 (normalized by $1/\zeta$ as it is given in Eq. (5)).

Additionally, in Figure 9b it is shown that in general grains with a volumetric rate of change larger than zero grow (curves 1 and 2), nearly equal to zero do not change (curves 3, 4 and 5) and smaller than zero shrink (curves 6 to 11) (compare also (Zöllner and Streitenberger 2007a).

In Figure 10a it appears that the simulated volumetric rate of change can be approximated by a quadratic polynomial in x (Zöllner and Streitenberger 2007b)

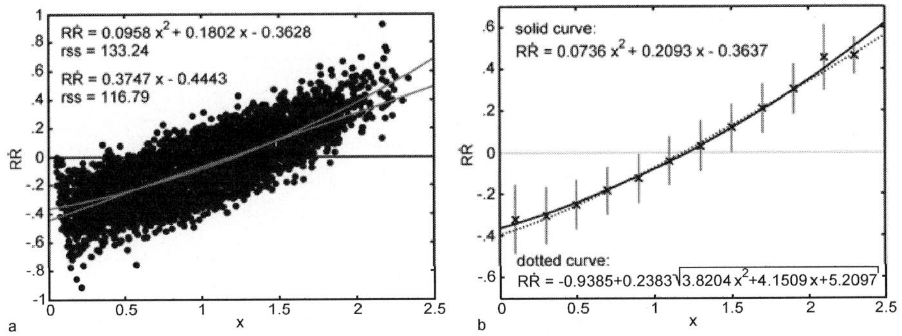

Fig. 10. a – Self-similar volumetric rate of change vs. relative grain size for all simulation data together with linear and quadratic least-squares fits; b - self-similar volumetric rate of change vs. relative grain size for data divided into size classes together with fit of Eq. (7) (solid curve) and plot of Eq. (6) with $s(x)$ from Eq. (3) as they both follow from fits to simulation data (dotted curve).

$$R\dot{R} = kH(x) = a_2 x^2 + a_1 x + a_0 . \tag{7}$$

This is consistent with the non-linear behaviour of the effective growth law resulting from the combination of Eqs. (3) and (6), which leads to an effective growth law in the form of Eq. (4), where contrary to Hillert's assumption $H(x)$ is a non-linear function, (Figure 10b).

4 Mean-Field Theory

In the statistical mean-field theory of grain growth [Zöllner (2006)] it is assumed that the growth of grains can be described by an average self-similar growth law, Eq. (4), for all grains of size R so that the grain size distribution function $F(R,t)$, characterized by the scaling form, Eq. (2), obeys the continuity equation

$$\frac{\partial F(R,t)}{\partial t} + \frac{\partial}{\partial R}\left(\dot{R}F(R,t)\right) = 0 . \tag{8}$$

In his pioneering work on grain growth (Hillert 1965) assumed a linear function for $H(x)$. Using stability arguments of the coarsening theory of (Lifshitz and Slyozov 1961; Wagner 1961) Hillert obtained his well-known grain size distribution function (Hillert 1965), which, however, never has been observed, neither experimentally nor by computer simulations.

Based on the parabolic approximation, Eq. (7), and the scaling assumption (2) for the grain size distribution, the integration of the continuity equation yields the following analytical expression for the normalized scaled grain size distribution function (Streitenberger and Zöllner 2006; Zöllner 2006),

$$f(x) = \frac{1}{x_c}\varphi\left(\frac{x}{x_c}\right), \tag{9}$$

$$\begin{aligned} \varphi(u) &= \frac{D[\gamma+\alpha\gamma]^{D/[2(1-\alpha\gamma)]}u}{\left[(1-\alpha\gamma)u^2 - \gamma u + \gamma + \alpha\gamma\right]^{1+D/[2(1-\alpha\gamma)]}} * \\ &\exp\left\{\frac{-D\gamma}{(1-\alpha\gamma)\sqrt{\Delta}}\left[\arctan\frac{-\gamma+2(1-\alpha\gamma)u}{\sqrt{\Delta}} - \arctan\frac{-\gamma}{\sqrt{\Delta}}\right]\right\} \\ x_c &= 1/\int u\varphi(u)du \\ \Delta &= 4\gamma(1+\alpha)(1-\alpha\gamma) - \gamma^2 \geq 0. \end{aligned} \tag{10}$$

Eq. (10) represents a two-parameter family of grain size distribution functions (Fig. 11a), which fulfil the requirement of volume conservation in conjunction with the existence of the D-th moment $<u^D>$ of the grain size distribution function if the parameters α and γ obey the conditions (Streitenberger and Zöllner 2006)

$$\begin{aligned} \gamma &\leq \frac{4(1+\alpha)}{4\alpha(1+\alpha)+1}, && \text{if } \alpha > 0 \\ \gamma &= 4, && \text{if } \alpha = 0 \end{aligned} . \tag{11}$$

For the limiting case $\Delta = 0$, that is for $4\gamma(1+\alpha)(1-\alpha\gamma)-\gamma^2 = 0$, Eq. (10) reduces to the one-parameter function (Fig. 11b)

$$\varphi(u) = \begin{cases} a \cdot u_0^a \cdot \exp(a) \cdot \dfrac{u}{(u_0 - u)^{a+2}} \exp\left(-\dfrac{a \cdot u_0}{u_0 - u}\right), & 0 \le u \le u_0 \\ 0, & \text{otherwise} \end{cases}, \qquad (12)$$

with $a = D(u_0 - 1)^2$, showing a finite cut-off at u_0 considered already in (Streitenberger 1998 and 2001, Zöllner and Streitenberger 2004 and 2006).

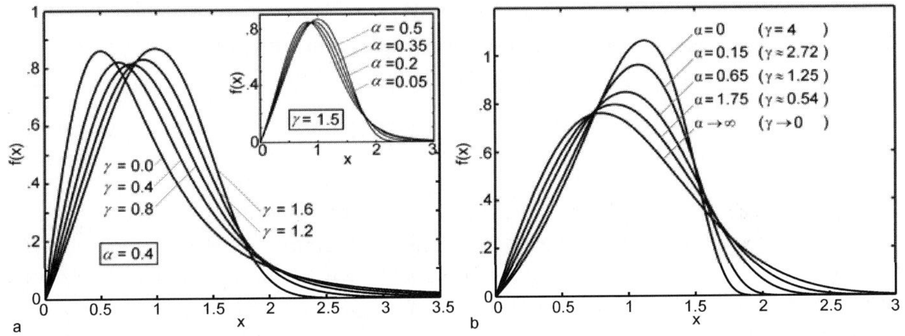

Fig. 11. Analytic grain size distribution functions: a – Eq. (10) rescaled to $f(x)$ for D = 3 and $\Delta > 0$; b – Eq. (12) for D = 3 and $\Delta = 0$.

5 Comparison of Simulation with Experimental Measurements and Analytical Theory

It is a well known problem that sections through a 3D grain ensemble yield smaller 2D grain sizes than the real 3D grain sizes (Ohser and Mücklich 2000). Parallel sections through grains yield different sizes and forms depending on place and orientation of the sectioning (compare Fig. 12).

Since most experimentally determined size distributions come from sectioning (Ohser and Mücklich 2000), for comparison 2D sections of the simulated 3D microstructure have to be used. In this case (Fig. 13a) experimental data of zone-refined iron, which shows normal grain growth, have been used determined by (Hu 1974). Despite some minor differences there is a good overall agreement between the 2D sectioning data from experiment and simulation (Zöllner and Strei-tenberger 2007c).

The comparison of 3D grain size distributions from simulation with experimental results for recrystallized and annealed pure polycrystalline iron obtained by serial sectioning (Zhang et al. 2004) is shown in Figure 13b, where additionally a fit of the theoretical expression Eq. (10) to our simulation data is plotted. It can be seen that the experimental size distribution of Zhang et al. shows some differences (Zöllner and Streitenberger 2007c). The reason why it is more peaked is unclear.

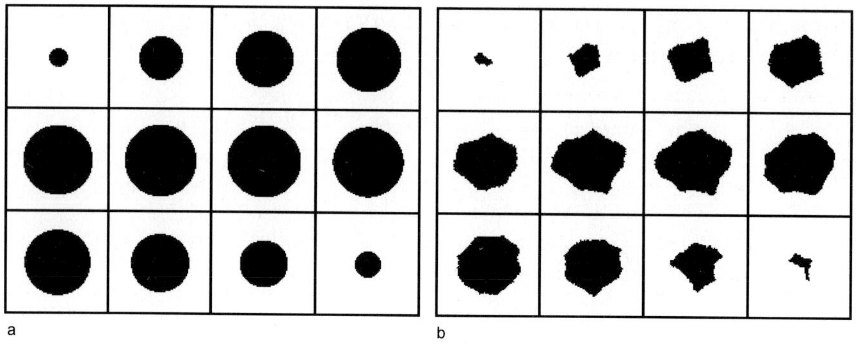

Fig. 12. Equally distanced sections through: a – a sphere; b – a grain

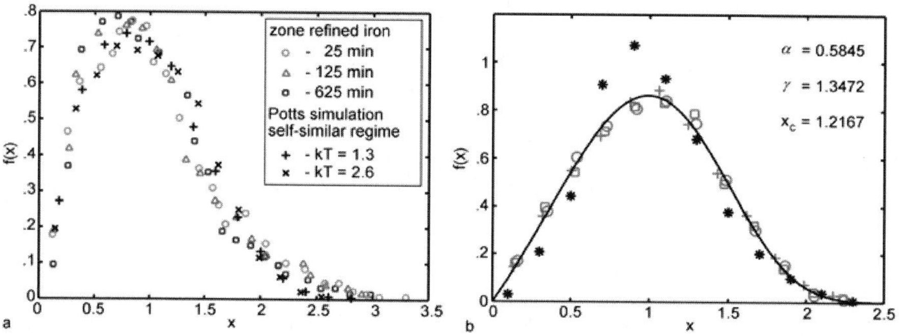

Fig. 13. Relative grain size distributions: a – of experimental data (Hu 1974) compared to 2D sections from Monte Carlo simulation; b – from experimental data (black stars) obtained by serial sectioning (Zhang et al. 2004) compared to simulated self-similar 3D size distribution (grey) with analytical fit, Eq. (10)

The least-squares fits of eqs. (10) and (12) to the simulation data are in very good agreement with the simulation results (Figure 14a). It can rather be seen that the one-parameter function is an approximation as good as the two parameter one.

The parameters α and γ can also be determined (Streitenberger and Zöllner 2006) from the parameters of the quadratic average growth law (7) fitted to the simulated microstructure (Fig. 10a). While α can be calculated immediately from $\alpha = a_2 x_c / a_1 = 0.6492$, γ is determined self-consistently by the scaling requirement that the scaled critical grain size $x_c = 1.22309$ of the simulated microstructure following from $\dot{R}(x_c) = 0$ has to be the same as $x_c = 1/\int u\varphi(u)du$ following from the grain size distribution function, (10), yielding $\gamma = 1.24876$. The resulting analytical grain size distribution function is in excellent agreement with the grain size distribution of the Monte Carlo Potts model simulation as shown in Figure 14b.

A comparison of the least-squares fit of Eq. (10) to the simulated size distribution with the analytical size distribution functions of (Hillert 1965) and (Louat 1974) are

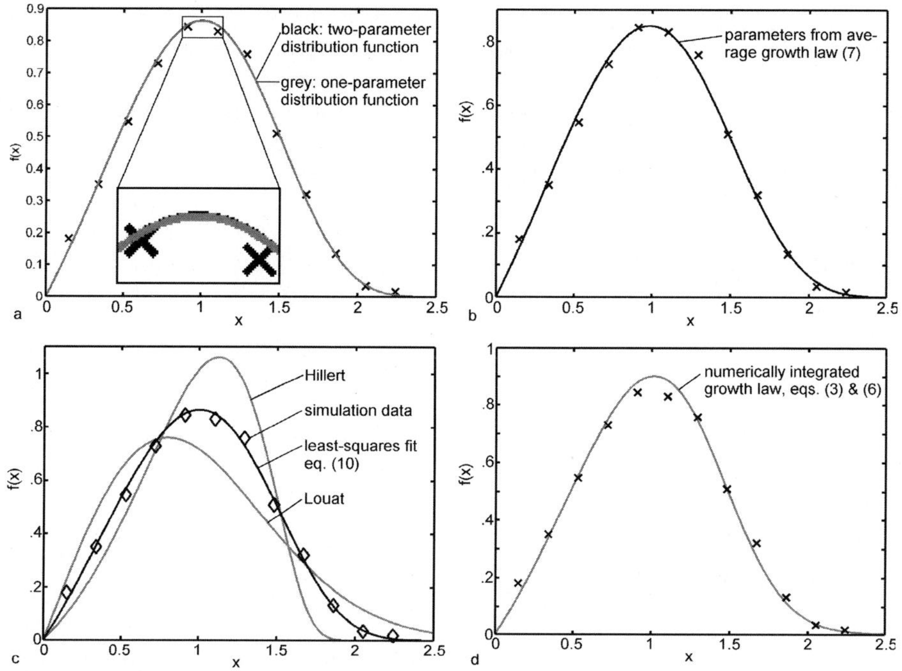

Fig. 14. Simulated grain size distribution together with: a – least-squares fit of (10) and (12); b – fit of (10) with parameters given by average growth law, Eq. (7); c – least-squares fit of (10), Hillerts (Eq. (10) for $\alpha = 0$) and Louats (Eq. (10) for $\alpha \to \infty$) size distribution; d – size distribution obtained by integration of average growth law (6) with Eq. (3).

shown in Figure 14c, where it is clear that the simulated distribution can be approximated very well by our theory but is quite different from the distributions of Hillert and Louat.

The average growth law $R\dot{R}$ can also be used free of the parabolic approximation, Eq. (7). Therefore, the non-linear growth law (6) is used as it has been fitted to the simulation results in Figure 9a. The function $s(x)$ is given in Figure 8a. Then the grain size distribution results from numerical integration of the growth law and is also in fair agreement with the results as can be seen in Figure 14d.

6 Conclusions

In the present work normal grain growth in three dimensions has been studied on the basis of large-scale Monte Carlo Potts model simulations, which enabled extensive statistical analyses of growth kinetics and topological properties of microstructures within the quasi-stationary coarsening regime. In particular, an average growth law could be derived from the simulation data, involving a quadratic dependency of the self similar grain volume change rate on the relative grain size x.

It has been shown that an adequate modification of the effective growth law allows a modification of the Lifshitz-Slyozov-Wagner procedure. Based on the effective growth law derived from simulation data an analytical size distribution function is derived, which is not only fully consistent with the requirement of total-volume conservation and the existence of a finite average grain volume but rather represents the simulation data of three-dimensional grain growth very well.

Additionally, 2D plane sections from simulated 3D grain structures were considered and compared with experimental data showing a very good agreement. The simulated size distribution shows – compared with an experimental grain size distribution for pure iron obtained by serial sectioning – also a fair agreement.

References

Abbruzzese, G., Lücke, K.: Statistical theory of grain growth: a general approach. Mater. Sci. Forum 204-206, 55–70 (1996)

Anderson, M.P., Srolovitz, D.J., Grest, G.S., Sahni, P.S.: Computer simulation of grain growth – 1. Kinetics. Acta Metall. 32, 783–791 (1984)

Anderson, M.P., Grest, G.S., Srolovitz, D.J.: Computer simulation of normal grain growth in three dimensions. Phil. Mag. B 59, 293–329 (1989)

Atkinson, H.V.: Theories of normal grain growth in pure single phase systems. Acta Metall. 36, 469–491 (1988)

Burke, J.E.: Some factor affecting the rate of grain growth in metals. Trans. Metall. Soc. AIME 180, 73–91 (1949)

Burke, J.E., Turnbull, D.: Recrystallization and grain growth. Progr. Met. Phys. 3, 220–292 (1952)

Glicksman, M.E.: Analysis of 3-d network structures. Phil. Mag. 85, 3–31 (2005)

Hilgenfeldt, S., Kraynik, A.M., Koehler, S.A., Stone, H.A.: An accurate von Neumann's law for three-dimensional foams. Phys. Rev. Lett. 86, 2685–2688 (2001)

Hillert, M.: On the theory of normal and abnormal grain growth. Acta Metall. 13, 227–238 (1965)

Holm, E.A., Glazier, J.A., Srolovitz, D., Grest, G.S.: The effect of lattice anisotropy and temperature on domain growth in the two-dimensional Potts model. Phys. Rev. A 43, 2662–2668 (1991)

Holm, E.A., Hassold, G.N., Miodownik, M.A.: On misorientation distribution evolution during anisotropic grain growth. Acta Mater. 49, 2981–2991 (2001)

Hu, H.: Grain growth in zone-refined iron. Can. Metall. 13, 275–286 (1974)

Huang, Y., Humphreys, F.J.: Subgrain growth and low angle boundary mobility in aluminum crystals of orientation {110}<001>. Acta Mater. 48, 2017–2030 (2000)

Hui, L., Guanghou, W., Feng, D., Xiufang, B., Pederiva, F.: Monte Carlo simulation of three-dimensional polycrystalline material. Mater. Sci. Eng. A 357, 153–158 (2003)

Ivasishin, O.M., Shevchenko, S.V., Vasiliev, N.L., Semiatin, S.L.: 3D Monte-Carlo simulation of texture-controlled grain growth. Acta Mater. 51, 1019–1034 (2003)

Kim, Y.J., Hwang, S.K., Kim, M.H., Kwun, S.I., Chae, S.W.: Three-dimensional Monte-Carlo simulation of grain growth using triangular lattice. Mater. Sci. Eng. A 408, 110–120 (2005)

Krill III, C.E., Chen, L.-Q.: Computer simulation of 3-D grain growth using a phase-field model. Acta Mater. 50, 3059–3075 (2002)

Lifshitz, I.M., Slyozov, V.V.: The kinetics of precipitation from supersaturated solid solutions. J. Phys. Chem. Solids 19, 35–50 (1961)

Louat, N.P.: On the theory of normal grain growth. Acta Metall. 22, 721–724 (1974)

Miodownik, M.A.: A review of microstructural computer models used to simulate grain growth and recrystallisation in aluminium alloys. J. Light Metals 2, 125–135 (2002)
Miyake, A.: Monte Carlo simulation of normal grain growth in 2- and 3-dimensions: the lattice-model-independent grain size distribution. Contrib. Mineral. Petrol. 130, 121–133 (1998)
Mullins, W.W.: Two-dimensional motion of idealized grain boundaries. J. Appl. Phys. 27, 900–904 (1956)
Mullins, W.W.: The statistical self-similarity hypothesis in grain growth and particle coarsening. J. Appl. Phys. 59, 1341–1349 (1986)
von Neumann, J.: Written discussion of grain shapes and other metallurgical applications of topology. In: Metal Interfaces, American Society for Metals, Cleveland OH, pp. 108–110 (1952)
Ohser, J., Mücklich, F.: Statistical analysis of microstructures in materials science. Wiley, Chichester (2000)
Read, T.W., Shockley, W.: Dislocation models of crystal grain boundaries. Phys. Rev. 78, 275–289 (1950)
Saito, Y.: Monte Carlo simulation of grain growth in three-dimensions. ISIJ Int. 38, 559–566 (1998)
Smith, C.S.: Grain shapes and other metallurgical applications of topology. In: Metal Interfaces, American Society for Metals, Cleveland OH, pp. 65–109 (1952)
Song, X., Liu, G.: A simple and efficient three-dimensional Monte Carlo simulation of grain growth. Scripta Mater. 38, 1691–1696 (1998)
Srolovitz, D.J., Anderson, M.P., Sahni, P.S., Grest, G.S.: Computer simulation of grain growth – II. Grain size distribution, topology and local dynamics. Acta Metall. 32, 793–802 (1984)
Streitenberger, P.: Generalized Lifshitz-Slyozov theory of grain and particle coarsening for arbitrary cut-off parameter. Scripta Mater. 39, 1719–1724 (1998)
Streitenberger, P.: Analytic model of grain growth based on a generalized LS stability argument and topological relationships. In: Gottstein, G., Molodov, D.A. (eds.) Recrystallization and grain growth, pp. 257–262. Springer, Berlin (2001)
Streitenberger, P., Zöllner, D.: Effective growth law from three-dimensional grain growth simulations and new analytical grain size distribution. Scripta Mater. 55, 461–464 (2006)
Streitenberger, P., Zöllner, D.: Topology based growth law and new analytical grain size distribution function of 3D grain growth. Mater. Sci. Forum 558-559, 1183–1188 (2007)
Sutton, A.P., Balluffi, R.W.: Interfaces in Crystalline Materials. Oxford Science Pub. (1995)
Thompson, C.V.: Grain growth and evolution of other cellular structures. Solid State Phys. 55, 269–316 (2001)
Wagner, C.: Theorie der Alterung von Niederschlägen durch Umlösen (Ostwald-Reifung). Z. Elektrochem. 65, 581–591 (1961)
Wakai, F., Enomoto, N., Ogawa, H.: Three-dimensional microstructural evolution in ideal grain growth - general statistics. Acta Mater. 48, 1297–1311 (2000)
Weygand, D., Bréchet, Y., Lépinoux, J., Gust, W.: Three-dimensional grain growth: A vertex dynamics simulation. Phil. Mag. B 79, 703–716 (1999)
Yu, Q., Esche, S.K.: A Monte Carlo algorithm for single phase normal grain growth with improved accuracy and efficiency. Comp. Mater. Sci. 27, 259–270 (2003)
Zhang, C., Suzuki, A., Ishimaru, T., Enomoto, M.: Characterization of three-dimensional grain structure in polycrystalline iron by serial sectioning. Metall. Mater. Trans. 35A, 1927–1933 (2004)
Zöllner, D.: Monte Carlo Potts Model Simulation and Statistical Mean-Field Theory of Normal Grain Growth. Shaker, Aachen (2006)
Zöllner, D., Streitenberger, P.: Computer Simulations and Statistical Theory of Normal Grain Growth in Two and Three Dimensions. Mater. Sci. Forum 467-470, 1129–1134 (2004)
Zöllner, D., Streitenberger, P.: Three Dimensional Normal Grain Growth: Monte Carlo Potts Model Simulation and Analytical Mean Field Theory. Scripta Mater. 54, 1697–1702 (2006)

Zöllner, D., Streitenberger, P.: Normal Grain Growth in Three Dimensions: Monte Carlo Potts Model Simulation and Mean-Field Theory. Mater. Sci. Forum 550, 589–594 (2007a)
Zöllner, D., Streitenberger, P.: Monte Carlo Potts model simulation and statistical theory of 3D grain growth. Mater. Sci. Forum 558-559, 1219–1224 (2007b)
Zöllner, D., Streitenberger, P.: New analytical grain size distribution in comparison with computer simulated and experimental data. In: Mücklich, F. (ed.) Fortschritte in der Metallographie. Praktische Metallographie Sonderband, vol. 39, pp. 97–102. DGM Werkstoff-Informationsgesellschaft mbH, Frankfurt (2007c)
Zöllner, D., Streitenberger, P.: Monte Carlo Simulation of Normal Grain Growth in Three Dimensions. Mater. Sci. Forum 567-568, 81–84 (2008)

Microstructural Influences on Tensile Properties of hpdc AZ91 Mg Alloy

D.G.L. Prakash[1] and D. Regener[2]

[1] Manchester Materials Science Centre, The University of Manchester
[2] Institut für Werkstoff- und Fügetechnik, Otto-von-Guericke-Universität Magdeburg

Abstract. This work is aimed to conclude the effect of section thickness, position and long term annealing (LTA) on the microstructure and its influences on the tensile properties such as yield strength (YS), ultimate tensile strength (UTS), ductility and fracture strain (FS) of high pressure die cast (hpdc) AZ91 Mg alloy. The variations in tensile properties with respect to different castings are explained using the microstructural quantities; average size, area fraction, number density and clustering tendency (CT) of microporosity, massive (β_m) continuous (β_c) and discontinuous (β_d) $Mg_{17}Al_{12}$ particles, and grain size. Evidence of different micro failure modes of the material are presented to explain the influence of different micro features on failure. The extend effects of shrinkage porosity and β ($\beta_{m,}$ $\beta_{c\ \mathrm{and}}$ β_d) particles on the deformation and failure of the material are discussed.

1 Introduction

Magnesium alloys are the lightest of all common metallic materials used in structural parts for the mass produced products. The significant growth in the consumption of magnesium and its alloys is due to high requirements in automobile and aeronautical industry. Most magnesium alloys are now produced by hpdc method and, specifically AZ91 accounts for more than 50 % of all high pressure die castings [1] and this is due to its good strength, ductility and castability [2].

Numerous studies have reported that the tensile properties of cast magnesium-based alloys depend on the section thickness of the castings [3-8]. Rodrigo et al. [7] found that the tensile strength increases with increasing section thickness while Sequeira et al. [4], Stich [8] and Schindelbacher [3] reported a reverse effect. Rodrigo et al. [7] and Sequeira et al. [4] found that the ductility increases with increasing section thickness while Stich [8] and Schindelbacher [3] found an opposite effect. Clearly there are discrepancies on the effect of section thickness of the material and in-depth studies are required for better understanding of this effect. However there is nothing documented on the effect of position on the tensile properties in the high pressure die casting.

Automotive components of Mg alloys may be exposed to moderate temperatures in the range of 60°C to 200°C. Various authors reported the presence of discontinuous (β_d) and continuous (β_c) [2, 9-11] precipitates in directly aged AZ91 casting. However, very limited information is documented on the quantitative characterization of different β phase. NMR spectroscopy is used recently for the bulk quantification of β phase [2, 12]. Present authors [13] quantitatively characterized the β_m (massive) and β_{c+d} (continuous + discontinuous) phases from the microstructure using an image processing technique and discussed the microstructural developments during annealing. Suman [9] performed LTA of AZ91 and AM60 castings at 120°C for 7, 30, 60 and 90 days and found that the YS increased (round tensile bar with 6.3 mm reduced section) by up to 40MPa and also stated that the YS decreases when the LTA time exceeds 60 days. Bowles et al. [2] performed a similar study of the same materials (120°C) for up to 208 days and obtained an increase in YS of 30MPa for 5mm thick castings and 6MPa for 2mm thick castings. Same authors [2] also stated the decrease in ductility for increasing LTA time. However, the role of microstructure on the properties with respect to the variations in LTA duration and temperature is not understood.

The different microstructural features of AZ91 alloy have complex geometry, their locations and arrangements are often non-uniform and usually strong spatial correlations exist. This multi length scale micro features cause multiple fracture micromechanisms, which, affect the fracture path and mechanical properties of this material. This also depends on the spatial arrangement, area fraction, size distribution and shape of different microstructural features. This highlights the requirement for a systematic microstructural quantitative analysis to obtain the variation in micro quantities with respect to the section thickness, position and LTA conditions of the castings and their relation to the tensile properties. Additionally, understanding the microscopic failure modes of the material is also equally important to predict the influence of the micro features on fracture behaviour, which is strongly related to the mechanical properties. Therefore, the aim of this work is to perform the same by quantitative characterization of microstructure and correlating the micro quantities with the tensile properties to conclude the effect of section thickness, position and LTA of hpdc AZ91 castings. Such an analysis is mandatory for the structural applications of this material.

2 Experimental Procedures

The as-cast thin (ACT) and as-cast thick plate (ACTP) castings of AZ91 alloy are produced using the cold chamber hpdc machine. All the castings are produced under the same conditions and more attention is not paid on the processing conditions as the scope of the present study is micro-macro property correlations. The chemical composition of the investigated hpdc AZ91 Mg alloy is 9.3% of Al, 0.12% of Mn, 0.79% of Zn, 0.02% of Si, 0.0007% of Cu, 0.0006% of Ni, 0.00046% of Fe and the remaining of Mg. The ACT casting is a flat as-cast tensile specimen with a rectangular cross section of 20 x 5mm with 50mm gauge length. The ACTP of the dimension 200x53x10mm are produced and the tensile specimens of a cross section of 10x10mm and 50mm gauge length are machined from edge and middle of the plates. A set of ACTP are further long term annealed (LTA) (a condition designated as T5) in air at temperatures of 150ºC and 200ºC for 1000hrs, and the tensile specimens are machined with the same dimension quoted for ACTP specimens. Uniaxial tensile test is performed on these specimens (around 20 in each case) at a constant strain rate of $10^{-4}s^{-1}$ in a computer controlled servohydraulic test machine at room temperature.

The specimens from each tensile test of different castings (ACT, ACTP-edge (ACTPE), ACTP-middle (ACTPM), LTA at 150ºC (LTA-150ºC) and LTA at 200ºC (LTA-200ºC)) are introduced to optical and scanning electron microscopy for microstructural analysis. A microstructural area of 25mm^2 (a quarter of the cross-section) from all castings is grabbed at 100X as continuous microstructural frames with an optical microscopy from the polished and unetched cross-section. This microstructural frames are used to create the microstructural montage and it is further introduced to image processing to quantify the gas and shrinkage microporosity. Nearest neighbour distance limit is used to separate the gas and shrinkage porosity and the detailed procedure is provided elsewhere [14-17]. The unetched cross-section is further etched to reveal the β_m phases. The microstructural area of 1.86mm^2 of all castings is grabbed from the edge and middle region of etched surface at 1000X for the quantification of β_m phase by image processing. In the case of LTA, optical and scanning electron microscopy is used to quantify β_m particles and the same location used to quantify β_m in optical microscopy is viewed in SEM at 4000X and the microstructures are grabbed for the quantitative characterization of β_{c+d} particle. The location of the β_m phase is computed as Cartesian coordinates by using the optical microstructures obtained for β_m quantification. These (X,Y) coordinates of β_m phase are used to identify the same phases in the SEM microstructures grabbed at 4000X to separate β_m phases from β_{c+d} phases. The separated β_m and β_{c+d} phases underwent for microstructural quantification studies and a montage of 0.74x10^{-2}mm^2 from the SEM (4000X) is used to quantify β_{c+d} phase. The clustering tendency (CT) of these features is explained by comparing nearest neighbour distance of them from the present montage with expected random arrangement and a detailed procedure is available elsewhere [14,18]. The values of CT below 1 indicate clustering and equal to 1 is random. The details of the quantification procedure of microporosity and β phase are presented elsewhere [14,18]. In addition, in-situ tensile analysis coupled with SEM is performed to understand the microscopic failure mode of the material.

3 Results and Discussion

This section explains the variations in macro properties, microstructural details, fracture behaviour, and micro-macro correlations with respect to the section thickness, position and LTA of the hpdc AZ91 magnesium alloy.

3.1 Macro Properties

The stress versus strain curves obtained from the tensile tests of the ACT, ACTPE, ACTPM and LTA castings are shown in Fig. 1. These alloys exhibit no yield point and the 0.2% proof strength is taken as an indication of the yield point. No necking is observed in the tested specimens and the different cases showed variations in hardening slope, plastic flow and fracture strain. A steep increase in hardening is observed in LTA-150°C material compared to other cases and LTA-200°C admits for the lower hardening slope. Fig. 2 gives the average YS (Fig. 2a), UTS (Fig. 2b), FS (Fig. 2c) and elongation (Fig. 2d) of ACT, ACTPE, ACTPM, LTA-150°C and LTA-200°C castings.

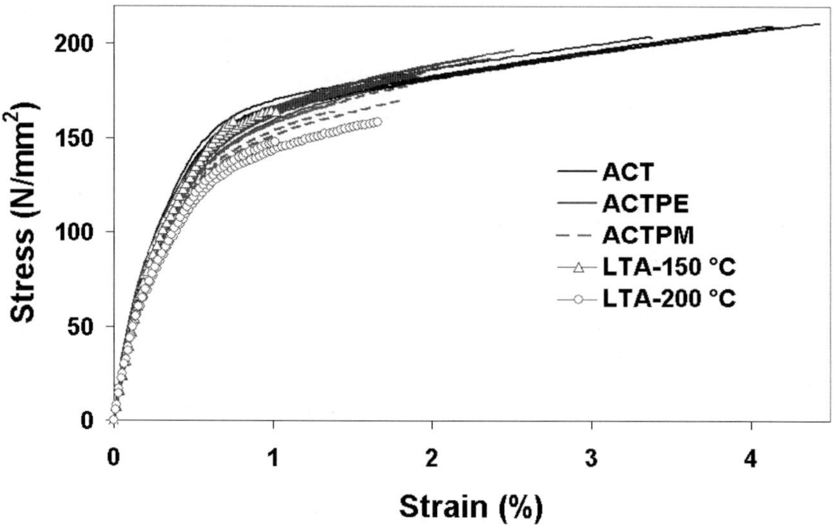

Fig. 1. Stress vs. strain results of some tensile experiments

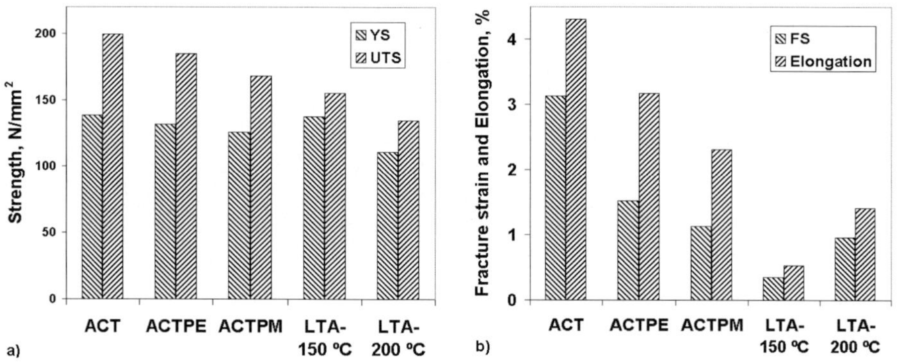

Fig. 2. Average YS (a), UTS (a), FS (b) and Elongation (b) of different castings

The ACT and LTA-150°C castings show better YS compared to others and LTA-200°C holds the least value. In the case of UTS, LTA castings have least properties compared to as-cast material. The YS and UTS are found to decrease significantly with increasing section thickness and the same is better for ACTPE compared to ACTPM castings. The average YS of LTA-150°C is high compared to ACTPE, ACTPM and LTA-200°C castings. As similar to YS, LTA-200°C declare low UTS values compared to LTA-150°C casting. The strain hardening increases significantly with decreasing section thickness and this also increases in ACTPE compared to ACTPM castings. This increasing strain hardening behaviour, high plastic flow and high FS provides better UTS for as-cast compared to LTA castings. The effect of section thickness, position and LTA on FS is high compared to the same on YS and UTS. The FS results show that the LTA castings show early failure compared to as cast material and ACT is superior among them. ACTP casting fails early compared to ACT castings indicating that the elongation of ACTP castings is low compared to ACT casting. The ACTPM fails early compared to ACTPE castings. LTA castings have low FS and ductility compared to as-cast material and among them LTA-150°C is the poor candidate.

The results conclude that strength and ductility are decreasing with increasing section thickness and the same is decreasing from ACTPM to ACTPE castings. Rodrigo et al. [7] reported a contradictory result that the tensile strength increases with increasing section thickness. However, Sequeira et al. [4], Stich [8] and Schindelbacher [3] state that the tensile strength decreases with increasing section thickness, which supports the present results. Further, the results of Stich [8] and Schindelbacher [3] support the present results in terms of ductility. Form et al [19] suggested that the local solidification time alters the microstructural development in as-cast material. This may lead to microstructural variation in ACT, ACTPE and ACTPM castings due to the variation in cooling rates during solidification. Additionally, microstructural variations are also expected due to LTA process. These microstructural variations would be responsible for the obtained changes in properties for different castings.

3.2 Microstructure

Metallographic examinations show that the hpdc castings show a distinct fine grained surface layer (chill zone) and a coarse grained interior. Fig. 3 is a typical microstructure of hpdc AZ91 alloy, which reveals microporosity, α-primary, α-secondary and β particles. Shrinkage and gas pores are the major processing defects in Mg alloys. High area fraction of gas and shrinkage pores in ACTP castings compared to ACT castings is obtained. A high number density of shrinkage pores compared to gas pores is observed in both ACT and ACTP castings; in particular, the number density of shrinkage porosity in ACTP casting is more than double of ACT casting. Size distribution of different microporosity (Fig. 4a) obtained from image processing validates these results. Significant microstructural variations between edge and middle region is shown in Fig. 4b, which represents the variations in pore density, area fraction and CT as a function of distance from one edge along the thickness direction of the ACT casting. This shows that the skin region of the casting has low area fraction, number density and clustering nature of pores in edge compared to middle of the castings.

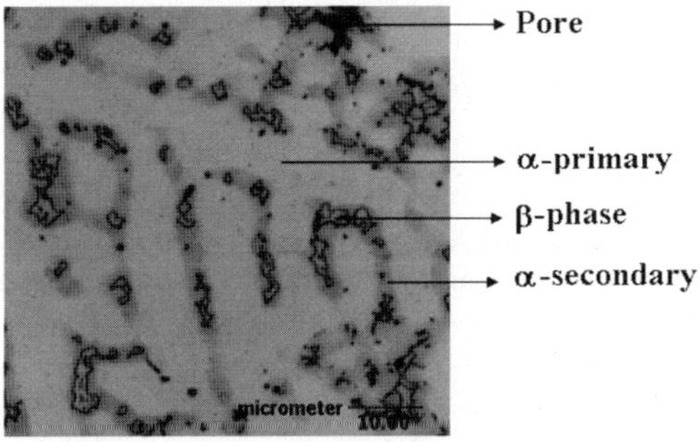

Fig. 3. Typical microstructure of hpdc AZ91 magnesium alloy

In similar to microporosity, morphological variations of β_m particles and grain size is also confirmed. The variations in average size, area fraction, CT and number density of β_m particles and grain size as a function of distance from edge to middle of the ACT casting is shown in Fig. 4c and Fig. 4d. Fig. 4c clearly indicates that there is no considerable variation in the area fraction of β_m particles from edge to middle of the casting. This plot also explains that the high number density of fine β_m particles arranged in edge compared to middle of the casting. The CT of β_m particles (Fig. 4c) increases from edge to middle of the casting. The grain size analysis (Fig. 4d) shows that it increases from edge to middle of the casting. Fine grains and fine β_m particle formation in edge is due to its high cooling rate during solidification. β_c and β_d precipitates are expected to precipitate in LTA castings. No microstructural changes except the size variation of β_m is detected in optical microscopy investigations using magnifications up to 1000X. However, these microstructural developments could be analysed by SEM microscopy at higher magnifications (above 2000X). Fig. 5 shows the SEM images of the LTA-150°C and LTA-200°C castings.

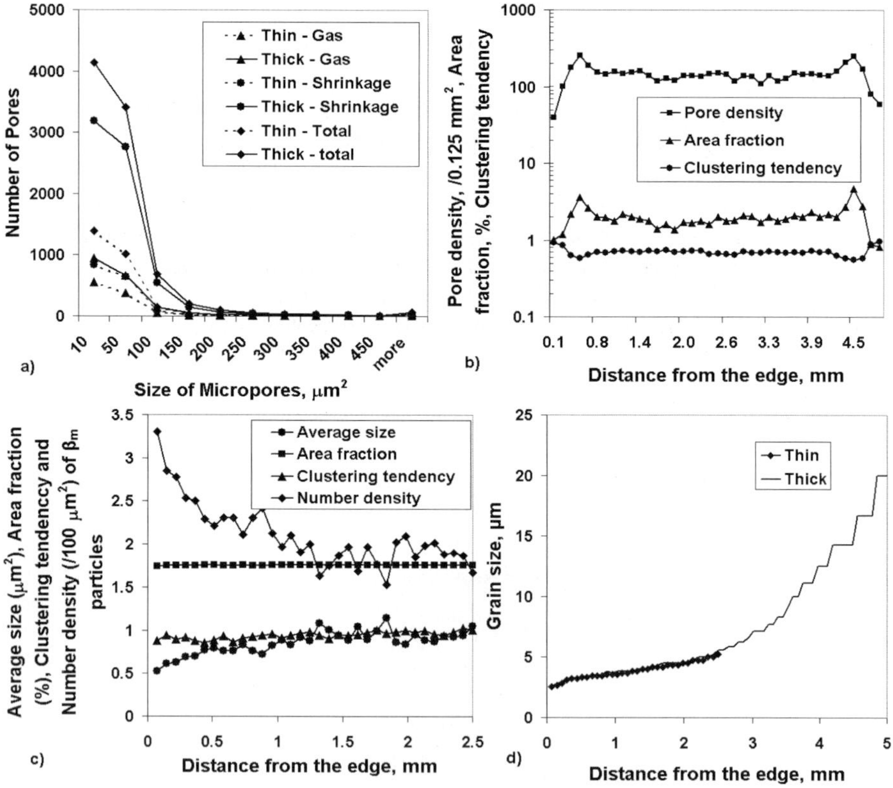

Fig. 4. a) Size distribution of microporosity and variation in micro quantities of b) micro porosity, c) β_m particles and d) grain size as a function of distance from the edge region

Both β_c (rod shape) and β_d (irregular lamellae/globular) precipitations are present in the microstructure of LTA-150°C casting (Fig. 5a) and particularly the closely spaced β_c precipitations have high number density and; around 80% of the matrix is invaded by both of these phases. Microstructure of LTA-200°C (Fig. 5b) shows high number density of β_d precipitation and low number density of β_c precipitations arranged in the eutectic region. Starink [20] investigated the ageing of Mg-Al alloys and reported that the β_c precipitation occurring throughout entire grains. However, the present examination shows that the β_c precipitation only occurs in the grain boundary/eutectic regions. The size of β_m phase is increasing with increasing LTA temperature. Area fraction of β_m and β_{c+d} phases in different LTA castings is shown in Fig. 6. The results show that the area fraction of β_m phase is very high in the LTA specimens compared to as cast specimens and it increases for increasing LTA temperature, particularly it is very high in LTA-200°C due to high diffusion of Al atoms towards the β_m phase and agglomeration of neighbouring particles.

These results explain that the nucleation rate of β_c and/or β_d phase increases as LTA temperature increases until a critical LTA temperature, which is around 150°C

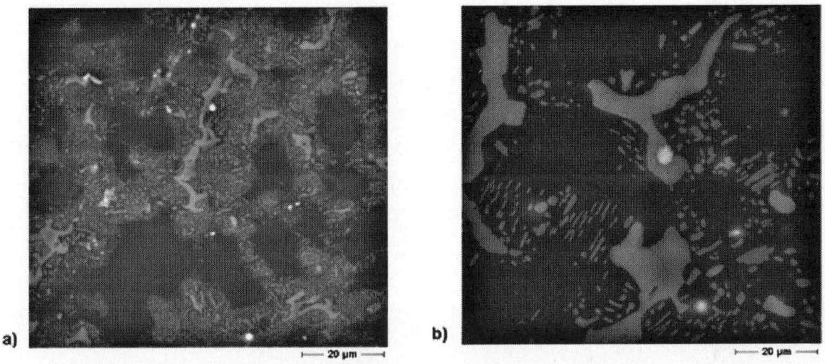

Fig. 5. SEM images shows the microstructural changes due to LTA a) 150°C b) 200°C

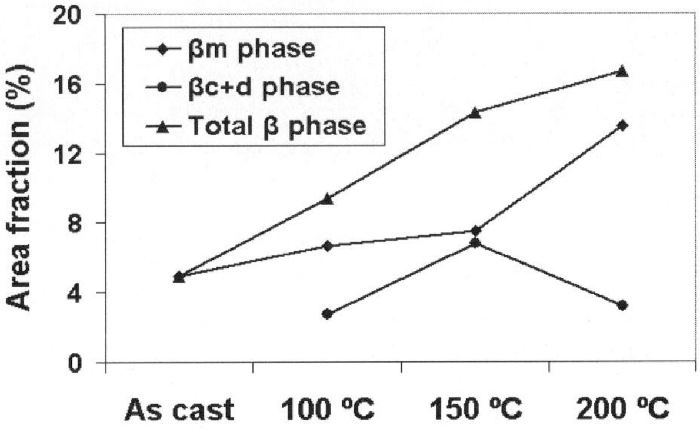

Fig. 6. Area fraction of β_m and β_{c+d} particles in as-cast and LTA castings

and there is a domination of agglomeration of neighbouring β_{c+d} particles beyond this critical LTA temperature. A similar trend is stated by Celotto [21]. Celotto [21] also stated that there is a dramatic increase in the number density of precipitates at temperatures lower than 150°C. The results also show that the discontinuous and continuous precipitation reactions occur competitively over a range of LTA temperatures.

3.3 Fracture Behaviour

The lack of independent slip systems of hpdc magnesium alloys causes less deformation due to the reduction of dislocation motion between the neighbouring grains which, leads to a grain boundary cleavage and grain size play an important role here. It is concluded that AZ91 alloy follows intergranular brittle failure and, cleavage and quasi-cleavage are the most common fracture modes [22]. The crack initiation and growth from shrinkage pores, damage of brittle β particles and grain boundary cleavage are found to be the primary failure modes of the present material from the in-situ tensile analysis [22]. Examples of these micro mechanisms are shown in Fig. 7. The predominant intergranular failure of hpdc AZ91 alloy is also due to the arrangement of shrinkage pores and β particles in intergranular region [22]. Regener et al. [23] also observed the β particle cracking in Mg alloys during in-situ tensile analysis. In addition, Yoo et al. [24] documented that damages in massive β-phase in the grain boundaries supports the grain boundary fracture. These results show the extent effect of shrinkage pores and β particles on failure of material [22]. The above explanations notice the greater influence of pores, β particles and grain size on fracture behavior of hpdc AZ91 alloy. Particularly, the size, shape and arrangement of these inhomogenities are important parameters which influence the mechanical properties by altering fracture behavior.

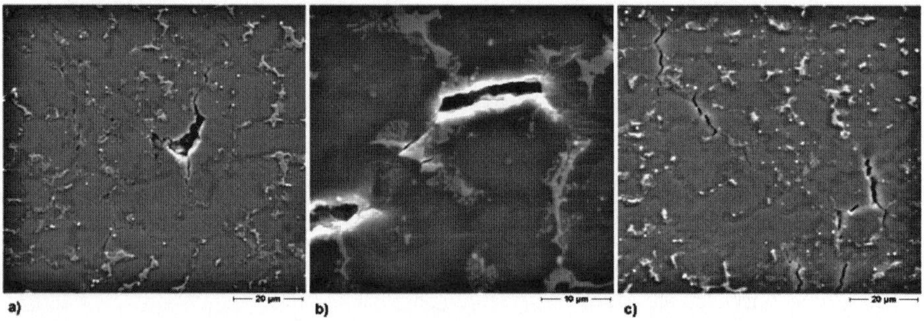

Fig. 7. Microscopic failure modes a) Crack initiation and growth from shrinkage pore b) damage of β particles and c) grain boundary fracture

3.4 Micro-Macro Interactions

The details of variation in microstructural quantities and tensile properties with respect to the section thickness, position and LTA is discussed in this section. The microstructural quantities of different castings and corresponding tensile properties are presented in Table 1. This confirms a marginal variation in area fraction, average

Table 1. Comparison of micro quantities and macro properties

Casting type	Area fraction, %		CT		Average size			UTS, N/mm²	YS, N/mm²	FS, %
	β_m / β_{c+d}	Pores	$\beta_{m/}$ β_{c+d}	Pores	$\beta_{m/}\beta_{c+d}$ (μm^2)	Pores (μm^2)	Grain (μm)			
ACT	1.5-2.7 /-	0.3-0.88	0.87-0.91 /-	0.64-0.75	1.34 /-	22.7	3.90	188-209	130-147	2.3-3.6
ACTPE	2.9-4.1 /-	1-1.27	0.65-0.70 /-	0.66-0.70	1.78 /-	25.9	5.91	171-190	128-135	1.41-1.9
ACTPM	3.1-5 /-	1.3-1.8	0.63-0.65 /-	0.59-0.63	2.18 /-	33.5	8.82	165-177	122-131	1-1.48
LTA-150 °C	6.3-8.1 /5.7-7.4	1-1.8	0.99-1.07 /0.38-0.43	0.59-0.70	3.35 /0.045	31.2	7.72	143-166	131-144	0.28-0.69
LTA-200 °C	12.5-14.9 /2.1-4.4	1-1.8	1-1.09 /0.84-0.9	0.59-0.70	5.73 /0.09	31.2	7.72	128-147	106-121	0.71-1.4

size and CT of pore, β_m and β_{c+d} particles, and average grain size with respect to the section thickness, position and LTA of the castings.

3.4.1 Section Thickness and Position

The area fraction of microporosity in ACT casting is less than 50% of ACTP castings and the same is around 26 % less in ACTPE compared to ACTPM. This increase in area fraction of porosity in ACTP castings, particularly in ACTPM region is due to the high probability of shrinkage pore formation during solidification. The average grain size and area fraction of β_m particles is almost double in ACTP castings compared to ACT castings due to the higher cooling rate of ACT castings during solidification. The drop in area fraction of β_m particles in ACTPE casting is around 13% compared to ACTPM castings. The average grain size in ACTPE is low compared to ACTPM castings and it is around 2/3rd of the latter. The drop in average size of microporosity and β_m particles in ACT compared to ACTP casting is around 27% and 32%, respectively. This is due to the high area fraction of skin region in ACT compared to ACTP castings. The high cooling rate in ACTPE compared to ACTPM section also cause fine β_m particles in ACTPE section and the average size is 18% low compared to ACTPM section. Around 23% drop in average size of microporosity is obtained in ACTPE casting compared to ACTPM due to the high probability of big shrinkage porosity on solidification. In the case of clustering nature, the increase of around 8% in β_m particles and 32% in microporosity is observed in ACTP compared to ACT castings.

The ACT castings are having better properties due to fine grain size, low area fraction and clustering nature of both microporosity and β_m phases, and the same is applicable to ACTPE when compares to ACTPM castings. The reduction in grain size of ACT castings is also due to high volume fraction of fine-grained hard skin, which improves the mechanical properties such as UTS and YS. The presence of soft [4, 25] and coarse grains in ACTPM section likely to restrict the twinning [26] may cause poor properties compared to ACTPE and ACT castings. Beside, transmission of slip across the boundaries is difficult in ACT and ACTPE castings due to the presence of high fraction of grain boundary region. The soft and coarse grains in ACTP, particularly the ACTPM castings cause early crack initiation and growth in grain boundaries and the grain boundary arrangements of shrinkage pores and β_m particles provide additional support to this process. In ACTP castings, the high area fraction and clustering nature of shrinkage pores and β_m phase particles and their arrangement in grain boundary regions may speed up the failure by early crack initiation and growth. Particularly, crack initiation and growth in micro level is strongly related to clustering nature of shrinkage pores and β_m particles which, gives a strong impact on FS quantities. This exactly reflects from the results of ACTP castings where high clustering nature of pores and β_m particles strongly reduces the values of FS by early failure and the same is applicable for ACTPM compared to ACTPE castings. β_m particles not only supports the fracture process by getting itself damaged [22] but also increases the strength of hpdc AZ91 alloy by the arrangement of high density of finer β_m in rim region.

In respect to tensile properties, porosity could have the effect of altering the stress field to initiate fracture, which affects crack propagation and reduces the effective load bearing nature of the material. When a tensile load is applied to a brittle alloy, the stress is concentrated around the pores and fracture begins at this point. Once the crack has formed, pores can be regarded as pre-existing elements of crack and hence lead to rapid crack propagation. Shrinkage porosity initiates fracture more rapidly by virtue of their sharper root radius; however, clustering of this pores leads to an easy link up between neighbouring pores which deteriorate the mechanical properties, especially to a loss of ductility.

3.4.2 Long Term Annealing

The LTA castings hold the similar morphology of microporosity of ACTP castings as this is the source of LTA castings. The quantification results confirm that the area fraction of β phase in as-cast material is low compared to LTA castings and it increases with increasing LTA temperature mainly due to the nucleation, diffusion and agglomeration reactions as explained. The area fraction of β_{c+d} particles in LTA-150°C castings is around two times of LTA-200°C castings. The high rate of fine and closely spaced β_c particle nucleation in LTA-150°C provides low average size, high CT and high area fraction for LTA-150°C castings. Compared to LTA-150°C case, LTA-200°C afford high average size, low CT and low area fraction of β_{c+d} particles due to agglomeration process. Besides, high average size, low CT (random arrangement) and high area fraction of β_m particles in LTA castings compared to as-cast material is observed due to the diffusion and agglomeration reactions. The increase in area fraction of β_m compared to as-cast material is around two times in LTA-150°C and around three times in LTA-200°C. The similar trend is obtained in average size of β_m particles.

With respect to the changes of micro quantities a marginal difference in UTS, YS and FS between as-cast and LTA castings is observed. Better YS of LTA-150°C castings compared to others is observed due to increased nucleation of β_c particles, which is a strengthener of this alloy. This is due to the fine β_c precipitates in the eutectic and grain boundary region resulting in the arrest of dislocation motion. The high number density of fine and low number density of bigger β_m particles in as-cast material compared to LTA-200°C would be the reason for better YS in as-cast material. This confirms that the strengthening ability of β particles decreases when its size increases. However, different trend is noticed in the case of UTS where as-cast material have better property due to high strain hardening, high plastic flow and resistance to failure compared to LTA castings. This is because of easy dislocation motion in as-cast material due to low number density and area fraction of β_m particles. Free from β_{c+d} particles is an additional supportive factor in this case. The high number density and area fraction of β particles in LTA castings compared to as-cast material cause a decrease in UTS due to the drop in fracture strain. However the high probability of arresting dislocation movement in LTA-150°C compared to LTA-200°C castings provides better UTS for LTA-150°C by a steep increase in hardening slope. Besides, high probability of damage in bigger brittle β_m particles of LTA-200°C castings softens the material, which decreases the UTS.

The FS values shows that the LTA-150°C fails early compared to other cases and the same is better in as-cast material. This depends on the fracture behaviour of the material and the important attribute is the grain boundary region as the material fails by intergranular brittle failure due to lack of dislocation motion from a grain to other. The arrangements of β particles in the grain boundary region also speed up the failure process [22]. The same weaken the grain boundary by increasing the area fraction of grain boundary which inhibits transmission of slip across the boundaries. These particles support the crack initiation and propagation in the grain boundary region due to the development of high stress concentration in the particles during deformation and this is severe in the case of LTA-150°C as the grain boundary region is almost occupied by the β_m and β_{c+d} particles. As-cast material holds better FS values compared to LTA-200°C due to the low size, area fraction and number density of the β_m particles. In the case of LTA-200°C, the probability of damage is high as it is having high number density of bigger β_m particles, which also speed up the failure process. These explanations also confirm the results of ductility of different castings.

4 Summary and Conclusions

The present study concludes that the mechanical properties (YS, UTS, FS and ductility) vary significantly with section thickness, section position and LTA conditions. The area fraction, average size and clustering tendency of microporosity, β_m and β_{c+d}, and grain size are quantified with respect to different conditions and their strong effects on tensile properties are confirmed. The improved properties with decreasing section thickness are confirmed. The low micro quantities of microporosity and β_m particles, and grain size provide better YS, UTS, FS and ductility for ACT castings compared to ACTP castings and for ACTPE section compared to ACTPM section. The higher area fraction of skin region provides better strength to the ACT material. The FS and ductility are more sensitive to the thickness of castings due to clustering nature of shrinkage pores and β_m particles and their intergranular arrangement is an additional influence. In addition to the reduction of FS, β_m phase also increases the strength of material. The area fraction of bulk β phase in as-cast material is low compared to LTA castings and it increases with increasing LTA temperature due to the nucleation of β_c and β_d particles and diffusion of Al atoms towards β particles during

LTA. As-cast material holds better UTS and ductility compared to LTA castings. LTA-150ºC offers vigorous nucleation of fine closely spaced β_{c+d} particles, which provides better YS compared to as-cast and LTA-200ºC castings and β_c particles act as a strengthener of LTA castings. The agglomeration of β particles takes place in LTA-200ºC castings provides poor YS and UTS compared to LTA-150ºC castings. The high plastic flow and FS provides better ductility for LTA-200ºC compared to the LTA-150ºC castings. The increase in size and area fraction of β particles in grain boundary region of LTA castings compared to as-cast material supports the failure process and leads to early failure in LTA castings.

References

[1] Mordike, B.L., Ebert, T.: Magnesium properties-applications-potential. Mater. Sci. Eng. A 302, 37–45 (2001)

[2] Bowles, A.L., Bastow, T.J., Davidson, C.J., et al.: The effect of low-temperature ageing on the tensile properties of high-pressure die-cast Mg-Al alloys. In: Magnesium Technology 2000, The Minerals, Metals and Materials Society, Nashville, USA, pp. 295–300 (2000)

[3] Schindelbacher, G., Rösch, R.: Mechanical properties of magnesium die casting alloys at elevated temperatures and microstructure independence of wall thickness. In: Mordike, B.L., Kainer, K.U. (eds.) Magnesium alloys and their applications, Wolfsburg, Germany, pp. 247–252 (1998)

[4] Sequeira, W.P., Dunlop, G.L.: Effect of section thickness and microstructure on the mechanical properties of high pressure die cast magnesium alloy AZ91D. In: Proceedings of the 3rd International magnesium conference, Institute of materials, Manchester, pp. 63–73 (1996)

[5] Orchan, W.L., Gruzleski, J.E.: Grain refinement, modification and melt hydrogen-their effects on microporosity, shrinkage and impact properties in A356 alloy. Trans. Am. Foundry Soc. 100, 415–424 (1992)

[6] Couture, A., Meier, J.W.: The effect of wall thickness on the tensile properties of Mg-Al-Zn castings. Trans. Am. Foundry Soc. 74, 164–173 (1966)

[7] Rodrigo, D., Murray, M., Mao, H., et al.: Effects of section size and microstructural features on the mechanical properties of die cast AZ91D and AM60B magnesium alloy test bars. In: SAE International congress and exposition, SAE International, Detroit, USA, pp. 785–789 (1999); Paper No.1999-01-0927

[8] Stich, A., Haldenwanger, H.G.: Dimension strategy for high-stress cast magnesium components. In: Magnesium 2000, 2nd International conference on magnesium science and technology, Dead Sea, Israel, pp. 27–34 (2000)

[9] Suman, C.: Heat treatment of magnesium die casting alloys AZ91D and AM60B. In: SAE International congress and exhibition, SAE technical paper series, Paper no. 890207 (1989)

[10] Basner, T.G., Evans, M., Sakkinen, D.J.: The effect of extended time aging on the mechanical properties of vertical vacuum cast aluminium-manganese and aluminium-rare earth magnesium alloys. In: Magnesium properties and applications for automobiles, society of automotive engineers, Detroit, USA, pp. 59–64 (1993)

[11] Sakkinen, D.J., Evans, M.: The effect of aging on magnesium die casting alloys. In: Proceedings of the 17th International die casting congress and exposition, Cleveland, USA, pp. 305–313 (1993)

[12] Song, G., Bowles, A.L., StJohn, D.H.: Corrosion resistance of aged die cast magnesium alloy AZ91D. Mater. Sci. Eng. A 366, 74–86 (2004)
[13] Leo Prakash, D.G., Regener, D., Vorster, W.J.J.: Effect of long term annealing on the microstructure of hpdc AZ91 Mg alloy: A quantitative analysis by image processing. Comput. Mater. Sci. (2008) doi:10.1016/j.commatsci.2008.01.040
[14] Leo Prakash, D.G., Regener, D.: 2D quantitative characterization of microstructural inhomogeneities in the pressure die cast AZ91 magnesium alloy. Prakt. Metallogr. 42, 555–575 (2005)
[15] Anson, J.P., Gruzleski, J.E.: The quantitative discrimination between shrinkage and gas microporosity in cast aluminum alloys using spatial data analysis. Mater. Charact. 43, 319–335 (1999)
[16] Anson, J.P., Gruzleski, J.E.: Effect of hydrogen content on the relative amounts of shrinkage and gas microporosity in a cast Al-7% Si foundry alloy. Trans. Am. Foundry. Soc 107, 135–142 (1999)
[17] Baudhuin, P., Leroy-Houyet, M.A., Quintart, J., et al.: Application of cluster analysis for characterization of spatial distribution of particles by stereological methods. J. Microsc. 115, 1–17 (1979)
[18] Leo Prakash, D.G., Regener, D.: Quantitative characterization of Mg17Al12 phase and grain size in HPDC AZ91 magnesium alloy. J. Alloys Compd. (2007) doi:10.1016/j.jallcom.2007.07.017
[19] Form, G.W., Ahearn, P.J., Wallace, F.J.: Mass effect on castings tensile properties. Trans Amer. Foundarymen's Soc. 67, 64–69 (1959)
[20] Starink, M.J.: Kinetic equations for diffusion-controlled precipitation reactions. J. Mater. Sci. 32, 4061–4070 (1997)
[21] Celotto, S.: TEM study of continuous precipitation in Mg-9%Al-I%Zn alloy. Acta Mater. 48, 1775–1787 (2000)
[22] Leo Prakash, D.G., Regener, D., Vorster, W.J.J.: Microscopic failure modes of hpdc AZ91HP Magnesium alloy under monotonic loading. Mater. Sci. Eng. A (2007) doi:10.1016/j.msea.2007.11.011
[23] Regener, D., Schick, E., Wagner, I., et al.: Strength and deformation behaviour of magnesium die casting alloys. Materialwiss. Werkst. 30, 525–532 (1999)
[24] Yoo, M.S., Kim, Y.C., Ahn, S., et al.: Tensile and creep properties of squeeze cast Mg alloys with various second phases. Mater. Sci. Forum, 419–424 (2003)
[25] Weiler, J.P., Wood, J.T., Klassen, R.J., et al.: Relationship between internal porosity and fracture strength of diecast magnesium AM60B alloy. Mater. Sci. Eng. A 419, 297–305 (2006)
[26] Westengen, H., Wei, L.Y., Aune, T., et al.: Effect of intermediate temperature aging on mechanical properties and microstructure of die cast AM alloys. In: Mordike, B.L., Kainer, K.U. (eds.) Magnesium alloys and their applications, Wolfsburg, Germany, pp. 209–214 (1998)

On Different Strategies for Micro-Macro Simulations of Metal Forming

A. Bertram[1], G. Risy[1], and T. Böhlke[2]

[1] Institut für Mechanik, Otto-von-Guericke-Universität Magdeburg
[2] Institut für Technische Mechanik, Universität Karlsruhe (TH)

Abstract. In metal forming processes, the accuracy of their simulations depends on the ability of the constitutive model to describe the relevant features of the material. For the inclusion of the texture-induced anisotropy, micro-macro models are favourable. However, the numerical effort must be drastically reduced for practical applications. A reduction of the number of crystallites on the macroscale will, unfortunately, result in an overestimation of the anisotropy. In this paper, three different methods are suggested which lead to a reduction of the numerical effort, for each of which this overestimation has been avoided by different means.

1 Introduction

During metal forming processes, large deformations occur which may lead to significant changes of the microstructure and, in particular, of the texture. In many cases, the initial material already possesses a texture, which will further evolve during additional deformations. In order to simulate such processes, the effects of both the initial and the evolving texture have to be taken into account. On the macro scale, this will cause a change of the elastic properties which is important for the springback behaviour, but also of the hardening and the anisotropy of the plastic flow. While the first is often negligible, the latter is of high technological importance since it results in typical effects like, e.g., the earing phenomenon.

Basically two different approaches to include these effects into the simulation can be found in the literature. Firstly, phenomenological models have been suggested to properly describe the anisotropy of the material. One of the oldest ones is perhaps that of von Mises [11] of an anisotropic yield criterion. In the sequel, many modifications and generalizations of this criterion have been suggested (see [13] for more references), but still a blossoming activity of inventing new yield criteria can be seen. Most of these approaches are capable to describe with more or less precision the typical forms of orthotropic or other yield loci, but hardly any of them takes into account that this anisotropy may change during the deformation, in particular under non-proportional processes. Accordingly, such approaches, although rather economical with respect to their computational costs, often turn out to be too limited in their capacities to describe the real material behaviour.

On the other hand, micro-macro simulations have recently become rather popular, since the mechanisms of the deformation on the microscopic level are much better understood than on the macro level. Crystal plasticity can be considered as a mature branch of material modelling, where issues like anisotropy and plastic spin are clearly understood. Various methods of homogenisation such as the Taylor model [10] or the Sachs model [9], self-consistent schemes, and representative volume elements (RVE) or unit cell approaches can be found in numerous investigations. All of them take into account a certain number of orientations that give rise to a crystallite orientation distribution function (CODF), which should be as close as possible to the real one. These approaches have in common that a rather detailed and realistic picture of the evolving texture can be described if only a large number of orientations is used, however, at the cost of enormous computation times. So these approaches are either able to describe the material behaviour rather realistically but lead to prohibitive costs for industrial applications, or they are drastically reduced by restrictive assumptions, but then they become less realistic. In particular, if only a small number of different orientations is taken into account, such models tend to overestimate the anisotropy. The Taylor model, e.g., although rather economic with respect to computation times, is known to even amplify this spurious effect. As a consequence, one is interested in the development of new models which lead out of this dilemma between needed precision and acceptable computational afford.

In the sequel, three methods are presented, which approach this problem in different ways, but with the same intention, namely to reduce the computational costs but to still introduce the evolution of the anisotropy at a realistic level. In fact, each of them contains a parameter which directly controls the anisotropy of the models and, thus, can be used for calibrating it in an optimal way.

The first method combines a polycrystal model with an isotropic one. Here, the stress tensor is composed of two parts, one of which results from a Taylor model containing the most important components and fibres of the CODF, while the other one is determined by an isotropic plasticity law of von Mises type, called *isotropic background*, which corresponds to a grey texture. The weighting factor of the two parts is a control parameter for the anisotropy.

The second method does not use singular orientations at each grain or integration point, but instead orientation components, which are constituted by von Mises-Fischer distributions. In contrast to sets of singular orientations, only a small number of such components is needed to model smooth CODFs. Moreover, by using the bandwidths of these components, the smoothness of the resulting CODF can be directly controlled. Large values will bring it closer to an isotropic one, small values to a more contoured distribution.

The third method is a two-scale model. On the micro level a simple Taylor model with a rigid-viscoplastic behaviour is used to simulate the evolution of the texture. From this information, a fourth order tensorial texture coefficient is determined, which contains the main information of the anisotropy. This tensor is used within a macrosopic model constituted by an elastic law, a flow rule, and a hardening rule.

All the three methods have in common that a scalar parameter controls the amount of the anisotropy. These three methods can be further combined if wanted. In the sequel we will describe each of them in more detail and give some results to demonstrate their effect on the anisotropy.

2 The Isotropic Background Model (IB)

The first model is an elastic-viscoplastic model based on discrete crystal orientations [3]. The macroscopic Kirchhoff stress is given by a superposition of the single crystal stresses. The model is enhanced by an isotropic constitutive equation modeling the isotropic part of the texture (IB - discrete Taylor model with isotropic background texture). More precisely, the IB model is modified by decomposing the stress tensor into two parts

$$\mathbf{T} = \nu \mathbf{T}_{\text{iso}} + (1-\nu)\, \mathbf{T}_{\text{cryst}}\,. \tag{1}$$

One part $\mathbf{T}_{\text{iso}}$ describes the isotropic effective viscoplastic behaviour due to a random texture. The other part $\mathbf{T}_{\text{cryst}}$ results from a the superposition of the crystal stresses. Consequently, we have two types of volume fractions. One type corresponds to the texture components, the other one describes how isotropic the microstructure is. The overestimation of anisotropy can be avoided by adapting the isotropic volume fraction ν.

3 The Continuous Taylor Model (CT)

The second model is an (elastic-)rigid-viscoplastic model based on a continuous CODF on the orientation space [2, 3] constituted by von Mises-Fischer distributions [5, 6] which permit an explicit modelling of the scattering around texture components. The von Mises-Fischer distribution is a central distribution. The scattering of a texture component can be described by a half-width value. If one amplifies the half-width values for all components by a joint factor β, then this value can be used as a parameter which controls the contour of the CODF and can, thus, be used to reduce its anisotropy. In contrast to the model of Raabe and Roters [7], the CT model directly incorporates this parameter for the calculation of the macroscopic stresses.

Both material models have been implemented into the finite element code ABAQUS [1] using the interface UMAT, and are applied to the simulation of deep drawing processes. Fig. 1 shows the experimental and the simulated earing profiles for the CT model. The starting texture for the model has been calculated from the experimental CODF section given in [4] for fcc aluminium. It can be seen that the CT model overestimates the earing height if the half-widths of the texture components are chosen in order to approximate the CODF in an optimal way (β=1). However, if the half-widths of the texture components are enlarged by a factor β=2 then the continuous model rather accurately predicts the earing profile. Thus, the modification of the half-widths allows to correct the predictions of the Taylor model which inherently overestimates the sharpness of the texture.

The predictions of the IB model is analysed in the case of a pure cube texture. In Fig. 2 the predicted earing profiles are shown for the IB model and the CT model. In the case of the discrete model, the volume fraction of the isotropic part has been varied in the range of 30% - 90%. In the case of the continuous model, the half-width has been varied in the range of 15° - 60°. It can be seen that the isotropic volume fraction of 30% corresponds approximately to a half-width of 15°. A volume fraction of 50%

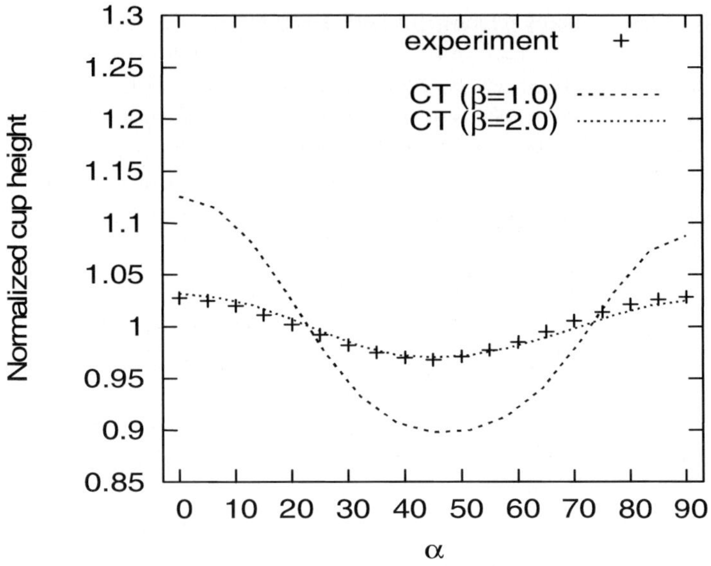

Fig. 1. Comparison of the earing profile calculated by the CT model with experimental data [4]

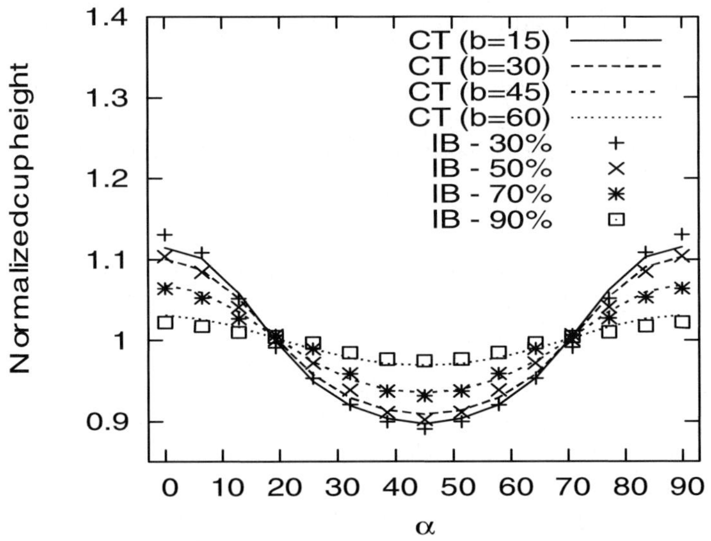

Fig. 2. Earing profile for a cube texture corresponding to different isotropic volume fractions in the IB model and different half-widths b [°] in the CT model

corresponds to a half-width of 30°. As a thumb rule, for half-width values larger than 10% one can use the fact that the isotropic volume fraction in the IB model is approximately given by $(10 + 240b/\pi)\%$ (b in rad). If one takes into consideration that even b should be increased by a factor of 2…3, then one has a rough estimate of the isotropic volume fraction in the IB model directly based on the CODF.

However, this estimate depends on the number of crystals involved in the IB model. The estimate given here can be considered as an upper bound. If more discrete orientations are used, the isotropic volume fraction should be smaller. Since for a small number of crystal orientations the discrete model is computational less expensive, this modification of the discrete Taylor model seems to be versatile.

4 The Two Scale Model (TS)

The third approach which we want to suggest combines the advantages of both a macroscopic and a microscopic approach [8, 14]. While the elastic law, the flow rule, and the hardening rule are formulated with respect to the macroscale, a 4th-order texture coefficient

$$V = \frac{1}{\sqrt{30}}\left(5\sum_{\alpha=1}^{N} v_\alpha \sum_{i=1}^{3} \mathbf{Q}_\alpha \mathbf{e}_i \otimes \mathbf{Q}_\alpha \mathbf{e}_i \otimes \mathbf{Q}_\alpha \mathbf{e}_i \otimes \mathbf{Q}_\alpha \mathbf{e}_i - \overset{2}{\mathbf{I}} \otimes \overset{2}{\mathbf{I}} - 2\overset{4}{\mathbf{I}}\right)$$

is used to capture the anisotropies on the macroscale. Here, N is the number of orientations with index α, v_α is its volume fraction, $\mathbf{Q}_\alpha$ its orientation in terms of an orthogonal tensor, e_i the crystallographic direction, $\overset{2}{\mathbf{I}}$ the second order identity tensor, and $\overset{4}{\mathbf{I}}$ the fourth order identity on the symmetric tensors after [12]. This texture coefficient is incorporated in the macroscopic elastic law, in the macroscopic yield criterion after [11], and in the associated macroscopic flow rule. Its evolution is determined by the use of a rigid-viscoplastic Taylor model. As a consequence, there is no need for an explicit modelling of the plastic spin on the macroscale. The rotation

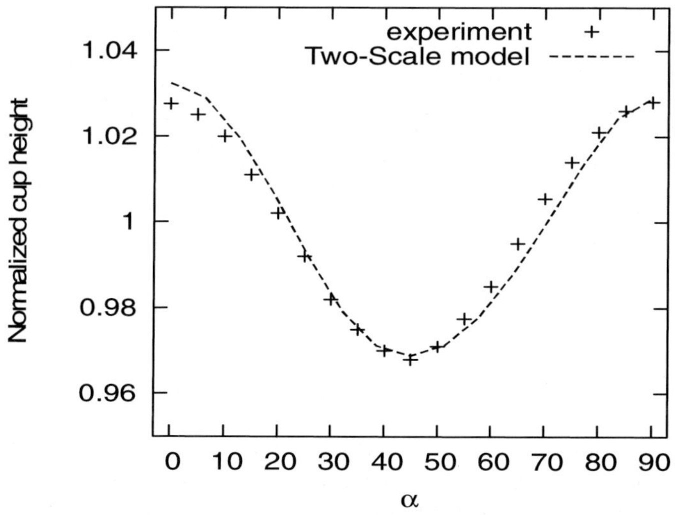

Fig. 3. Comparison of the earing profile calculated by the TS model with experimental data [4]

of the crystal lattice vectors in relation to the material is taken into account by the micro-mechanical model. The macroscopic anisotropy results from a specific orientation distribution on the microscale which changes with large inelastic deformations.

The predictions of the Two Scale model for the earing profile is shown in Fig. 3 together with the experimental data [4]. The starting texture for both models has been again calculated from the experimental CODF section given in [4]. The influence of the 4th-order anisotropy tensor in the macroscopic flow rule is controlled by the scalar factor η [8]. When choosing $\eta = 0.02$, there is a good agreement between the two scale approach and the experimental results.

Compared to classical micro-macro models, the computation of the macroscopic stress is much simpler and faster. Since the texture evolves slowly compared to the yield stress, an update of the texture coefficient is not required in each time step. Furthermore, even if only a small number of crystal orientations is used, the anisotropy is not necessarily overestimated since the discrete orientations enter the model through the 4th-order texture coefficient specifying the quadratic flow rule.

5 Conclusions

We presented three approaches for micro-macro simulations of metal forming processes. In each of them a different method is applied to reduce the spurious overestimation of the anisotropy of the Taylor model and other micro-macro models with a reduced number of orientations. By comparing these three methods from their results, some conclusions can be drawn. As expected, both the CT model and the IB model over-predict the earing behaviour drastically, in particular if only a small number of grains is taken into account. For the CT model, a larger half width parameter of the von Mises-Fischer distributions leads to a clear reduction of the earing anisotropy. The result is a reduction of the contour of the texture, which also helps to compensate the over-prediction due to the Taylor assumption. By an appropriate calibration of this parameter, the results show a good agreement with the experimental findings.

The same effect has been accomplished in the DT model by appropriately calibrating the weight of the isotropic background. Some coarse rules for estimating the values of these factors for the two models could be given.

The TS model is perhaps the most promising one of the three, since it not only reduces the anisotropy to a realistic amount, but also leads to a small computational effort.

References

[1] ABAQUS/Standard. Hibbitt, Karlsson & Sorensen, Inc. (2008)

[2] Böhlke, T., Risy, G., Bertram, A.: A texture component model for anisotropic polycrystal plasticity. Comp. Mat. Science 32, 284–293 (2005)

[3] Böhlke, T., Risy, G., Bertram, A.: Finite element simulation of metal forming operations with texture based material models. Modelling Simul. Mater. Sci. Eng. 14, 1–23 (2006)

[4] Engler, O., Kalz, S.: Simulation of earing profiles from texture data by means of a visco-plastic self-consistent polycrystal plasticity approach. Materials Science and Engineering A 373, 350–362 (2004)

[5] Eschner, T.: Texture analysis by means of modelfunctions. Textures and Microstructures 21, 139–146 (1993)

[6] Matthies, S.: Standard functions in texture analysis. Phys. Stat. Sol. B 101, K111–K115 (1980)

[7] Raabe, D., Roters, F.: Using texture components in crystal plasticity finite element simulations. Int. J. Plast 20, 339–361 (2004)

[8] Böhlke, T., Risy, G., Bertram, A.: A micro-mechanically based quadratic yield condition for textured polycrystals. ZAMM 88 5, 379–387 (2008)

[9] Sachs, G.: Zur Ableitung einer Fließbedingung. Z. Verein dt. Ing. 72, 734–736 (1928)

[10] Taylor, G.: Plastic strain in metals. J. Inst. Metals 62, 307–324 (1938)

[11] Mises, R.v.: Mechanik der plastischen Formänderung bei Kristallen. Z. angew. Math. Mech. 8(3), 161–185 (1928)

[12] Böhlke, T., Bertram, A.: The evolution of Hooke's law due to texture development in polycrystals. Int. J. Solids Struct. 38(52), 9437–9459 (2001)

[13] Bertram, A.: Elasticity and Plasticity of Large Deformations, 2nd edn. Springer, Berlin (2008)

[14] Risy, G.: Modellierung der texturinduzierten plastischen Anisotropie auf verschiedenen Skalen. Ph.D. thesis, Otto von Guericke Universität Magdeburg (2007)

Simulation of Texture Development in a Deep Drawing Process

V. Schulze[1], A. Bertram[1], T. Böhlke[2], and A. Krawietz[3]

[1] Institut für Mechanik, Otto-von-Guericke-Universität Magdeburg
[2] Institut für Technische Mechanik, Universität Karlsruhe (TH)
[3] Technische Fachhochschule Berlin

Abstract. In this paper the effect of the texture development during the deep drawing of a ferritic steel is studied with a reduced crystal plasticity model. This model has been developed for the application in the forming and springback simulation for industrial applications. Based on a specific optimisation scheme, the number of crystals used at each integration point of the finite elements is reduced to less than 100. Even with such a low number of crystals, it is possible to predict the qualitative development of the texture during the deep drawing process.

1 Introduction

In the development of tools for deep drawing processes, the application of finite element based simulations has become a standard procedure. While the prediction of the strain distribution can be performed with significant precision using conventional phenomenological material models, these models lack to predict the development of the anisotropy of the material during the metal forming operation. The latter is of high interest for the prediction of stresses in the material. With this information, the accuracy of the springback simulation as well as the prediction of the failure of the material can be improved. In this study, a micro-mechanical approach is used to model the material behaviour. Typical examples of phenomenological material models used in commercial finite element codes are the Hill-48 model implemented in PamStamp and the similar Barlat-89 model in LS-Dyna (Hallquist 1998; Hill 1948; Barlat and Lian 1989). These models are able to approximate the anisotropy of the yield locus by three parameters (*r*-values) that can be identified by tension tests in three directions. These relatively simple models have been extended to more sophisticated models, e.g., by Barlat et al. (2005). All these models have the assumption in common that the anisotropy of the material remains constant during the deformation process.

Conventional ferritic steels consist of agglomerates of body-centred cubic crystals. While the deformation of these crystals can be modelled fairly easily, the complex macroscopic behaviour is the result of the orientation distribution, morphology, and interaction of the grains. Therefore, such a micro-mechanical material model has the advantage that the development of the material behaviour is much less restricted compared to phenomenological models. However, the improvement in the material description definitely requires an increased computational effort, which depends on the material modelling of the crystals, the type of the micro-macro-transition, and the number of crystals taken into account.

There are many approaches that are used for the micro-macro-transition, such as the relaxed constraint (RC) models (Honneff and Mecking 1978), self-consistent (SC) models (Kröner 1961), and finite element models, in which a representative volume element is used in the homogenisation schemes (Bronkhorst et al. 1992; Böhlke et al. 2007). A major drawback of these more detailed approaches is that they are either modelled for specific load cases (RC-models) or lead to a large additional computational effort compared with the rather simple Taylor model. Consequently, the latter methods are typically limited to very detailed simulations of small material regions. A special aspect of crystal plasticity models for anisotropic materials is the need to approximate the initial texture of the material under consideration. The field of the texture approximation has been the topic of intensive studies itself (see, e.g., Böhlke et al. 2006b; Risy 2007; Böhlke et al. 2008). The number of crystals needed for this approximations has an important impact on the computational effort. In many cases it is possible to approximate a measured texture by a group of texture components, which can be described by a mean orientation and a corresponding scatter parameter. Many studies use special techniques to calibrate these components by the measured texture (see, e.g., Toth and Van Houtte 1992; Böhlke et al. 2006a).

In order to improve the accuracy of finite element simulations of forming operations and springback, we use a crystal plasticity model for bcc crystals, which has been implemented in a user subroutine in the finite element code LS-Dyna (Hallquist 1998). The rate-independent crystal plasticity model uses the pencil glide assumption to reduce the computational effort on the micro scale. The homogenization is based on the Taylor assumption. Special emphasis is given to an approximation of the crystallographic texture by a low number of crystal orientations without overestimating the mechanical anisotropy. Therefore, the number of crystals is minimized by a special approximation procedure in combination with an isotropic background model. The results from this model are compared with experimental measurements as well as with the results of the standard Barlat-89 model (Barlat and Lian 1989).

Notation. Throughout the text, a direct tensor notation is preferred. The scalar product and the dyadic product are denoted by $\boldsymbol{A} \cdot \boldsymbol{B} = \mathrm{tr}(\boldsymbol{A}^{\mathrm{T}}\boldsymbol{B})$ and $\boldsymbol{A} \otimes \boldsymbol{B}$, respectively. A linear mapping of 2nd-order tensors by means of a forth-order tensor $\mathbb{C}$ is written as $\boldsymbol{A} = \mathbb{C}[\boldsymbol{B}]$. Traceless tensors (deviators) are designated by a prime, e.g., $\boldsymbol{A}'$. A superimposed bar indicates that the quantity corresponds to the macroscale.

2 Constitutive Equations

In the following the multiplicative decomposition of the deformation gradient $\boldsymbol{F} = \boldsymbol{F}_{\mathrm{e}}\boldsymbol{F}_{\mathrm{p}}$ into an elastic part $\boldsymbol{F}_{\mathrm{e}}$ and a plastic part $\boldsymbol{F}_{\mathrm{p}}$ will be used, see e.g., Krawietz 1986; Bertram 1999, 2008). It is further assumed that a linear relation between a generalized stress and a corresponding generalized strain measure can be used for the formulation of the elastic law, since the elastic strains are small. For this study, the St.

Venant-Kirchhoff law formulated in terms of quantities with respect to the undistorted configuration

$$\boldsymbol{S}_e = \mathbf{C}[\boldsymbol{E}_e] \text{ with } \boldsymbol{S}_e = \det(\boldsymbol{F}_e)\, \boldsymbol{F}_e^{-1}\, \boldsymbol{\sigma}\, \boldsymbol{F}_e^{-\mathrm{T}} \tag{1}$$

the 2nd Piola-Kirchhoff stress tensor, $\boldsymbol{\sigma}$ the Cauchy stress tensor, $\boldsymbol{E}_e = (\boldsymbol{C}_e - \boldsymbol{I})/2$ Green's strain tensor, and $\boldsymbol{C}_e = \boldsymbol{F}_e^{\mathrm{T}} \boldsymbol{F}_e$ the right (elastic) Cauchy-Green tensor. For the modeling of the plastic flow, an evolution equation for the plastic part of the deformation gradient is used

$$\dot{\boldsymbol{F}}_p \boldsymbol{F}_p^{-1} = \sum_{\alpha \in A}^{N} \dot{\gamma}_\alpha \boldsymbol{d}_\alpha \otimes \boldsymbol{n}_\alpha \tag{2}$$

with the slip rate $\dot{\gamma}_\alpha$, the slip direction $\boldsymbol{d}_\alpha$, and the slip plane normal $\boldsymbol{n}_\alpha$ of the slip system α. A denotes the set of active slip systems. N is the total number of slip systems. The yield condition in each glide system is given by a scalar equation depending on the weighted shear stress τ_α and the critical resolved shear stress τ_α^C after Schmid

$$\phi_\alpha\left(\tau_\alpha, \tau_\alpha^C\right) = \left|\tau_\alpha\right| - \tau_\alpha^C = 0$$

The weighted shear stress τ_α is determined by the projection of the weighted Mandel stress tensor $\boldsymbol{Z}_e = \boldsymbol{C}_e \boldsymbol{S}_e / \rho_0$ into the slip system $\tau_\alpha = \boldsymbol{Z}_e \cdot \boldsymbol{d}_\alpha \otimes \boldsymbol{n}_\alpha$ with ρ_0 being the mass density in the reference placement.

Ferritic steels consists of body-centred cubic crystals (bcc). The primary glide systems of these crystals are the {110}<111>, {112}<111>, and {123}<111> systems. In a conventional rate-independent crystal model, all these glide systems have to be tested for admissible combinations that satisfy the yield condition and the consistency condition. A systematic testing sequence would be cumbersome. In the context of pencil glide for given slip directions $\boldsymbol{d}_\alpha$ corresponding to the lattice directions <111>, only the slip plane normals $\boldsymbol{n}_\alpha$ have to be determined. For given stress state

$$\boldsymbol{n}_\alpha = \frac{1}{\mu} \boldsymbol{s}_\alpha, \quad \boldsymbol{s}_\alpha = \left(\boldsymbol{I} - \boldsymbol{d}_\alpha \otimes \boldsymbol{d}_\alpha\right) Z_e^{\mathrm{T}} \boldsymbol{d}_\alpha, \quad \mu = \left\|\boldsymbol{s}_\alpha\right\| \tag{3}$$

determines a shear vector for each slip direction $\boldsymbol{d}_\alpha$ (Schulze et al. 2007). Since the vector $\boldsymbol{s}_\alpha$ is orthogonal to the glide direction $\boldsymbol{d}_\alpha$, one derives the result

$$\tau_\alpha = \boldsymbol{s}_\alpha \cdot \boldsymbol{n}_\alpha = \left\|\boldsymbol{s}_\alpha\right\| = \mu .$$

The hardening is modelled by a phenomenological approach based on an accumulated slip in each slip system. We assume that the critical weighted shear stress τ_α^C depends on a hardening parameter ξ_α, defined by

$$\dot{\xi}_\alpha = (1 - q)\gamma_\alpha + q \sum_\beta \dot{\gamma}_\beta , \tag{4}$$

where q is the ratio of self and latent hardening. The hardening is modelled by the ansatz of Swift

$$\tau_\alpha^C = A_1(1 + A_2\xi_n)^n . \tag{5}$$

Since the model is meant for industrial applications, the micro-macro transition is performed by the Taylor model (Taylor 1938). The Taylor model typically overestimates the anisotropy of a material. This drawback of this model is even increased, if only a small number of discrete crystals is used. Therefore, an isotropic background model is needed. Here an isotropic von Mises plasticity model with fictitious volume fraction is used (hybrid model) in order to model the grey texture.

3 Model Identification and Finite Element Simulation

The material DX53D+Z used in this study is a mild ferritic deep drawing steel. The identification of the crystal plasticity model is performed in two stages. In the first step, the initial texture of the material is approximated. With the fixed initial orientation of the crystals in the model, the elastic and plastic constants are approximated using macroscopic measurements in a second step. Details concerning the parameter identification are given in Schulze et al. (2007). The approximation of the texture is based on a mixed integer quadratic approximation scheme using sharp components with the same scatter to approximate a given texture (Böhlke et al. 2006a). The advantage of this method is that it can be applied to arbitrary crystal and texture classes, the existence of an error bound for the approximation, and the user-independence of the approximation results.

With this approximation, the initial orientation of the crystals as well as their respective volume fraction can be determined in one optimisation procedure. For the verification of the material model, the deep drawing of a circular cup is performed. Subsequent to this process, the cups are cut into slices or rings, and these rings are opened so that the springback of the part can be evaluated (Rohleder 2002). Furthermore, samples are taken from the cups in order to determine the texture after the deformation. The surface texture of the material is measured by conventional X-rays.

Table 1. Material parameters of DX53 D+Z

	E in GPa	G in GPa	ν	q	A_1 GPa	A_2	n	ξ_0	v_{iso}
A16M	114.88	94.99	0.3302	1.4	0.0299	410.090	0.275	0.0330500	0.56
A32M	114.35	94.82	0.3320	1.4	0.0292	409.970	0.270	0.0330500	0.54
A48M	111.60	95.00	0.3415	1.4	0.0271	400.147	0.278	0.0326575	0.54
A64M	113.97	94.71	0.3304	1.4	0.0291	414.920	0.270	0.0330500	0.52
A80M	114.30	94.81	0.3307	1.4	0.0292	409.960	0.270	0.0330400	0.52
A96M	113.75	94.63	0.3334	1.4	0.0292	409.940	0.270	0.0330400	0.50

The simulation of the deep drawing process uses the meshed CAD-geometry of the tools. The tool elements are rigid bodies and used to simulate the contact between the tool and the blank. The blank is modelled by under-integrated Belytschko-Lin-Tsay elements (Hallquist 1998), which are routinely used in industrial computations. In the simulation, a friction coefficient of $\mu = 0{,}075$ is used. This value has been determined in separate measurements. The penalty parameter is equal to the suggested value for LS-Dyna (Hallquist 1998). The material parameters are given in Tab. 1. Details of the identification scheme can be found in (Schulze et al. 2007).

4 Results and Discussion

The results of the strain distribution in the rolling direction for DX53D+Z are given in Figs. 1 and 2. The strain cuts show a distinct increase in the accuracy of the prediction of the major and minor strains compared with the reference model.

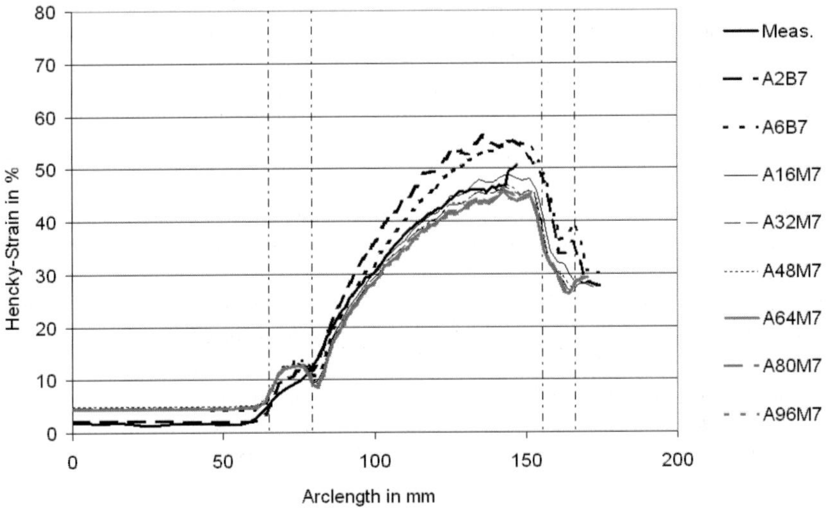

Fig. 1. Major strain in rolling direction

Considering the prediction of the earing, the result is improved by the application of the crystal plasticity model in combination with the von Mises model for the background. Only the model with 16 crystals is unable to reproduce the shape of the earing after the deformation (Fig. 3-4). For the models with 32 or more crystals, the result is in good agreement with the measurements. The mean error of the flange draw-in is reduced from more than 3.3mm in the reference model to less than 1.6mm in the crystal models. The earing height predicted by the Barlat models is in the range of more than 10mm while it is less than 4 mm with the hybrid models with more than 32 crystals, and, therefore, within the range of the measurements (3.6mm ± 0.7mm). The

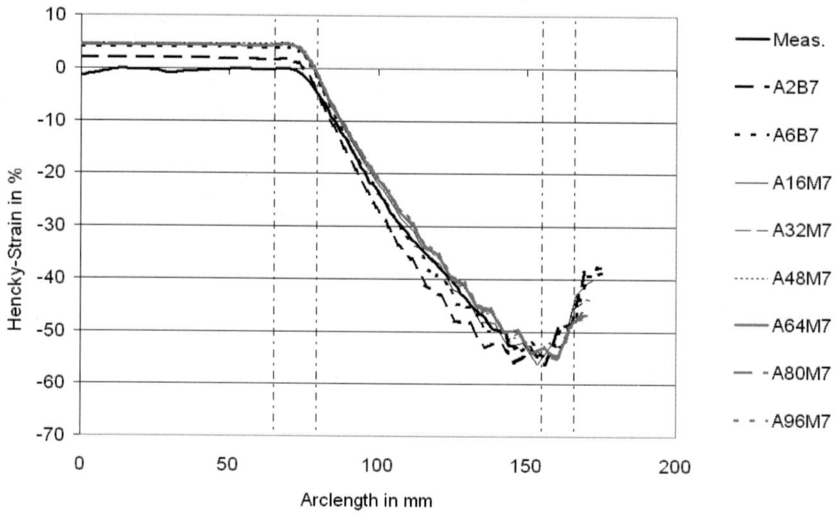

Fig. 2. Minor strain in rolling direction

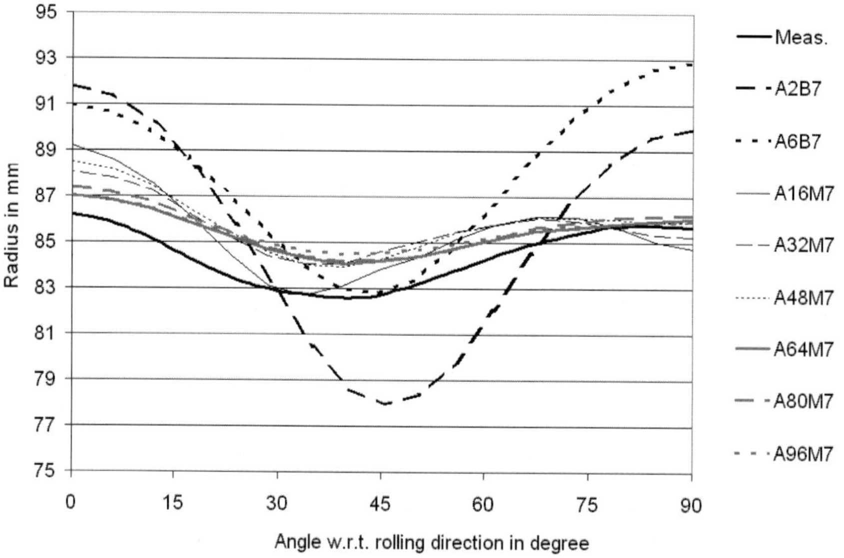

Fig. 3. Earing profile for models with isotropic background

measurements of the texture of DX53 after the deep drawing shows that the fibre structure of the texture has been transformed into a dominant component. Due to symmetry reasons, this component appears twice in the Euler space representation. The odf-plots are given in Fig. 5(a) in cuts in the φ_2-direction.

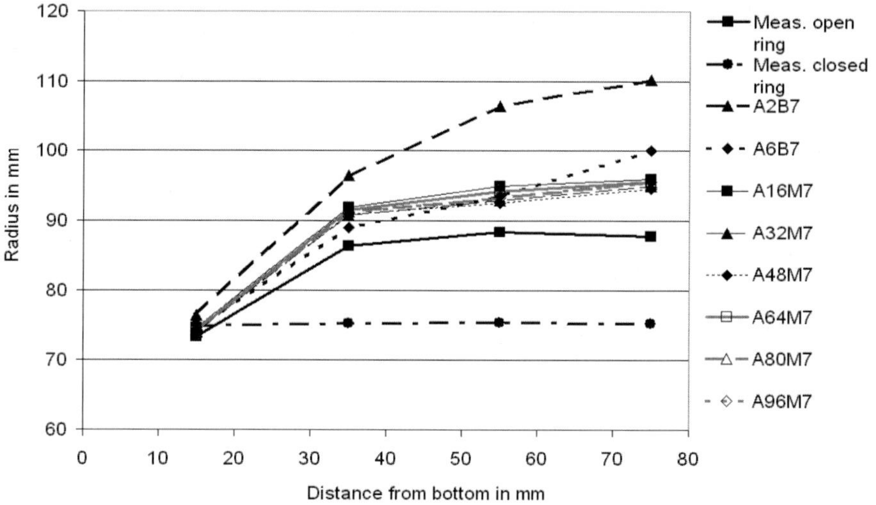

Fig. 4. Springback results: Barlat model (7 IP), crystal model with the isotropic von Mises component (7 IP)

The simulation clearly shows the development of the crystallite orientation. The results of this development are given in Fig. 5-6. The single crystals have been approximated with single components by a half-width scatter of 6°. It can be observed that the discretisation with only 16 crystals and the von Mises model as the isotropic background corresponds well to the measured texture. The model is able to reproduce the orientation of the dominant component. However, the intensity of the texture is overestimated. This is a typical behaviour of a Taylor-type model.

The results of the springback measurement and simulation are given in Fig. 3-4. The measurement shows a reduction of the diameter of the lowest ring (4) after the opening. All other rings have an increasing diameter, with the maximum value at the second highest ring (2).

The crystal plasticity models with 7 integration points in thickness direction are in good correlation with the measurements. The diameter reduction of the lowest ring is predicted well. In contrast, the Barlat model with the exponent of m = 2 fails to predict this behaviour. All simulations with the Barlat model and the hybrid model predict the maximum diameter at the highest ring (1), but the differences between ring 1 and 2 are minor if a crystal plasticity model.

Discussion. The evaluation of the model behaviour during a deep drawing process and a subsequent cutting and springback operation shows that the number of crystals has only a minor influence on the simulation accuracy. The results of the strain distribution are nearly equal for all configurations under consideration. The application of the crystal plasticity model slightly improves the accuracy of the strain prediction compared to the Barlat model.

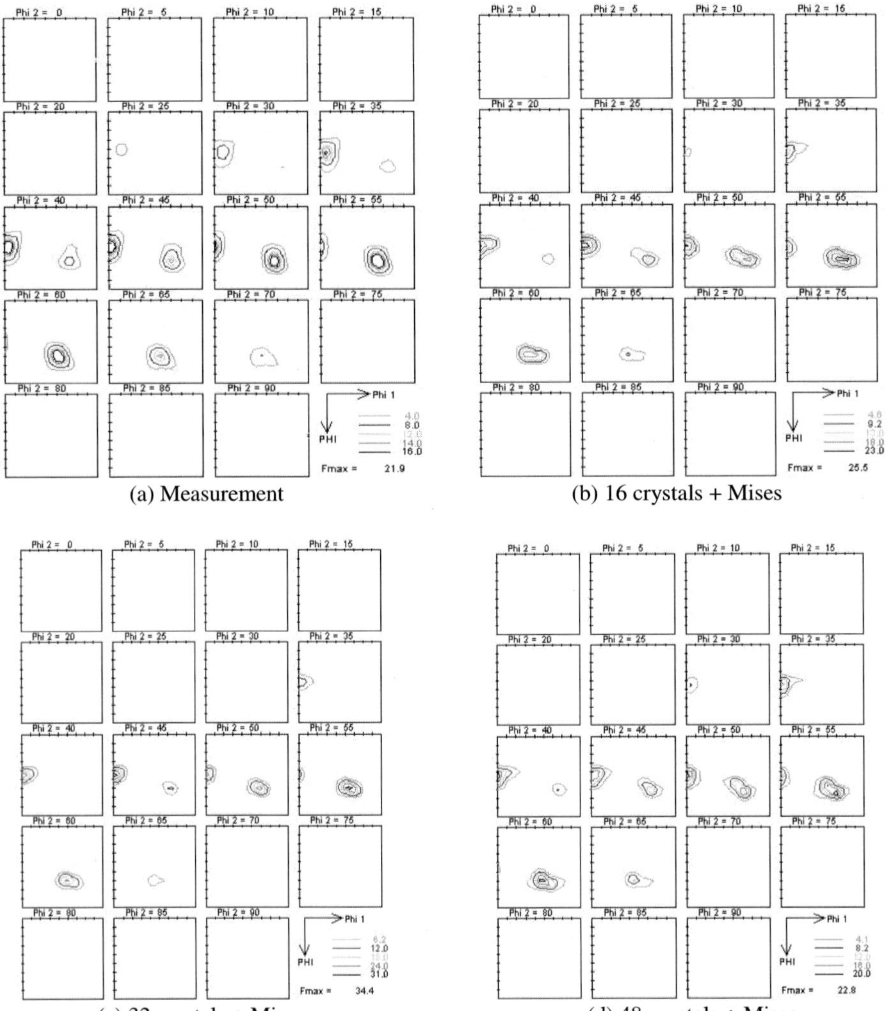

(a) Measurement (b) 16 crystals + Mises

(c) 32 crystals + Mises (d) 48 crystals + Mises

Fig. 5. Textures after the deep drawing in rolling direction, measurement and 16-48 crystals with von Mises background

A more interesting result is gained from the earing evaluation. There, the crystal plasticity model with the von Mises background model is able to improve the predictions significantly up to the value of the measurement errors.

The evaluation of the texture development shows that the initial fibre evolves during the deep drawing process into a structure that is dominated by a few components.

The simulation is able to reproduce the development of the primary components of the texture. The orientation of these components is reproduced well, while the intensity of the texture is overestimated, which is a typical behaviour of a Taylor-type model. Considering the springback results, it can be stated that the crystal plasticity

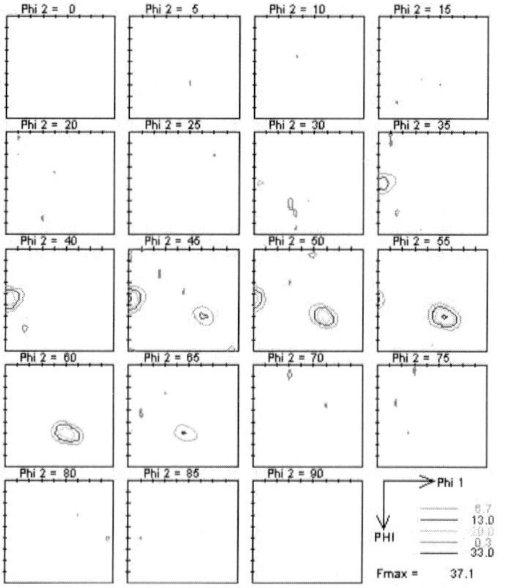

(a) 64 crystals + Mises

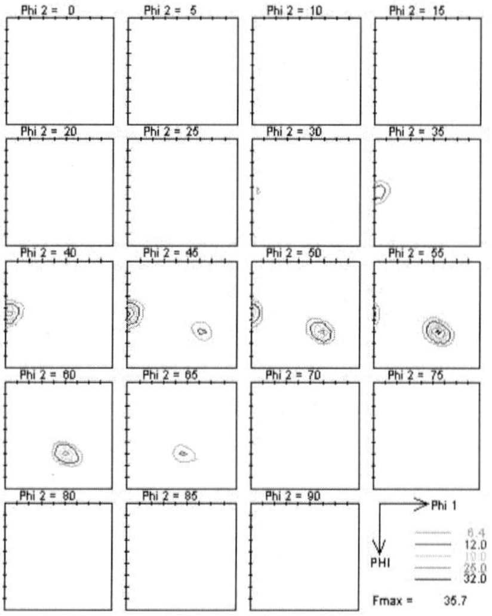

(b) 80 crystals + Mises

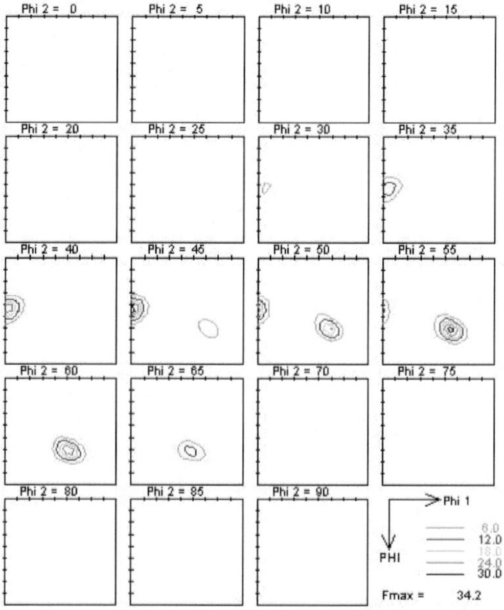

(c) 96 crystals + Mises

Fig. 6. Textures after the deep drawing in rolling direction, 64-96 crystals with von Mises background

model is able to correctly reproduce the qualitative development of the springback. While the Barlat model with m = 2 does not predict the closing of the lowest ring. In other studies (Schulze 2006; Schulze et al. 2007) it is shown that the closing predicted by the Barlat model with m = 6 is also predicted for other materials, that open the lowest ring. Only the new material model is able to predict this material dependent behaviour correctly.

The overall error of the conventional model is also higher than the one of the crystal plasticity model. The number of crystals does not have a significant impact on the simulation accuracy. Even with only 16 crystals, the qualitative agreement of the numerical predictions is acceptable. Taking all these results into consideration, the crystal plasticity model is able to improve the simulation accuracy for the given case with only 32 crystals and the von Mises component for the approximation of the isotropic background.

5 Summary and Conclusions

The study shows how crystallographic information can be incorporated into a continuum mechanical modelling of sheet metal forming. Based on a specific optimisation scheme, a low-dimensional description of the texture is obtained. The model is able to increase the simulation accuracy for a typical deep drawing process with a subsequent springback evaluation. The predictions for the earing profile, the strain distribution in the sheet, and the texture evolution during the deep drawing process are in good agreement with the measurements, if at least 32 crystals are used. For the model identification, a texture measurement is needed in addition to the conventional tension tests in three directions of the blank. Therefore, the measurement effort is not increased dramatically. Even with such a reduced modelling, the computational effort compared with the Barlat model is increased by two orders of magnitude. Therefore, such a model will be available in the near future only with parallel computing as well as for mid size problems.

References

Barlat, F., Lian, J.: Plastic behavior and stretchability of sheet metals. Part I: A yield function for orthotropic sheets under plane stress conditions. Int. J. Plast. 5, 51–66 (1989)

Barlat, F., Aretz, H., Yoon, J.W., Karabin, M.E., Brem, J.C., Dick, R.E.: Linear transformation-based anisotropic yield functions. Int. J. Plast. 21, 1009–1039 (2005)

Bertram, A.: An alternative approach to finite plasticity based on material isomorphisms. Int. J. Plast. 15(3), 353–374 (1999)

Bertram, A.: Elasticity and Plasticity of Large Deformations. Springer, Heidelberg (2005, 2nd ed (2008)

Böhlke, T.: Crystallographic texture evolution and anisotropy. Simulation, modeling, and applications. Shaker (2001)

Böhlke, T., Haus, U.-U., Schulze, V.: Crystallographic texture approximation by quadratic programming. Acta Mat. 54, 1359–1368 (2006a)

Böhlke, T., Risy, G., Bertram, A.: Finite element simulation of metal forming operations with texture based material models. Modelling Simul. Mater. Sci. Eng. 14, 365–387 (2006b)

Böhlke, T., Glüge, R., Klöden, R., Skrotzki, W., Bertram, A.: Finite element simulation of texture evolution and Swift effect in NiAl under torsion. Modelling Simul. Mater. Sci. Eng. 15, 619–637 (2007)

Böhlke, T., Risy, G., Bertram, A.: micro-mechanically based quadratic yield condition for textured polycrystals. ZAMM 88(5), 379–387 (2008)
Bronkhorst, C.A., Kalidini, S.R., Anand, L.: Polycristalline plasticity and the evolution of crystallographic texture in FCC metals. Philosophical Transactions of the Royal Society London A 341, 443–477 (1992)
Hallquist, J.O.: LS-DYNA theoretical manual. LSTC (1998)
Hill, R.: A theory of the yielding and plastic flow of anisotropic metals. Proc. Roy. Soc. London A 193, 281–297 (1948)
Honneff, H., Mecking, H.: A method for the determination of the active slip systems and orientation changes during single crystal deformation. In: Proceedings of the 5th Conference on Textured Materials (ICOTOM), vol.1, pp. 265–275 (1978)
Krawietz, A.: Materialtheorie. Springer, Heidelberg (1986)
Kröner, E.: Zur plastischen Verformung des Vielkristalls. Acta Metallurgica 9, 155–161 (1961)
Risy, G.: Modellierung der texturinduzierten plastischen Anisotropie auf verschiedenen Skalen. Dissertation, Universität Magdeburg (2006)
Rohleder, M.: Simulation rückfederungsbedingter Formabweichungen im Produktentstehungsprozeß von Blechformteilen. Shaker (2002)
Schulze, V.: Anwendung eines kristallplastischen Materialmodells in der Umformsimulation. Dissertation, Universität Magdeburg (2006)
Schulze, V., Bertram, A., Böhlke, T., Krawietz, A.: Texture-Based Modeling of Sheet Metal Forming and Springback. Preprint 3-2007, Continuum Mechanics, Institute of Engineering Mechanics, Karlsruhe Institute of Technology (2007)
Taylor, G.I.: Plastic strain in metals. J. Inst. Metals 62, 307–324 (1938)
Toth, L., Van Houtte, P.: Discretization techniques for orientation distribution functions. Text. Microstruct. 19, 229–244 (1992)

Modelling and Simulation of the Portevin-Le Chatelier Effect

C. Brüggemann[1], T. Böhlke[2], and A. Bertram[1]

[1] Institut für Mechanik, Otto-von-Guericke-Universität Magdeburg
[2] Institut für Technische Mechanik, Universität Karlsruhe (TH)

Abstract. During deformations of an Al-Mg alloy (AA5754) dynamic strain aging occurs in a certain range of temperatures and strain-rates. A manifestation of this phenomenon, usually referred to as the Portevin-Le Chatelier (PLC) effect, consists of the occurrence of strain localisation bands accompanied by discontinuous yielding. The PLC effect is due to dynamic dislocation-solute interactions and results in negative strain-rate sensitivity of the flow stress. The PLC effect is detrimental to the surface quality of sheet metals and also affects the ductility of the material. Since the appearance of the PLC effect strongly depends on the tri-axiality of the stress state, three-dimensional finite element simulations are necessary in order to optimise metal forming operations. We present a geometrically non-linear material model which reproduces the main features of the PLC effect. The material parameters are identified by experimental data from tensile tests. Special emphasis is put on the prediction of the critical strain for the onset of the PLC effect and the statistical characteristics of the stress drop distribution.

1 Physical and Mechanical Characteristics of the PLC Effect

The material instability called PLC effect is a special type of dynamic strain aging, which is observed during deformation processes of, e.g., Al, Fe, Cu, and Ni base alloys. This effect describes the discontinuous yielding in strain localisation or shear bands (Fig. 1 left). While the deformation process initially begins in a stable manner, instantaneously shear bands appear in dependence of the deformation rate and temperature. For special alloys the critical strain for the onset of the PLC effect is minimal for special strain rates, while for smaller or greater strain rates the PLC effect occurs later in process, which is called inverse and normal behaviour, respectively (Fig. 1 right). Strain localisations produce a typical serrated material response in the stress strain curves of tensile tests. In stress driven tests steps are observed in the stress-strain curve. However, for strain rate driven tests the serrations have different statistical characteristics in dependence of the applied strain rate, and so one typically distinguishes three different types (Fig. 1 middle)

A: high strain rate - stochastic development of bands over the hole specimen
B: medium strain rate - hopping bands
C: low strain rate - the PLC-band goes continuously through the specimen

As a consequence, the material quality is reduced, which is manifested in a loss of ductility and a reduction of the surface quality of the deformed part. In Figure 1 (left)

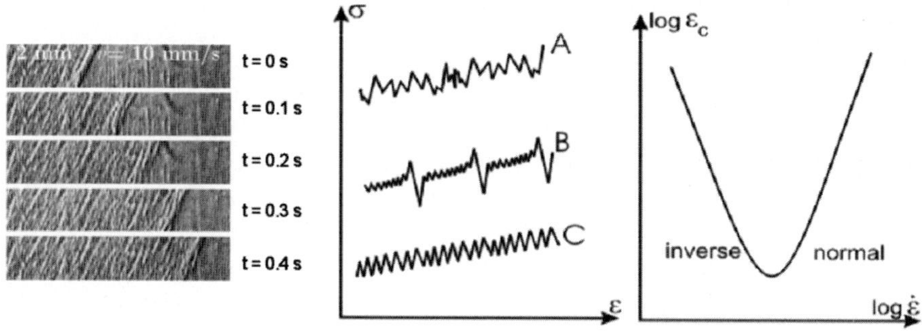

Fig. 1. Localized deformation (left, Franklin et al. 2000); typical material response (middle); inverse behaviour of the critical strain (right)

we see that the waviness and roughness of the deformed specimen is increased, so expensive post processing is needed to ensure product quality. In the case of the inverse and normal behaviour of the critical strain, however, the production process can be optimised by avoiding these critical strain rates, where the onset of PLC effect is earliest in process.

2 Experimental Findings

To deal with the PLC effect, tensile tests have been performed. Polycrystalline flat specimens (gage length 21mm, width 5.1mm, thickness 1.55mm) have been

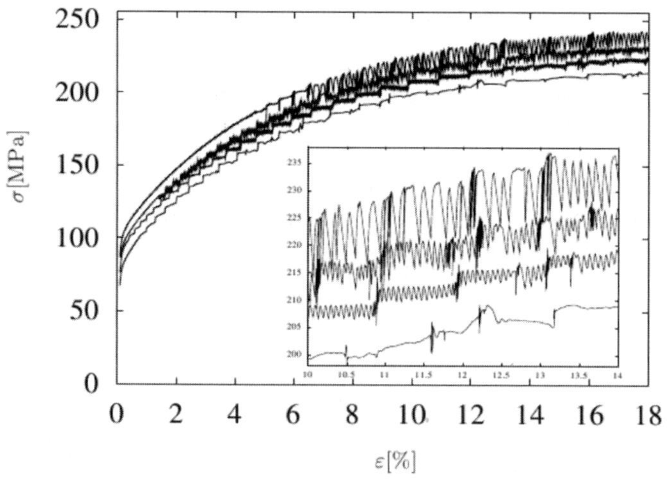

Fig. 2. Comparison between the experimental data of tensile tests for different plastic strain rates. The curves are deliberately shifted along the ordinate axis for clarity. The strain rate from the upper curve downwards are as follows: $2{\cdot}10^{-5}\mathrm{s}^{-1}$ [+15MPa], $2 \cdot 10^{-4}\mathrm{s}^{-1}$ [+10MPa], $8 \cdot 10^{-4}\mathrm{s}^{-1}$ [+5MPa], $6 \cdot 10^{-3}\mathrm{s}^{-1}$ [+0MPa].

manufactured from an Al-Mg alloy (AA5754 - Mg 2.6-3.6%). All samples are cut from sheet materials with the tensile axis chosen in the rolling direction. After grinding and polishing, the specimens are annealed for two hours at 400°C and afterwards quenched in water. Tensile tests are performed at room temperature with four different imposed strain rates between $2{\cdot}10^{-5}s^{-1}$ and $6{\cdot}10^{-3}s^{-1}$. The experimental stress-strain curves are shown in Figure 2. With higher strain rates, lower normalized stress-strain curves result, and also the stress drop distribution is depending on applied strain rate, which can be seen in the enlargement.

The three typical statistical characteristics of the stress drop distribution can also be distinguished. For the high strain rate, small stress drops are dominating. For the medium strain rates, two different stress drop amplitudes are dominating, and for the low strain rate, bigger stress drop amplitudes are prevailing. More details about the statistical analysis of stress strain curves are given in for example Lebyodkin et al. (1969), Lebyodkin et al. (2000), Lebyodkin et al. (2001) and Kubin et al. (2002).

3 Micromechanical Mechanisms

Free dislocations become temporarily arrested at obstacles like forest dislocations (Fig. 3). During the waiting time t_w at the obstacles, solute atoms migrate towards the dislocations, condensing in a cloud around the dislocation and producing additional pinning. The pinning force necessary to move the dislocation increases in time t_a and the system ages. As soon as the stress becomes large enough, the pinning breaks down and the dislocation flashes over a distance ω to stop again. Steady pinning and unpinning of the free dislocations from the solute atoms produces the PLC effect.

A detailed interpretation and the theoretical background of the PLC effect can be found in Penning (1972), Kubin and Estrin (1996), as well as Rizzi and Hähner (2004).

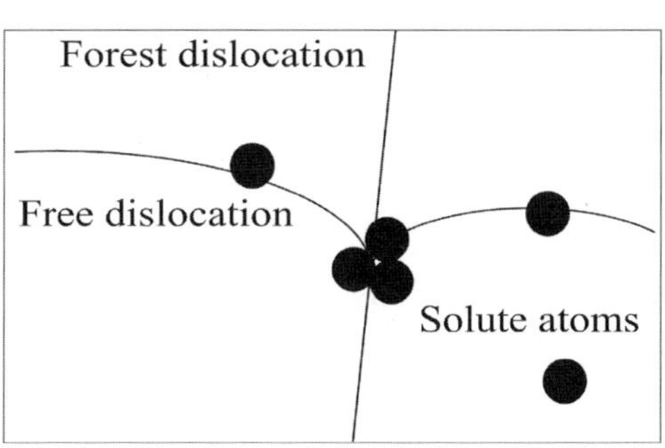

Fig. 3. Interaction of dislocations and solute atoms

4 Modelling of the PLC Effect

In the following, an isotropic elastic viscoplastic material model is formulated which considers large viscoplastic deformations but only small elastic deformations. For the elastic deformations the St. Venant-Kirchhoff law is given by

$$\boldsymbol{\tau} = \mathbb{C}\left[\mathbf{E}_e^A\right], \quad \mathbb{C} = \lambda \cdot \mathbf{I} \otimes \mathbf{I} + 2G\mathbb{I}^S, \quad \mathbf{E}_e^A = \tfrac{1}{2}\left(\mathbf{I} - \mathbf{B}_e^{-1}\right). \tag{1}$$

The Kirchhoff stress tensor $\boldsymbol{\tau}$ is connected to the elastic Almansi strain tensor $\mathbf{E}_e^A$ by the isotropic stiffness tensor $\mathbb{C}$. λ and G denote the Lamé coefficients and $\mathbf{B}_e = \mathbf{F}_e \mathbf{F}_e^{\mathrm{T}}$ the elastic left Cauchy Green tensor. The kinematics are described by

$$\mathbf{L} = \dot{\mathbf{F}}\mathbf{F}^{-1}, \quad \mathbf{D} = \tfrac{1}{2}\left(\mathbf{L} - \mathbf{L}^{\mathrm{T}}\right), \tag{2}$$

where $\mathbf{L}$ is the velocity gradient and $\mathbf{F}$ the deformation gradient. $\mathbf{D}$ denotes the symmetric part of the velocity gradient. The viscoplastic behaviour is given by a von Mises type flow rule for large plastic but small elastic deformations

$$\mathbf{D}_p = \frac{3\dot{\varepsilon}_p}{2\sigma}\boldsymbol{\tau}', \quad \mathcal{L}_v(\mathbf{B}_e) = \dot{\mathbf{B}}_e - \mathbf{L}\mathbf{B}_e - \mathbf{B}_e\mathbf{L}^{\mathrm{T}}, \tag{3}$$

$$-\frac{1}{2}\mathcal{L}_v(\mathbf{B}_e) = sym\left(\mathbf{L}_p\mathbf{B}_e\right), \quad sym\left(\mathbf{L}_p\mathbf{B}_e\right) \approx \mathbf{D}_p,$$

where

$$\dot{\varepsilon}_p = \sqrt{\frac{2}{3}}\left\|\mathbf{D}_p\right\|, \quad \sigma = \sqrt{\frac{3}{2}}\left\|\boldsymbol{\tau}'\right\|, \quad \boldsymbol{\tau}' = \boldsymbol{\tau} - \frac{1}{3}tr(\boldsymbol{\tau})\,\boldsymbol{I} \tag{4}$$

and the prime denotes the symmetric and traceless part of a tensor.

On the micro scale the PLC effect is produced by additional pinning of dislocations temporarily arrested at localized obstacles (forest dislocation junctions) by diffusing solutes. This mechanism is represented by a scalar flow rule for the equivalent plastic strain rate originally provided by Zhang et al. 2001, modified to adapt it to the material behaviour of the above described Al-Mg alloy. The equivalent plastic strain rate is given by

$$\dot{\varepsilon}_p = \dot{\varepsilon}_o \exp\left(\frac{\sigma - \sigma_d - P_1(\varepsilon_0 + \varepsilon_p)C_s}{S}\right), \tag{5}$$

where $\dot{\varepsilon}_0$, ε_0 and P_1 are constants. σ_d describes the strain hardening and S the strain rate sensitivity effect of the flow stress

$$\sigma_d = d_1 + d_2\left(1 - \exp\left(-\frac{\varepsilon_p}{d_3}\right)\right), \quad S = s_1 + s_2\sqrt{\varepsilon_p}. \tag{6}$$

The concentration of solute atoms at pinned dislocations is

$$C_s = \left(1 - \exp\left(-P_2 t_a^n\right)\right) C_m \tag{7}$$

and depends on the accumulated plastic strain ε_p and the aging time t_a which is an internal variable

$$\dot{t}_a = 1 - \frac{t_a}{1_w}, \qquad t_w = \frac{\Omega}{\dot{\varepsilon}_p}, \qquad \Omega = \omega_1 + \omega_2 \varepsilon_p^{\beta}. \tag{8}$$

The aging time t_a is the time available to solute atoms to diffuse to dislocations. The rate of change of t_a depends on t_a and the waiting time t_w, the time a dislocation spends at localized obstacles before it unpins.

In order to solve the differential equations, the backward Euler method is used with the assumptions that the initial conditions are $\varepsilon_p(t = 0) = 0$, $t_a\,(t = 0) = 0$. The material parameters are fitted to the experiments by using the hardening behaviour of the

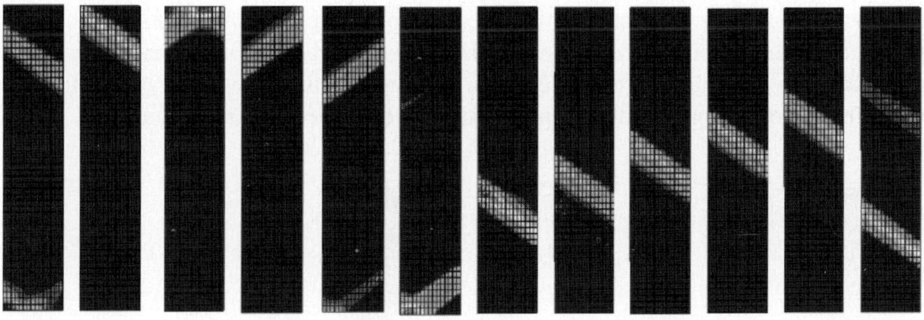

Fig. 4. FE simulation of PLC band propagation

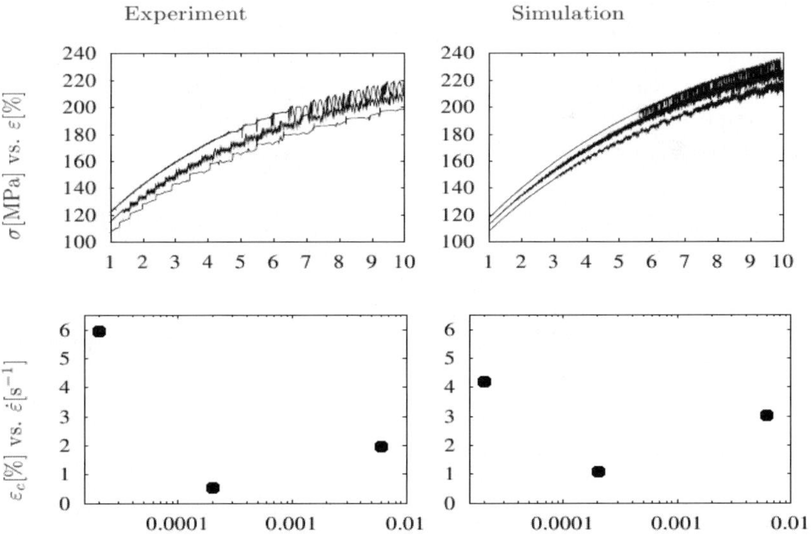

Fig. 5. Comparison of experimental findings with numerical results: σ vs. ε. The curves are deliberately shifted along the ordinate axis for clarity. The strain rate from the upper curve downwards are as follows: $2{\cdot}10^{-5}\text{s}^{-1}$ [+10MPa], $2 \cdot 10^{-4}\text{s}^{-1}$ [+5MPa], $6 \cdot 10^{-3}\text{s}^{-1}$ [+0MPa] (top) and ε_c vs. $\dot{\varepsilon}$ (bottom).

normalized stress strain curves for the hardening parameters. For the parameters describing the aging behaviour the stability analysis of the strain rate sensitivity of the stress is used. The material parameters of the above model for the Al-Mg alloy (*AA*5754) are listed in the table below.

5 Comparison of Experimental Findings with Simulation

In Figure 5 and 6 the experimental findings are compared to the numerical results.

The hardening behaviour is reproduced well. The different types of serrations observed experimentally are recovered by the simulation. The comparison of the statistical behaviour (Fig. 6), such as the stress drop distribution function shows acceptable agreement. The shift from type A for high strain rates, where small stress drops magnitudes are dominating, to type B for medium strain rates, to type C for small strain rates, where bigger stress drop magnitudes occur, can be reproduced.

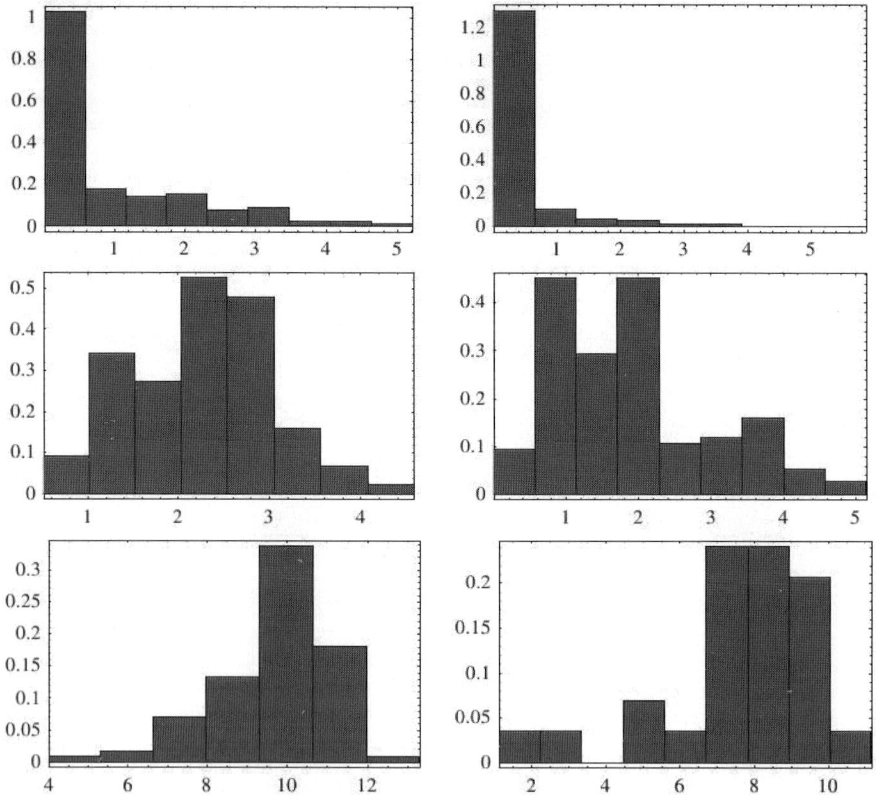

Fig. 6. Comparison of the stress drop distribution function (probability *P* vs. stress drop amplitude $\Delta\sigma$) for experimental (left) and simulation (right) results, for the stress strain curves of three different imposed strain rates, $6{\cdot}10^{-3}s^{-1}$ (top), $2{\cdot}10^{-4}s^{-1}$ (middle), $2{\cdot}10^{-5}s^{-1}$ (bottom)

Another characteristic feature is the critical strain ε_c , which is shown in Figure 5 (bottom). With the current material model this effect can also be reproduced in general.

To ensure that the results are not depending on the finite element mesh size, simulations have been performed with three different numbers of elements, 10, 30 and 150 (Fig. 7). The results show good agreement concerning the hardening behaviour and the onset of jerky flow, but the statistical behaviour of the serrations shows differences, in particular for small strain rates.

Table 1. Material parameters of an AlMg alloy

parameter	value	parameter	value
P_1	325.0	d_1 (MPa)	75.0
P_2 ($s^{1/3}$)	4.0	d_2 (MPa)	165.0
β	1.0	d_3	7.0 10^{-2}
ω_1	$1.0 \cdot 10^{-6}$	E (MPa)	70000
ω_2	$1.0 \cdot 10^{-4}$	ν	0.3
s_1 (MPa)	0.41	ρ (kg/m^3)	2650
s_2 (MPa)	2.91	$\dot{\varepsilon}_0$ (s^{-1})	2.3 10^{-7}
n	1/3	ε_0	0.001

6 Further Numerical Results Compared to Experimental Findings in the Literature

In the literature other experimental results concerning the PLC effect in different deformation processes of the alloy Al-Mg are listed. For example, stress rate driven tests are performed by Kovacs et al (2000). The test results with a stress rate of 0.5MPa/s is reproduced by the simulation and shows a comparable hardening behaviour and the typical steps for stress rate driven tests, but the onset of the PLC effect is predicted to late by the simulation. One possible reason for this may be the different Mg concentration and the different solution heat treatment temperatures. Also, compression test results can be found for an Al-Mg alloy in Horvath et al (2007). Tests are performed at three different stress rates, 0.02, 0.2 and 3MPa/s. Simulation results can reproduce the hardening behaviour in general (Fig. 7, right).

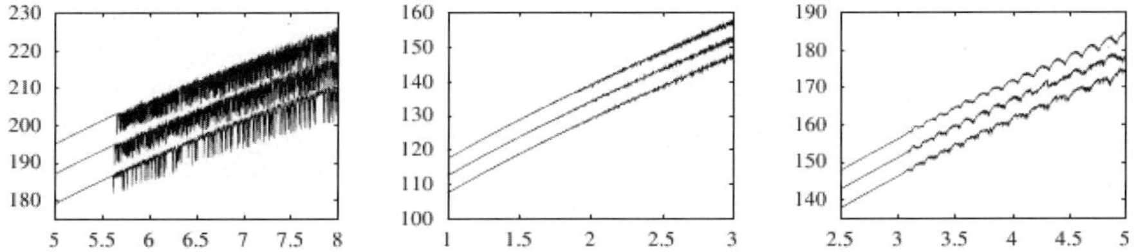

Fig. 7. σ vs. ε comparison of numerical results produced ($2 \cdot 10^{-5} s^{-1}$ (right), $2 \cdot 10^{-4} s^{-1}$ (middle), $6 \cdot 10^{-3} s^{-1}$ (right)) by using 10 (bottom), 30 (middle) and 150 (top) elements in the FE model. The curves are deliberately shifted along the ordinate axis for clarity.

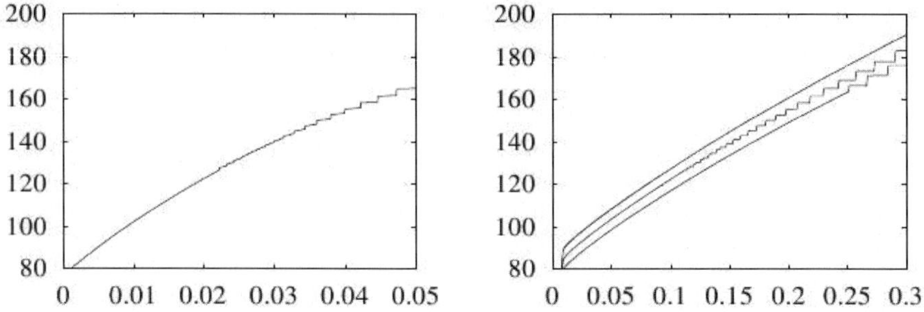

Fig. 8. Examples of numerical results σ vs. ε for stress rate driven (left) and compression tests with different stress rates (right). The curves are deliberately shifted along the ordinate axis for clarity.

After the literature and also in the simulations, the different tested stress rates have hardly any influence on the hardening behaviour. And the typical steps for stress driven tests are shown too. The differences for the onset of serrations may occur in the compression tests also due to different Mg concentrations and different solution heat treating temperatures.

7 Conclusions

FE simulations have been performed with a 3D isotropic, geometrically nonlinear PLC model. Material parameters have been identified, and the results of the simulations compared with experimental findings. The comparison of experiment with simulation shows good agreement for the strain hardening behaviour and the statistical behaviour of the stress drops for all strain rates. Also the inverse effect of the onset of instability could be reproduced by this material model. Simulation results of experimental data in the literature are shown.

Acknowledgments. The authors gratefully thank Y. Estrin (TU Clausthal, Institut für Werkstoffkunde und Werkstofftechnik) and M. Lebedkin for the possibility of performing the deformation tests and many helpful discussions.

References

[1] Lebyodkin, M., Dunin-Barkowskii, L., Lebedkina, T.: Statistical and multifractal analysis of collective dislocation processes in the Portevin-Le Chatelier effect Physical. Mesomechanics 4, 9 (2001)

[2] Kubin, L., Fressengeas, C., Ananthakrishna, G., In Nabarro, F., Duesbery, M.: Collective Behaviour of Dislocations in Plasticity. Dislocations in Solids 102 (2002)

[3] Zhang, S., McCormick, P., Estrin, Y.: The morphology of Portevin-Le Chatelier bands: finite element simulation for Al-Mg-Si. Acta. Mater. 49, 1087 (2001)

[4] Horvarth, G., Chinh, N., Gubicza, J., Lendvaih, J.: Plastic instabilities and dislocation densities during plastic deformation in Al-Mg alloys. Mat. Sci. Eng. A445(446), 186 (2007)

[5] Kovacs, Z.s., Lendvai, J., Vörös, G.: Localized deformation bands in Portevin-LeChatelier plastic instabilities at a constant stress rate. Mat. Sci. Eng. A 279, 179 (2000)
[6] Franklin, S.V., Mertens, F., Marder, M.: Portevin-Le Chatelier effect. Physical Review E 62, 8195 (2000)
[7] Bross, S., Hähner, P., Steck, E.A.: Mesoscopic simulations of dislocation motion in dynamic strain ageing alloys. Computational Materials Science 26, 46 (2003)
[8] Marciniak, Z., Kuczynski, K.: Limit Strains in the processes of stretch-Forming sheet metal. Int. J. Mech. Sci. 9, 609 (1967)
[9] Lebyodkin, M., Brechet, Y., Estrin, Y., Kubin, L.: Statistical behaviour and strain localization patterns in the Portevin-Le Chaterlier effect. Acta mater 44, 4531 (1996)
[10] Lebyodkin, M., Dunin-Barkowskii, L., Brechet, Y., Estrin, Y., Kubin, L.P.: Spatiotemporal dynamics of the Portevin-Le Chatelier effect: experiment and modelling. Acta mater. 48, 2529 (2000)
[11] Rizzi, E., Hähner, P.: On the Portevin-Le chatelier effect: theoretical modeling and numerical results. Int. J. Plast. 20, 121 (2004)
[12] Penning, P.: Mathematics of the Portevin-Le Chatelier effect. Acta Metall. 20, 1169 (1972)
[13] Estrin, Y., Kubin, L.: Spatial Coupling and propagative plastic instabilities. Continuum models for Materials with Microstructure 365 (1995)

Plastic Deformation Behaviour of Fe-Cu Composites

Y. Schneider[1], A. Bertram[1], T. Böhlke[2], and C. Hartig[3]

[1] Institut für Mechanik, Otto-von-Guericke-Universität Magdeburg
[2] Institut für Technische Mechanik, Universität Karlsruhe (TH)
[3] Institut für Werkstoffphysik und -technologie, TU Hamburg-Harburg

Abstract. Two-phase composites, which consist of spherical polycrystalline α-iron and copper particles, are studied mechanically under large plastic deformation. Such polycrystals are produced from mixtures of iron and copper powders by powder metallurgy. Due to the significant difference of the yield stress in the iron and the copper phase in which the slip system geometry is also dissimilar, a high heterogeneity and anisotropy characterize the plastic deformation behaviour. In such bcc-fcc polycrystals, the harder phase shows higher stresses while the softer phase undergoes a larger deformation. To successfully predict the mechanisms of the plastic deformation for a certain grain, effects of the major factors should be taken into account like, e.g., the microscopic interaction, the influence of neighbouring grains, the phase volume fraction, the morphology, and the initial crystallographic texture. In this work, an elasto-viscoplastic material model is applied in axisymmetric finite element simulations, whereas the macroscopic material behaviour is established based on constitutive equations of the single crystal. In the simulations, real two-dimensional microstructures are selected as cross-sections in the axisymmetric model. The material parameters are identified from the experimental data in compression tests. Numerical predictions include the flow behaviour, the crystallographic texture, and the local strain in Fe-Cu composites. In particular, a quantitative study is performed for the mean value of the local strain in both phases, which shows a good agreement with the experimental result for the Fe17-Cu83 composite under tension. Numerical predictions and experimental measurements are compared for the flow behaviour and the texture in both Fe17-Cu83 and Fe50-Cu50 composites.

1 Introduction

Multiphase metals are widely applicable in the automobile and aerospace industries, since they can show good ductility, enhanced strength at elevated temperature, and improved corrosion resistance. During the deformation process in such polycrystals, the microstructure and its evolution are essential for the determination of the macroscopic mechanical behaviour. The main features of the microstructure include the morphology, the arrangement, and the orientation of grains. The volume fraction of each phase and the interaction among grains are also important factors which influence the macroscopic material properties.

In such two-phase polycrystals, the harder phase shows higher stresses than the softer phase (Raabe et al. 1995; Soppa et al. 1998). To investigate the influence of local mechanical interactions among the phases on the plastic behaviour in such polycrystals, α-Fe-Cu composites as model materials have been studied in this work.

Commentz et al. (1999); Commentz (2000); Hartig and Mecking (2005) and Daymond et al. (2005) investigated for the first time the complex plastic deformation of this type of composite. Here, we introduce a mechanical approach based on finite elements to numerically study properties of iron-copper composites and compare the results with experimental data (Commentz 2000). We incorporate two-dimensional (2D) realistic morphologies, the internal interphase, and the crystallographic texture in these calculations to predict the micro- and macro-mechanical properties of Fe-Cu composites under simple tensile and compressive loads until large plastic strains.

This work is structured as following. The production of samples and processes of experimental tests are briefly introduced in Section 2. In Section 3, we summarize the constitutive equations for single crystals. The material behaviour of polycrystals is established based on the equations of the single crystal. Section 4 indicates the numerical predictions, which concern local and global deformation properties of Fe-Cu composites, and are compared with experimental results.

Notation. In the present work, 2nd- and 4th-order tensors are presented as $\boldsymbol{A}$ and $\mathbb{C}$, respectively. $\boldsymbol{A}^{-1}$, $\boldsymbol{A}^{\mathrm{T}}$ and $\dot{\boldsymbol{A}}$ indicate the inverse, the transpose and the material time derivative of the tensor $\boldsymbol{A}$. A linear mapping between 2nd-order tensors is written as $\mathbb{C}[\boldsymbol{A}]$. $\boldsymbol{A} \otimes \boldsymbol{B}$ is the dyadic product of the tensors $\boldsymbol{A}$ and $\boldsymbol{B}$. $\|\boldsymbol{A}\|$ denotes the Frobenius norm of tensor $\boldsymbol{A}$.

2 Experiment

Iron-copper composites are produced from mixtures of iron and copper powders by powder metallurgy. Such powders have a purity higher than 99.9% and consist of spherical polycrystalline particles with a diameter less than 63μm. Such a production of iron-copper polycrystals follows three steps: the mixing, the pre-compression, and the final compression. The porosity of the composite is less than 1vol.%, and the samples are named after the volume fraction of the composition. Figure 1 shows the microstructure of the aforementioned Fe-Cu polycrystals (after composition) where the darker phase represents the iron. The stress-strain behaviour is studied by compressive tests which are performed on cylindrical samples (height: 9mm, diameter: 6mm) at room temperature with a constant strain rate $\dot{\varepsilon} = 10^{-4} s^{-1}$. The experiments have been performed by Commentz (2000). Detailed information concerning the experiment can be found in Commentz *et al.* (1999) and Commentz (2000). The $\sigma - \varepsilon$ curves are friction corrected with a friction coefficient $\mu = 0.235$. After the compression test (90% logarithmic plastic strain), the sample is ground and polished until its middle plane being parallel to the top and bottom surface is laid open. The texture measurement is accomplished on this middle surface.

Pole figures are measured for three reflections, namely {200}, {211} and {220} for the iron phase and {200} {220} {311} for the copper phase, by scanning the hexagonal grid (Matthies and Wenk 1992). The measured data are further processed in a 5×5 grid. A local resolved strain measurement is performed with a scanning electron

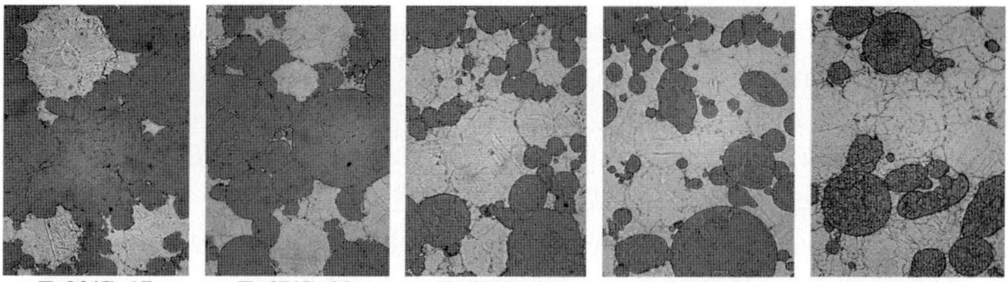

Fe83/Cu17 Fe67/Cu33 Fe50/Cu50 Fe33/Cu67 Fe17/Cu83

Fig. 1. Microstructures of Fe-Cu polycrystals with different phase volume fractions (Commentz 2000)

microscopic method (Commentz 2000) on a tensile sample of the Fe17-Cu83 composite. The measured cut-out is extracted from the middle plane (through the tension direction and the transverse direction) of the unloaded sample, and has a rectangular geometry with a dimension of $640\times480\mu m^2$ and $160\times120\mu m^2$ for the undeformed case and at large deformations, respectively.

3 Polycrystal Modelling

3.1 Constitutive Model

We denote the deformation gradient by $\boldsymbol{F}$. In the formulation of the elastic law, the 2nd Piola Kirchhoff stress tensor $\boldsymbol{S}$ is assumed to be a function of the right Cauchy Green tensor $\boldsymbol{C} = \boldsymbol{F}^{\mathrm{T}}\boldsymbol{F}$

$$\boldsymbol{S}=\boldsymbol{h}_p(\boldsymbol{C}), \tag{1}$$

where $\boldsymbol{h}_p$ is the actual elastic law. During the plastic deformation, $\boldsymbol{h}_p$ can be related to a reference law $\boldsymbol{h}_0$ by the isomorphy condition

$$\boldsymbol{h}_p(\boldsymbol{C}) = \boldsymbol{P}\,\boldsymbol{h}_0(\boldsymbol{P}^{\mathrm{T}}\boldsymbol{C}\boldsymbol{P})\,\boldsymbol{P}^{\mathrm{T}}, \tag{2}$$

where $\boldsymbol{P}$ is the plastic transformation (Bertram 2005). $\boldsymbol{F}_e$ can be defined as

$$\boldsymbol{F}_e := \boldsymbol{F}\boldsymbol{P}. \tag{3}$$

This leads formally to the same decomposition suggested, e.g., by Lee (1969), if we identify $\boldsymbol{F}_p$ by $\boldsymbol{P}^{-1}$, i.e. $\boldsymbol{F} = \boldsymbol{F}_e\boldsymbol{F}_p$. $\boldsymbol{F}_e$ indicates the elastic distortion, the dilatation and the rotations which also accounts for rigid body rotations. $\boldsymbol{F}_p$ consists of crystallographic slips along the slip systems $(\boldsymbol{d}_\alpha , \boldsymbol{n}^\alpha)$. The plastic incompressibility implies $\det(\boldsymbol{F}_p) = 1$. Since the elastic strains in our materials are small, we assume a linear relation between the 2nd Piola-Kirchhoff stress

$$\boldsymbol{S}_e = \det(\boldsymbol{F}_e)^{-1}\,\boldsymbol{F}_e^{-1}\,\boldsymbol{\sigma}\,\boldsymbol{F}_e^{-T}$$

with the Cauchy stress $\boldsymbol{\sigma}$ and Green's strain $\boldsymbol{E}_e^G$ in the undistorted configuration

$$\boldsymbol{S}_e = \mathbb{C}[\,\boldsymbol{E}_e^G\,] \text{ with } \boldsymbol{E}_e^G = \frac{1}{2}(\boldsymbol{C}_e - \boldsymbol{I}) \tag{4}$$

with a (constant) elasticity tensor $\mathbb{C}$, the elastic right Cauchy-Green tensor $\boldsymbol{C}_e = \boldsymbol{F}_e^T \boldsymbol{F}_e$, and the 2nd-order unit tensor $\boldsymbol{I}$. A tilde indicates that a quantity is formulated with respect to the undistorted configuration which is characterized by the fact that corresponding symmetry transformations are elements of $SO(3)$. The Kirchhoff stress tensor $\boldsymbol{\tau}^K$ is given by $\boldsymbol{\tau}^K = \boldsymbol{F}_e \boldsymbol{S}_e \boldsymbol{F}_e^T$.

The flow rule is taken from finite crystal visco-plasticity theory which specifies the time evolution of $\boldsymbol{P}$ in terms of the shear rate $\dot{\gamma}_\alpha$ and the Schmid tensors $\tilde{\boldsymbol{M}}^\alpha$ ($\alpha = 1 \ldots N$). The shear rates $\dot{\gamma}_\alpha$, the Schmid resolved shear stresses τ_α, and the Schmid tensors $\tilde{\boldsymbol{M}}^\alpha$ are given as

$$\dot{\gamma}_\alpha = \dot{\gamma}_0 \operatorname{sgn}(\tau_\alpha) \left| \frac{\tau_\alpha}{\tau_\alpha^C} \right|^m, \quad \tau_\alpha = \boldsymbol{C}_e \boldsymbol{S}_e \cdot \tilde{\boldsymbol{M}}^\alpha \approx \boldsymbol{S}_e \cdot \tilde{\boldsymbol{M}}^\alpha,$$

$$\tilde{\boldsymbol{M}}^\alpha = \tilde{\boldsymbol{d}}_\alpha \otimes \tilde{\boldsymbol{n}}^\alpha \tag{5}$$

respectively (Hutchinson 1976). The constant $\dot{\gamma}_0$ is the reference shear rate. A certain slip system α is specified by the slip direction $\tilde{\boldsymbol{d}}_\alpha$ and the slip plane normal $\tilde{\boldsymbol{n}}^\alpha$. For a given $\boldsymbol{L} = \dot{\boldsymbol{F}} \boldsymbol{F}^{-1}$, the flow rule can be formulated in terms of $\boldsymbol{F}_e$

$$\dot{\boldsymbol{F}}_e \boldsymbol{F}_e^{-1} = \boldsymbol{L} - \boldsymbol{F}_e \tilde{\boldsymbol{k}}\left(\boldsymbol{T}_e', \tau_\alpha^C\right) \boldsymbol{F}_e^{-1},$$

$$\tilde{\boldsymbol{k}}\left(\boldsymbol{T}_e', \tau_\alpha^C\right) =: \sum_{\alpha=1}^{N} \dot{\gamma}_\alpha \left(\boldsymbol{T}_e', \tau_\alpha^C\right) \tilde{\boldsymbol{M}}^\alpha, \tag{6}$$

where $\boldsymbol{T}_e = \boldsymbol{F}_e^T \tau^K \boldsymbol{F}_e^{-T}$ is the Mandel stress tensor. For $\boldsymbol{F}(0) = \boldsymbol{I}$, the initial condition of the differential Eq. (6) is given as $\boldsymbol{F}_e(0) = \boldsymbol{Q} \in SO(3)$, where $\boldsymbol{Q} = \boldsymbol{g}_i(0) \otimes \boldsymbol{e}_i$ is the initial orientation of the single crystal with the lattice vectors $\boldsymbol{g}_i$ and the orthonormal basis $\boldsymbol{e}_i$.

It is a reasonable assumption that the slip systems of fcc materials harden iso-tropically (Kocks and Mecking 2003) such that only one critical resolved shear stress τ^C appears in Eq. (6). For simplicity and limited by the experimental data, this concept is also applied to the iron phase of the Fe-Cu composites. Slip systems of copper and iron are chosen as <110>{111} and <111>{110} (Hartig and Mecking 2005), respectively.

The materials under consideration have been submitted to monotonous deformations (simple tension and compression) at room temperature. Under such conditions, the plastic deformation is characterized by the accumulation of dislocations in the crystal lattice. The Kocks-Mecking hardening rule, which emphasizes the mechanisms of the dislocation growth, the accumulation and the annealing is a suitable rule to be applied for materials in this work (Kocks and Mecking 2003).

The hardening rule used in the present work is

$$\dot{\tau}^C = \Theta_0 \left(1 - \frac{\tau^C}{\tau_c^\upsilon \left(\tau_\alpha, \tau^C \right)} \right) \dot{\gamma}\left(\tau_\alpha, \tau^C \right) , \dot{\gamma}\left(\tau_\alpha, \tau^C \right) = \sum_{\alpha=1}^{N} \left| \dot{\gamma}_\alpha \left(\tau_\alpha, \tau^C \right) \right|$$

$$\tau_c^\upsilon = \tau_{c0}^\upsilon \left| \frac{\dot{\gamma}\left(\tau_\alpha, \tau^C \right)}{\dot{\gamma}_0^*} \right|^{\frac{1}{n}} . \tag{7}$$

In Eq. (7), Θ_0, τ_{c0}^υ and $\dot{\gamma}_0^*$ are input material parameters which can be identified from experiments. N is the number of slip systems. $\dot{\gamma}_0^*$ is a material constant which is in agreement with the order of magnitude expected from the dislocation theory. n denotes the stress exponent. Detailed information on work-hardening can be found in Kocks (1976) and Kocks and Mecking (2003).

3.1.1 Homogenisation of Stresses

The above constitutive equations are suitable for a single crystal. Materials in this work are assumed to be free of pores and cracks. To describe the transition from the micro to the macro variables, we apply the representative volume element (RVE). Macro fields are determined through homogenizing the corresponding micro fields by appropriate averages over the RVE. The effective Kirchhoff stress tensor is given by the volume average over the reference volume V

$$\overline{\tau}^K = \frac{1}{V} \int_V \tau^K dV . \tag{8}$$

3.2 Finite Element Simulation

To predict the deformation behaviour of Fe-Cu polycrystals, an axisymmetric model is used in the finite element simulation which is well applicable for complex structures, inhomogeneities, and anisotropies. Here, the cross-sections of axisymmetric models are identified from the cut-outs of real microstructures, see Figure 2 and 3. In order to emphasize the interaction among the grains, the FE net is generated by the public domain software OOF (NIST, 2003). The numerical simulations are performed by ABAQUS (ABAQUS/Standard (2003)). The volume fraction of the iron phase is 22% and 49% in the simulation for the mentioned two composites. The element type applied in the simulation is CGAX3H which belongs to the generalized axisymmetric solid element of ABAQUS. Homogeneous boundary conditions are used. Schneider (2008) gives more information about the FE modelling and the applied material parameters.

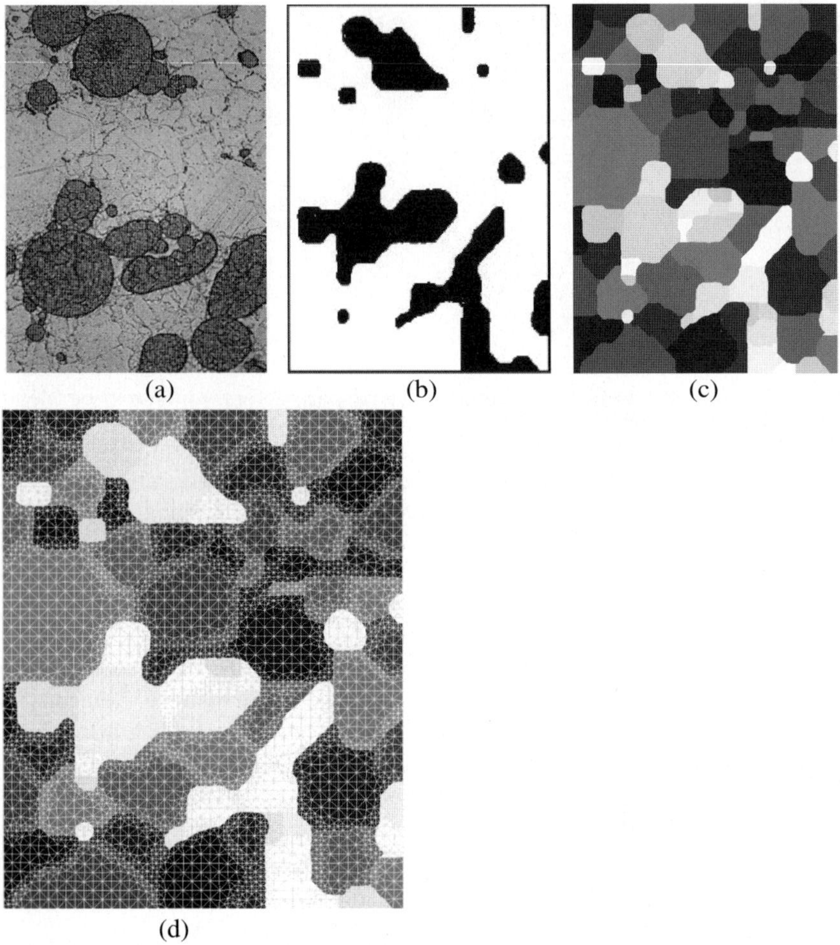

Fig. 2. Real microstructure (a), identified microstructure with the darker phase Fe (b), identified grains (c), finite element net with the refined mesh on the grain boundary (d) of the Fe17-Cu83 composite

4 Results and Discussion

4.1 Stress-Strain Flow Behaviour

Since the microscopic stress-strain behaviour can be modified by the anisotropy of the grain orientation distribution, the number of grain orientations should be representatively enough to predict the stress-strain curves which are comparable with the experimental ones.

In Figure 4, the stress-strain curve is averaged from 18 calculations with different initial grain orientations for the Fe17-Cu83 composite. Since, as already mentioned in

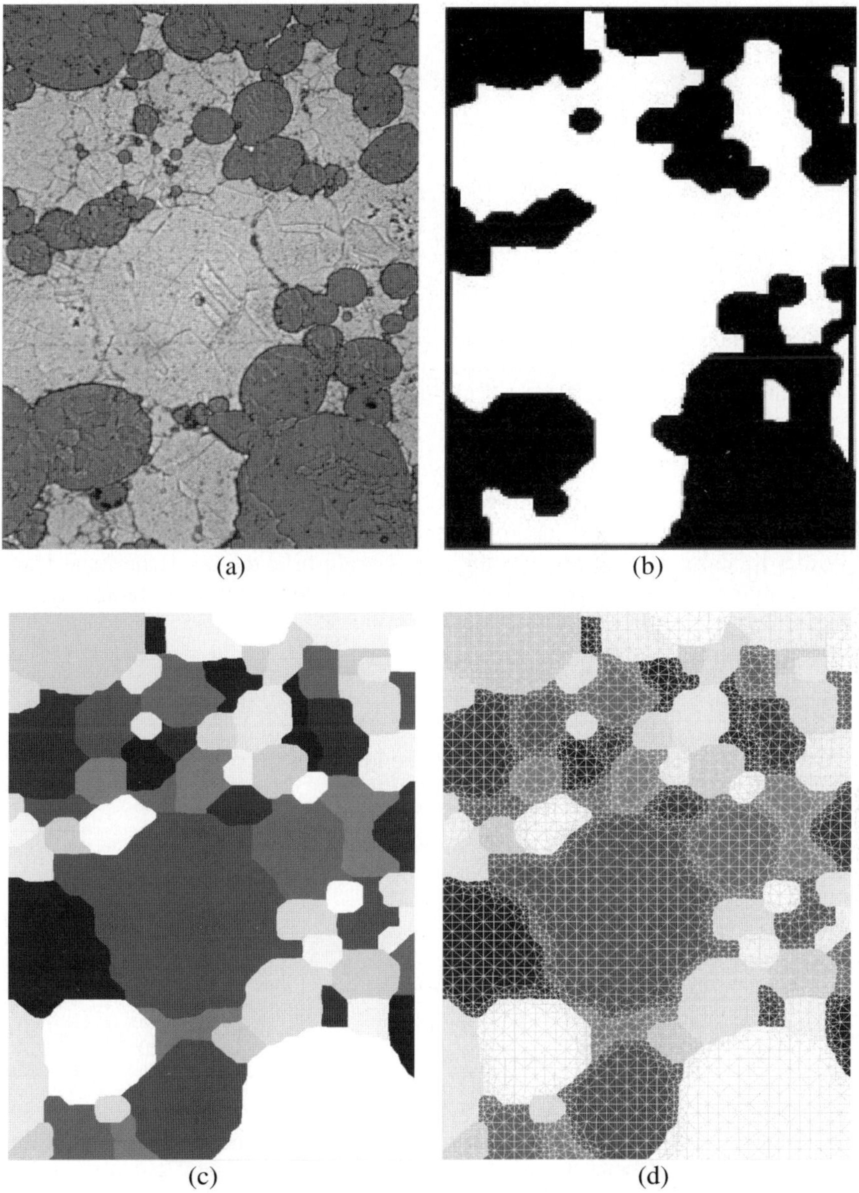

Fig. 3. Real microstructure (a), identified microstructure with darker phase Fe (b), identified grains (c), finite element net with the refined mesh on the grain boundary (d) of the Fe50-Cu50 composite

Section 3.2, the volume fraction of the Fe phase is about 22% in the axisymmetric simulation, it is reasonable that the simulated stress-strain curve is located between the experimental ones of the Fe33-Cu67 and the Fe17-Cu83 composite.

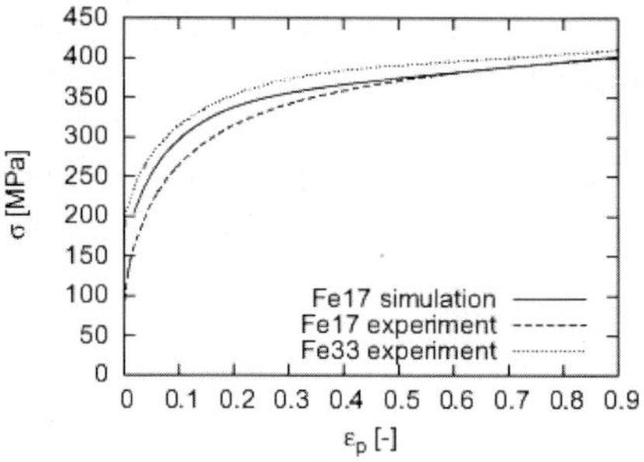

Fig. 4. Stress-strain curves of simulation and experiment for Fe17-Cu83 composite till ε_p=90% under simple compression load

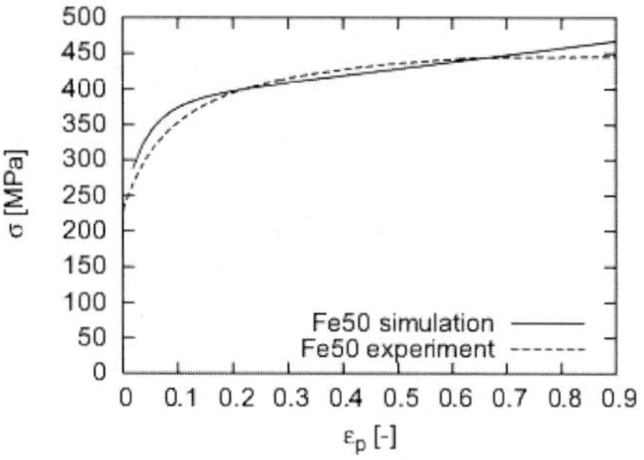

Fig. 5. Stress-strain curves of simulation and experiment for Fe50-Cu50 composite till ε_p=90% under simple compression load

Figure 5 compares the stress-strain curves of the experiment and the FE prediction which is an average of 22 simulations with different initial grain orientations. In the Fe50-Cu50 case, the simulated curve matches also well the experimental one.

In order to show the influence of the initial grain orientation on the local flow behaviour, Figure 6 and 7 exhibit the normalized stress-strain curves obtained by

$$\sigma_N = \frac{\sigma_{\text{max/min}}(t)}{\sigma_{\text{aver}}(t)} \,, \quad \sigma_{\text{aver}} = \frac{\sum_{i=1}^{n} \sigma_i}{n} \qquad (9)$$

with n=18 for the Fe17-Cu83 and n=22 for the Fe50-Cu50 composite.

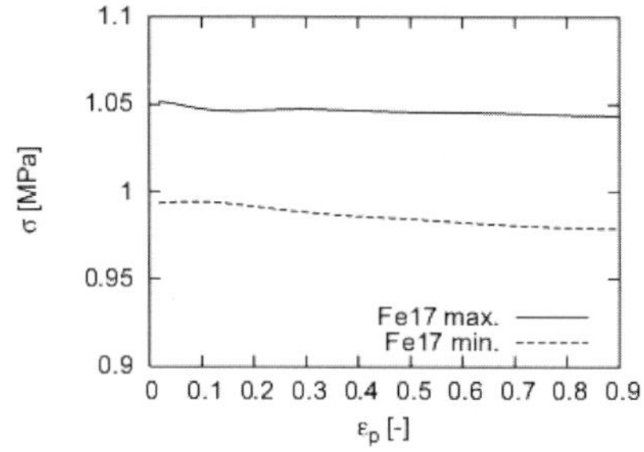

Fig. 6. Minimum and maximum stress-strain curves among 18 simulations for the Fe17-Cu83 composite; stress is normalized by averaged stress (Figure 4)

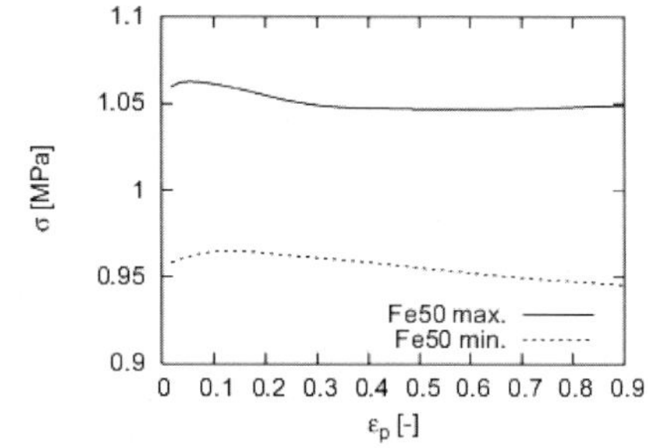

Fig. 7. Minimum and maximum stress-strain curves among 22 simulations for the Fe50-Cu50 composite; stress is normalized by averaged stress (Figure 5)

σ_N presents the normalized stress. The maximum stress and the minimum stress are given by the simulated stress-strain curves which are located at the highest and the lowest position among all the simulations, correspondingly. The deviation from the averaged stress-strain curve is about 1% to 2% for the minimum stress-strain curve (dashed line in Figure 6) of the Fe17-Cu83 composite while it is about 5% for the maximum stress-strain one. This 5% deviation remains true for the maximum and minimum stress-strain curves for the Fe50-Cu50 composite (Fig. 7).

Different orientation distributions of grains may cause 5% fluctuation for the local stress-strain curves, if a relatively small number of grains is considered in the microstructure in the axisymmetric simulation. According to Cheong and Busso (2006), such number of grains is between 10 to 100.

Fe-Cu composites exhibit an increased stiffness (compared to the pure copper) and ductility (compared to the pure iron). The phase-stress partition, i.e., the softer phase transfers the load to the harder phase, is the reason for the former property. Figure 8 and 9 show the stress flow in each phase for the Fe17-Cu83 and the Fe50-Cu50 composite, respectively. σ_{phase} and σ present the stress for the Fe or the Cu phase and the total stress, whereas both stresses are the averaged values from 3 simulations. These three simulations are chosen from the 18 (Fe17-Cu83) and the 22 (Fe50-Cu50) calculations. For a given strain, the stresses predicted by two of these 3 simulations give the largest and the smallest stress, respectively. The location of the third stress-strain curve is nearest to the averaged (stress strain) curve among the 18 (Fe17-Cu83) and the 22 (Fe50-Cu50) calculations. We define σ_{phase}/σ as the normalized phase stress. At the beginning of the yielding, the strength of the iron phase decreases fast, and that of the copper phase exhibits the reverse effect for both composites (Figure 8 and 9), whereas the rate of this decrease or increase is higher for the Fe17-Cu83 composite than for the Fe50-Cu50 composite. This trend diminishes with the increase of the plastic strain.

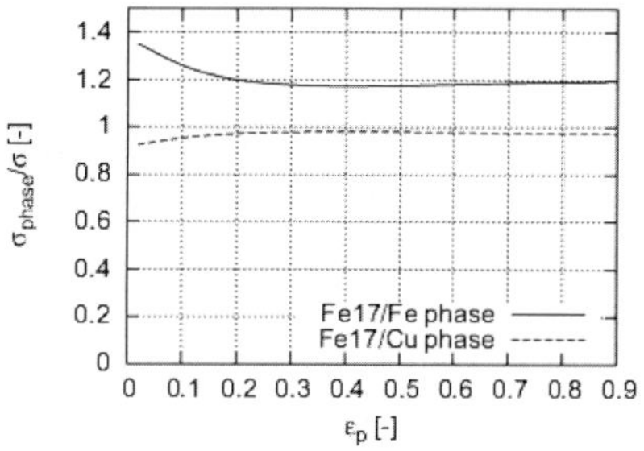

Fig. 8. Normalised stress-strain curves of Fe and Cu phases for Fe17-Cu83 composite until plastic strain ε_p=90%

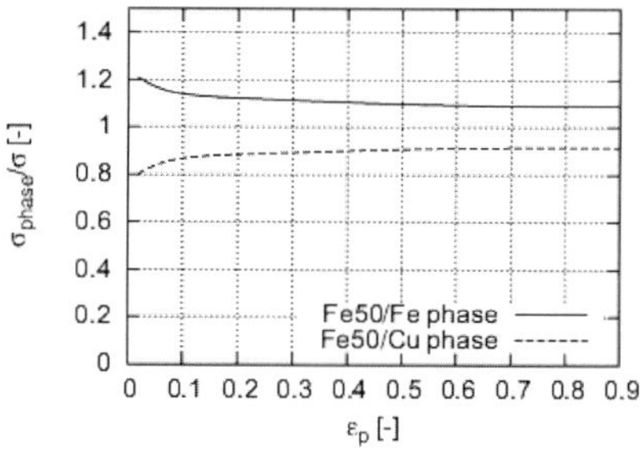

Fig. 9. Normalised stress-strain curves of Fe and Cu phases for Fe50-Cu50 composite until plastic strain ε_p=90%

At last, the normalized phase stress converges to a certain value which varies according to the ratio of (Fe:Cu) phase volume fractions. At the beginning of the yielding of the composite, it is observed that the iron phase starts to yield partly in experiments (Commentz 2000). On the other hand, some iron phase is still in the elastic-plastic transition up to higher macroscopic strains (Commentz *et al.* 1999). This means that the plastic deformation of the iron phase takes place step by step. If the harder phase yields, the load will be transferred back to the softer phase (Daymond *et al.* 2005). As a result, the stress of the iron phase decreases, while the stress of the copper phase shows the opposite effect with an increase of the plastic strain. The observations in the experiment prove that the numerical results in Figure 8 and 9 are reasonable for the phase stress flow behaviour.

4.2 Crystallographic Texture

The crystallographic textures are presented as inverse pole figures, which are obtained from 18 and 22 calculations for Fe17-Cu83 and Fe50-Cu50 composites, respectively. The iron-phase texture presented as the standard inverse pole figure is indicated in Figure 10 for both the simulation and the experiment at 90% plastic strain under compression load.

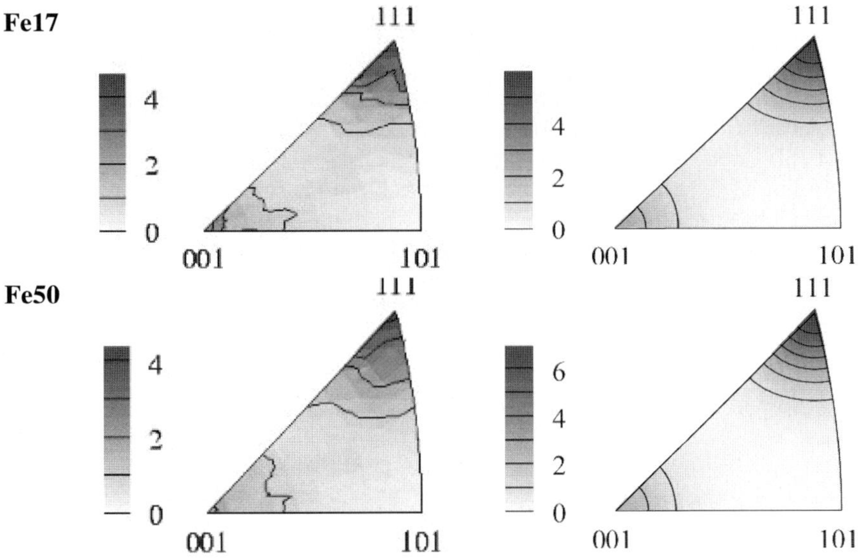

Fig. 10. Texture (inverse pole figure) of Fe phase from axisymmetric FE experiments (left) and simulations (right) at ε_p=90% for Fe17-Cu83 and Fe50-Cu50 composite

Two fibre components, <100>-fibre and <111>-fibre are observed, which are typical for compressively deformed α-Fe (Kocks *et al.* 1998). Such a fibre texture is well captured by the simulations for both the Fe17-Cu83 and the Fe50-Cu50 composite. The prediction also demonstrates the difference of the fibre intensity due to the phase volume variation, even though the maximum value of the fibre intensity is slightly higher in the Fe50-Cu50 than in the Fe17-Cu83 composite. While the experiment shows approximately the same fibre intensity. Since each inverse pole figure includes more than 600 orientations which are initially assigned randomly to the grains, the effect of the local grain orientations on the texture can be neglected. In this case, the local interaction and the phase arrangement are two major factors influencing the texture evolution in the simulation. Due to the limited number of grains in both real microstructures considered, the influence of the local interaction on the texture is a minor factor. The local morphology and the restricted confinement (homogeneous

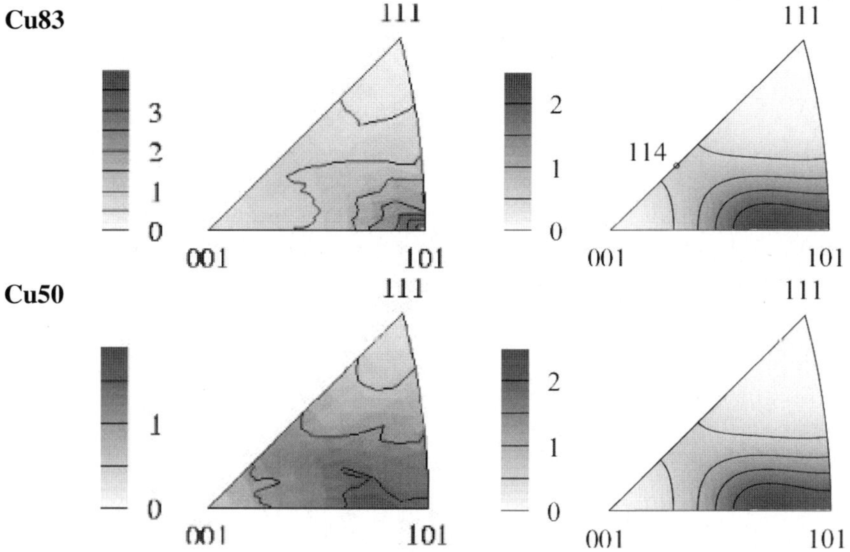

Fig. 11. Texture (inverse pole figure) of Cu phase from axisymmetric FE experiments (left) and simulations (right) ε_p=90% for Fe17-Cu83 and Fe50-Cu50 composite

boundary condition, in particular for the small cut-outs of the real microstructure) could be reasons for the sharper textures of the simulations than the real ones. The particle distribution of the Fe50-Cu50 composite may result in a higher fibre-intensity than that of the Fe17-Cu83 one, since such distributions lead to an even stiffer material structure in which four large iron particles are located on the structure boundaries.

The simulated and the experimental texture of the copper phase is given in Figure 11 for the Fe17-Cu83 and the Fe50-Cu50 composite. In the simulated textures, the pronounced <110>-fibre develops to the <411> direction. A rather soft texture of the Cu phase is presented by the measurement for the Fe50-Cu50 composite, which is not well simulated by the FE model. Besides the above mentioned reasons, i.e., the local interaction, the limited material structure, and the boundary conditions, the present model may not be able to catch the local change of the activation of the slip systems rapidly enough due to the isotropic hardening assumption.

4.3 Local Strain Distribution

The anisotropy of the constituent grains causes high heterogeneities in the local strain field when the polycrystalline structure is under load. We analysed the local strain for both the iron and the copper phase at a 19.8% tensile plastic strain. Figure 12 presents the histogram of the local plastic strain for the iron and the copper phase in the simulation (left column) and the experiment (right column, Commentz (2000)), where LD, TD, and LD/TD indicate the loading, the transverse, and the shear direction, respectively. Both the distribution and the mean value of the plastic strain match the reality well for the iron phase in all the mentioned directions. The strain of the iron phase behaves more heterogeneously in the loading direction (LD) than in the shear direction

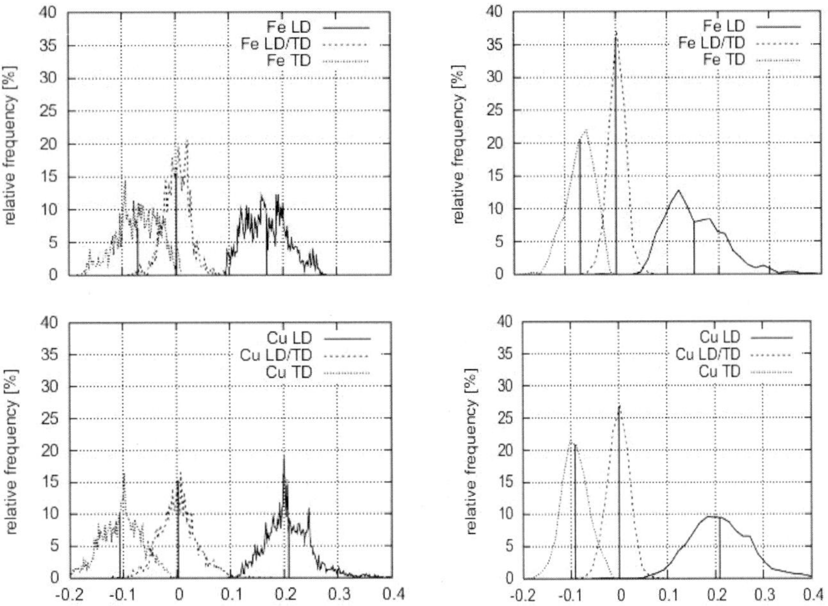

Fig. 12. Strain distribution of the Fe and the Cu phase in the axisymmetric FE simulation (left column) and the experiment (right column) for the Fe17-Cu83 composite at the tensile strain ε_p=19.8%

(LD/TD) due to the wider range and the larger oscillation of the distribution curve in the LD direction, while this non-homogeneity lies in the middle for the transverse direction (TD).

The mean value of the plastic strain is 16.99% in the loading direction for the simulation and 15.3% for the experiment. Both values are much smaller than the mean value of the composite (19.80%). These mean values are -7.1% and -7.0% for the transverse direction in the numerical and the experimental predictions, correspondingly. There is no obvious deviation of the mean value from the total one for the shear direction. Generally, the (absolute) mean value of the plastic strain for the harder iron-phase is less than that of the composite.

For the copper phase, the experiment (in Figure 12) clearly presents a wider range of the strain distribution in the loading direction than in the other two directions, i.e., there is more inhomogeneity in the LD direction. This property is well predicted by the numerical simulation. In the loading direction, the numerical curve even captures the second peak shown at approximately 27% plastic strain in the experiment. The mean value of the copper-phase plastic strain is 20.8% in the simulation and 20.9% in the experiment. The corresponding results for the TD directions are -10.6% for the FE prediction and -9.1% in the experiment, respectively. Like in the iron phase, this value is still approximately zero for the shear direction (LT/TD). Contrary to the harder phase, the mean value of the plastic strain for the Cu phase is higher than that of the total composite. This corresponds to the general conclusion for the two-phase polycrystal mechanical behaviour, i.e., the softer phase burdens more deformation than the harder one.

5 Summary

In order to understand the mechanical behaviour of the Fe-Cu composites and, particularly, the coupling of the microscopic and the macroscopic deformation behaviour under large plastic deformations, axisymmetric simulations have been performed by the finite element software ABAQUS. The elasto-viscoplastic material model has been applied. Material structures are modelled based on real microscopic cut-outs, in which regions near grain boundaries are finer meshed than other parts. Two composites, Fe17-Cu83 and Fe50-Cu50, are taken as representative microstructures which are investigated in detail. From the investigations of the local deformation, the flow behaviour, the texture, and the distribution of the strain, we draw the following conclusions.

In the case that the local stress is influenced only by the initial grain orientations in a given microstructure, i.e., all other conditions are kept identical in the simulations, a deviation of 5% to 10% can be observed for the local stress normalized by the macroscopic stress.

The stress-strain behaviour is well captured for both Fe-Cu composites mentioned before. The stress of each phase is sensitive to the amount of the plastic deformation at the early stage of the yielding. The stress in the iron phase decreases fast, and that in the copper phase shows the reverse effect.

The simulated texture of the iron phase describes well the fibre type texture and the (maximum) fibre intensity variation according to the volume change of phases. The predicted texture of the Cu phase also captures the experimental fibre texture in the Fe17-Cu83 composite. Due to the limited number of grains in the microstructure for the simulation, the Fe50-Cu50 composite presents a higher fibre intensity than the Fe17-Cu83 composite, while no obvious difference is shown in reality.

Concerning the distribution of the plastic strain (Fe17-Cu83 composite), the mean value of the strain in both the harder phase and the softer phase presents a deviation from the total mean value in the normal and the transverse direction. The copper phase undergoes larger deformations than the iron phase in the composite. These properties are well predicted by the axisymmetric simulation.

The mean value of the strain is quantitatively well predicted for both the iron and the copper phase.

Acknowledgement. This project has been supported by the Deutsche Forschungsgemeinschaft (DFG) under the grant BE 1455/10.

References

[1] Bertram, A.: Elasticity and Plasticity of Large Deformations - an Introduction, 2nd edn. Springer, Heidelberg (2005, 2nd edn. 2008)
[2] Cheong, K., Busso, E.: Effects of lattice misorientations on strain heterogeneities in FCC polycrystals. 54, 671–689 (2006)

[3] Commentz, B.: Plastische Verformung von zweiphasigen Eisen-Kupfer-Verbundwerkstoffen. In: Dissertation. Technische Universität Hamburg-Harburg. Shaker Verlag (2006)
[4] Commentz, B., Hartig, C., Mecking, H.: Micromechanical interaction in two phase iron-copper polycrystals. Comp. Mat. Sci. 16, 237–247 (1999)
[5] Daymond, M.R., Hartig, C., Mecking, H.: Interphase and intergranular stress generation in composites exhibiting plasticity in both phases. Acta. Mater. 53, 2805–2813 (2005)
[6] Hartig, C., Mecking, H.: Finite element modelling of two-phase Fe-Cu polycrystals. Comp. Mat. Sci. 32, 370–377
[7] Hutchinson, J.: Bounds and self-consistent estimates for creep of polycrystalline materials. Proc. R. Soc. Lon. A 348, 101–127 (1976)
[8] Kocks, U.: Laws for work-hardening and low-temperature creep. J. Eng. Mater. Techn (ASME) 98, 75–85 (1976)
[9] Kocks, U., Mecking, H.: Physics and phenomenology of strain hardening: the FCC case. Progr. Mat. Sci. 48, 171–273 (2003)
[10] Kocks, U., Tome, C., Wenk, H.: Texture and Anisotropy: Preferred Orientation in Polycrystals and Their Effect on Materials Properties. Cambridge Univ. Pr., Cambridge (1998)
[11] Matthies, S., Wenk, H.: Optimization of texture measurements by pole figure coverage with hexagonal grids. Phys. Stat. Sol. (a) 133, 253–257 (1992)
[12] NIST: OOF: Object-Oriented Finite Element Analysis of Real Material Microstructures Working Group. ppm2oof1.1.24. NIST: National Institute of Standards and Technology (2003), http://www.ctcms.nist.gov/oof
[13] Raabe, D., Heringhaus, F., Hangen, U., Gottstein, G.: Investigation of a Cu-20 mass% Nb in situ composite Part I: fabrication. microstructure and mechanical properties. Z.Metallkd 86(6), 405–415 (1995)
[14] Schneider, Y.: Simulation of the Deformation Behaviour of Two-phase Composites. Dissertation, Fakultät für Maschinenbau, Otto-von-Guericke-Universität Magdeburg (2008)
[15] Soppa, E., Amos, D., Schmauder, S., Bischoff, E.: The influence of second phase and/or grain orientations on deformation patterns in a Ag polycrystal and in Ag/Ni composites. Comp. Mater. Sci. 13, 168–176 (1998)
[16] ABAQUS/Standard, Hibbitt, Karlsson & Sorensen, Inc. (2003)

Regularisation of the Schmid Law in Crystal Plasticity

S. Borsch[1] and M. Schurig[2]

[1] Institut für Mechanik, Otto-von-Guericke-Universität Magdeburg
[2] Federal Institute for Materials Research and Testing (BAM), Berlin

Abstract. Plastic deformations of crystals are governed by a multitude of slip systems resulting in a yield locus with corners and an irregular flow rule. To overcome these difficulties, a regularization is needed resulting in a single equation that gathers the contributions of the distinct slip systems. We investigate two complementary approaches (regularization of the yield locus, and the plastic potential, respectively). The main result is a coincidence of the approaches only for uniform slip system activity which is the case in polyslip for similar hardening states.

1 Introduction

In crystals, plastic deformation is mainly determined by dislocation movement along glide planes. The gross effect of many glide events can be modelled by the slip rates of the crystallographic slip systems. These can be seen as independently acting. Accordingly, the multi-surface theory of plasticity has been well established (Koiter 1953). It uses the formalism of the theory of plasticity for each single slip system, adding up the effect. Accordingly, a combined flow rule that contains in its symmetric and anti-symmetric parts both the plastic strain as well as the plastic spin.

In the rate-independent theory, a system of equation results, which in its solution gives the slip rates. But the decomposition of an arbitrary volume preserving deformation into a number of shear modes is in general not unique. Accordingly, this problem can be ill-posed. A solution can be found by use of a generalized inverse, see (Anand and Kothari 1996; Miehe et al. 1999; Knockaert et al. 2000).

Another solution is the adoption of a rate-dependent theory, resulting in a constitutive law for the slip rates, usually by means of a power law. The ambiguity is hereby resolved. Rate-insensitive behaviour is often modelled by a high exponent (Peirce et al. 1983; Asaro 1983; Kocks et al. 1998).

It is also possible to combine the multiple yield surfaces of the slip systems in a single yield surface of a regularized theory. Instead of using the first slip system to be active as indicator for the onset of yield (a kind of Maximum or L_∞-norm) which results in a multi-facet compound yield locus. Mollica and Srinivasa (2002) use an L_ψ-norm to combine several segments of a yield function for sheet metals into a single yield function.

Such a way has been gone independently by Gambin (1991) and Arminjon (1991). A single yield function that can be interpreted as a kind of L_ψ-norm is used, resulting in areas where more than one slip system is simultaneously active. This is not restricted to corners, like in the intersection of the slip surfaces of multiple mechanisms, but in a strict sense nearly everywhere.

Besides the regularization of the yield locus, also the plastic potential that determines the flow rule, can be regularized in a very similar fashion. This has been proposed by Kiryk and Petryk (1998).

Both approaches with an associated flow rule result in models that either approximate the elastic domain or the flow rule. A third approach that approximates both at the cost of ending with a non-associated flow rule has been proposed by Schurig and Bertram (2003). In this paper, we discuss these three approaches.

Tensors are denoted by boldface capital letters (**A**), forth order tensors like $\mathbb{C}$. The partial derivative with respect to **A** is written as $\partial_{\mathbf{A}}$, the tensor product as $\mathbf{A} \otimes \mathbf{B}$, the composition of tensors has no extra symbol, $\mathbf{AB}$. The mapping of a tensor by a fourth order tensor is written as $\mathbb{C}[\mathbf{A}]$, a dot symbolizes the scalar product of the according vectors or tensors, $\mathbf{A} \cdot \mathbf{B}$.

2 Framework of Non-hardening Crystal Plasticity

In the context of elastoplastic materials with isomorphic elastic ranges, see Bertram (1999), we use an elastic reference law in the undistorted state for the Kirchhoff stress tensor **JT** and the right Cauchy-Green tensor $\mathbf{C} = \mathbf{F}^{\mathrm{T}}\mathbf{F}, \mathrm{J} = \sqrt{\det \mathbf{C}}$, where **F** is the deformation gradient, $\tilde{\mathbb{C}}$ is a constant fourth order tensor containing the single crystal constants,

$$\mathbf{JT} = \mathbf{FP}\,\tilde{\mathbb{C}}\left[\frac{1}{2}\left(\mathbf{P}^{\mathrm{T}}\mathbf{CP} - \mathbf{1}\right)\right]\left(\mathbf{FP}\right)^{\mathrm{T}} \tag{1}$$

The unimodular inelastic transformation **P** contains plastic effects and enters the elastic reference law only through the elastic transformation $\tilde{\mathbf{F}} := \mathbf{FP}$. There exists an imbedding into theories based on multiplicative decomposition of the deformation gradient **F** since the identity

$$\mathbf{F} = \tilde{\mathbf{F}}\mathbf{P}^{-1} \tag{2}$$

holds (Bertram 2005).

The ratio $\left|\frac{\tau_\alpha}{\tau_\alpha^c}\right|$ indicates whether or not a certain slip system is active. τ_α is the resolved shear stress of a slip system α and τ_α^c its critical value. A pull-back operation by $\tilde{\mathbf{F}}$ leads to $\tilde{\mathbf{C}} = \tilde{\mathbf{F}}^{\mathrm{T}}\tilde{\mathbf{F}}, \tilde{\mathbf{S}} = \tilde{\mathbf{F}}^{-1}\mathbf{T}\tilde{\mathbf{F}}^{-\mathrm{T}}$ and

$$\tau_\alpha = \tilde{\mathbf{C}}\tilde{\mathbf{S}} \cdot \tilde{\mathbf{M}}_\alpha \tag{3}$$

where $\tilde{\mathbf{M}}_\alpha = \tilde{\mathbf{d}}_\alpha \otimes \tilde{\mathbf{n}}^\alpha$ is the (constant) Schmid tensor with slip direction $\tilde{\mathbf{d}}_\alpha$ and slip plane normal $\tilde{\mathbf{n}}^\alpha$ fixed to the crystal lattice. The hardening state is described by the vector $\mathrm{g} = \left\{\tau_1^c, \ldots, \tau_N^c\right\}$, and the dual hardening variable z.

If a set of N independent mechanisms potentially contribute to plastic flow, an additive relation is usually assumed for the rates (Koiter 1953)

$$\mathbf{P}^{-1}\dot{\mathbf{P}} = \sum_{\alpha=1}^{N} \hat{\mathbf{\Pi}}_\alpha \qquad \dot{z} = \sum_{\alpha=1}^{N} \hat{z}_\alpha \tag{4}$$

For each mechanism, a convex yield criterion

$$\phi_\alpha\left(\tilde{\mathbf{C}}\tilde{\mathbf{S}}, g\right) \leq 0 \tag{5}$$

is used. (5) are weak inequality constraints for (4). The elastic domain is defined by the intersection,

$$\mathcal{E} := \left\{\tilde{\mathbf{C}}\tilde{\mathbf{S}} \mid \phi_\alpha\left(\tilde{\mathbf{C}}\tilde{\mathbf{S}}, g\right) \leq 0, \alpha = 1 \ldots N\right\}. \tag{6}$$

The intersection of convex domains is also convex (see, Boyd and Vandenberghe (2004)). For simplicity, we assume non-hardening plasticity with constant g. The contributions to the flow rule are obtained from the normality to a plastic potential,

$$\hat{\mathbf{\Pi}}_\alpha = \lambda_\alpha \partial_{\tilde{\mathbf{C}}\tilde{\mathbf{S}}} \psi_\alpha. \tag{7}$$

A further restriction can be obtained by assigning $\psi_\alpha = \phi_\alpha$, an assumption often made that comes out of the maximum dissipation principle (Bishop and Hill 1951a,b), which is widely used in the literature (Chin and Mammel 1969; Lubliner 1990, Section 3.2.2; Neff 2000; Nguyen 2000; Miehe et al. 2002).

It is an equivalent variational formulation of the consistency condition and the normality rule. Thus it is a geometrical and not a thermodynamical principle as the name seems to indicate. The problem can be formulated as follows,

$$\begin{aligned} &\mathbf{P}^{-1}\dot{\mathbf{P}} = -\sum_{\alpha=1}^{N} \lambda_\alpha \partial_{\tilde{\mathbf{C}}\tilde{\mathbf{S}}} \psi_\alpha \\ &\dot{z} = -\sum_{\alpha=1}^{N} \lambda_\alpha \partial_g \psi_\alpha \\ &\lambda_\alpha \phi_\alpha = 0 \quad \text{if} \quad \alpha = 1 \ldots N, \\ &\lambda_\alpha \geq 0 \quad \text{if} \quad \alpha = 1 \ldots N, \\ &\phi_\alpha \leq 0 \quad \text{if} \quad \alpha = 1 \ldots N. \end{aligned} \tag{8}$$

Eq. (8) is a differential-algebraic system of equations (DAE) for the flow and the hardening rule. Using the maximum dissipation principle, it represents the Kuhn-Tucker conditions for the resulting unrestricted variational problem. In this case, $(8)_1$ contains the normality rule. Note that the minus sign in the flow rule (8) results from the definition of $\mathbf{P} = \mathbf{F}_p^{-1}$. The set of active inelastic mechanisms, is called the active set

$$\mathcal{A} := \left\{\alpha \mid \phi_\alpha = 0, \lambda_\alpha > 0\right\}. \tag{9}$$

By differentiation, the DAE can be solved

$$\lambda_\beta = \begin{cases} \sum_{\alpha\in\mathcal{A}} \left(g^{-1}\right)_{\beta\alpha} \mathbf{P} * \tilde{\mathbf{A}}_\alpha \cdot \dot{\mathbf{E}} & \beta \in \mathcal{A} \\ 0 & \beta \notin \mathcal{A} \end{cases} \tag{10}$$

using the consistency matrix

$$g_{\alpha\beta} = \frac{\partial \dot{\phi}_\alpha}{\partial \lambda_\beta} = \tilde{\mathbf{A}}_\alpha \cdot \tilde{\mathbf{C}}\ \partial_{\tilde{\mathbf{C}}\tilde{\mathbf{S}}} \psi_\beta + a_\alpha \partial_g \psi_\beta \tag{11}$$

and the definitions

$$\begin{aligned} \tilde{\mathbf{A}}_\alpha &= 2\mathrm{sym}\left(\partial_{\tilde{\mathbf{C}}\tilde{\mathbf{S}}} \phi_\alpha \tilde{\mathbf{S}}\right) + \tilde{\mathbb{C}}\left[\tilde{\mathbf{C}}\ \partial_{\tilde{\mathbf{C}}\tilde{\mathbf{S}}} \phi_\alpha\right] \\ a_\alpha &= \partial g \phi_\alpha \frac{\partial g}{\partial z}. \end{aligned} \tag{12}$$

Insertion in the flow rule (-)- -yields

$$\mathbf{P}^{-1}\dot{\mathbf{P}} = -\sum_{\alpha,\beta\in\mathcal{A}} \left(g^{-1}\right)_{\beta\alpha} \left(\partial_{\tilde{\mathbf{C}}\tilde{\mathbf{S}}} \psi_\beta\right) \otimes \mathbf{P} * \tilde{\mathbf{A}}_\alpha \left[\dot{\mathbf{E}}\right], \tag{13}$$

whereas the hardening rule (-)- -results in

$$\dot{z} = -\sum_{\alpha,\beta\in\mathcal{A}} \left(g^{-1}\right)_{\beta\alpha} \left(\partial_g \psi_\beta\right) \otimes \mathbf{P} * \tilde{\mathbf{A}}_\alpha \left[\dot{\mathbf{E}}\right], \tag{14}$$

In numerical solution procedures, the determination of $\mathcal{A}$ imposes some difficulties as the Index-2 DAE problem can be ill posed due to a non-uniqueness of the solution of the algebraic part. One possible remedy for that problem is to use a regularization.

Instead of using the multi-faceted elastic domain of (6), a single surface can be introduced, see (Gambin 1991; Arminjon 1991), (interacting slip systems model),

$$\Phi_{\text{iss}}\left(\tilde{\mathbf{C}}\tilde{\mathbf{S}}, g\right) = \frac{1}{n+1} \sum_\alpha \left(\phi_\alpha\left(\tilde{\mathbf{C}}\tilde{\mathbf{S}}, g\right) + 1\right)^n - 1. \tag{15}$$

The same can be formally done for the plastic potentials, see (Rice 1970; Kiryk and Petryk 1998), (viscoplastic potential model),

$$\psi_{\text{vp}}\left(\tilde{\mathbf{C}}\tilde{\mathbf{S}}, g\right) = \frac{1}{n+1} \dot{\gamma}_0 \sum_\alpha g_\alpha \left(\phi_\alpha\left(\tilde{\mathbf{C}}\tilde{\mathbf{S}}, g\right) + 1\right)^n - \omega_c. \tag{16}$$

A representation of such a regularization is shown in Fig. 1.

Both can be put in (13) if only a single mechanism is used. Of particular interest is the maximum dissipation case that results if for both, Φ and ψ the same choice out of Φ_{iss} and ψ_{vp} is used. Then by comparison with the multi-mode equations, results for the λ_α can be obtained as in Schurig (2006).

$$\lambda_\alpha^{\text{iss}} = \lambda \frac{1}{g_\alpha} (\phi_\alpha + 1)^n$$
$$\lambda_\alpha^{\text{vp}} = \lambda \dot{\gamma}_0 \frac{1}{g_\alpha} (\phi_\alpha + 1)^n \tag{17}$$

The constant $\dot{\gamma}_0$ can be dropped by suitable normalization. $(17)_1$ explicitly contains the absolute value of the hardening value, which is physically questionable. This deficiency is avoided in $(17)_2$, but at the price that possibly the elastic domain is not properly represented.

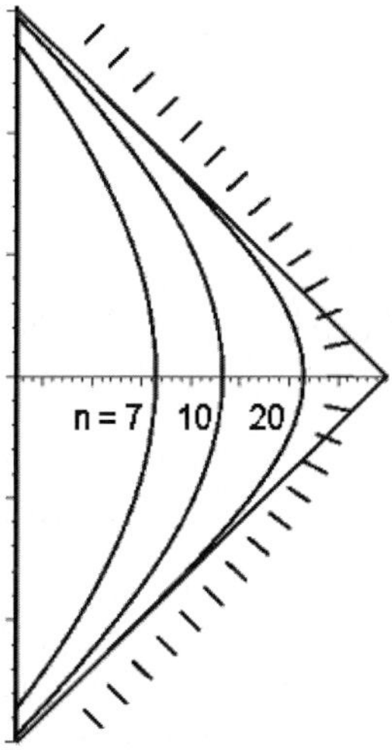

Fig. 1. Approximation of the multi mode yield locus by a single regularized function in a corner

A detailed study of Eq. (17) follows in the next section.

3 The Effect of Regularization in Polyslip Situations in Two Dimensions

3.1 Analytical Examination with a Reduction on Two Dimensions

Here, we present a comparison of the regularizations of Gambin's interacting slip systems (ISS) as well as the viscoplastic potential (VP) for a single time increment, starting out of the undistorted state $\mathbf{P} = 1$. We consider a material with two slip systems activated. A slip system, indexed α, is characterized by a pair of orthogonal vectors: the vector of the slip direction $\mathbf{d}_\alpha$, and the vector of the slip plane $\mathbf{n}^\alpha$. The orientation of both slip systems in the plane is fully described by the angle ψ, contrary in sign for each glide plane: $\psi_1 = -\psi_2$ (Fig. 2, right),

$$\tilde{\mathbf{d}}_1 = \begin{pmatrix} \sin(\psi) \\ \cos(\psi) \end{pmatrix} \quad \tilde{\mathbf{d}}_2 = \begin{pmatrix} -\cos(\psi) \\ \sin(\psi) \end{pmatrix} \quad \tilde{\mathbf{n}}^1 = \begin{pmatrix} -\sin(\psi) \\ \cos(\psi) \end{pmatrix} \quad \tilde{\mathbf{n}}^2 = \begin{pmatrix} \sin(\psi) \\ \cos(\psi) \end{pmatrix} \tag{18}$$

The Schmid tensors $\tilde{\mathbf{M}}_\alpha = \tilde{\mathbf{d}}_\alpha \otimes \tilde{\mathbf{n}}^\alpha$ can be expressed as

$$\begin{aligned} \tilde{\mathbf{M}}_1 &= \mathrm{sym}\begin{pmatrix} -\cos(\psi)\sin(\psi) & \cos^2(\psi) \\ -\sin^2(\psi) & \sin(\psi)\cos(\psi) \end{pmatrix} \\ \text{and } \tilde{\mathbf{M}}_2 &= \mathrm{sym}\begin{pmatrix} -\cos(\psi)\sin(\psi) & -\cos^2(\psi) \\ \sin^2(\psi) & \sin(\psi)\cos(\psi) \end{pmatrix} \end{aligned} \tag{19}$$

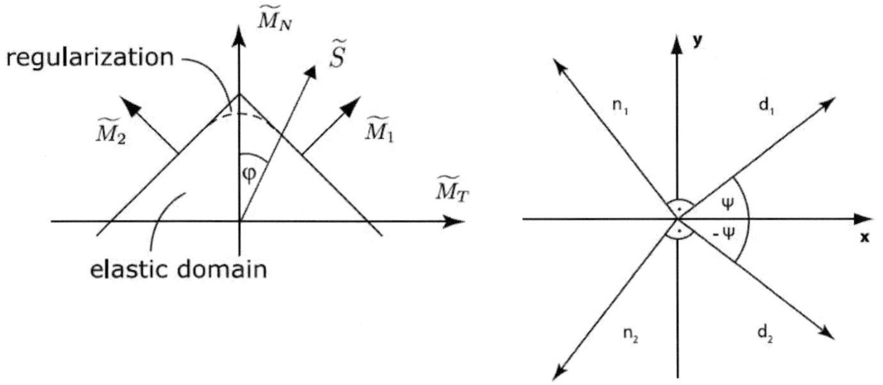

Fig. 2. Left: stress direction in 2D; right: orientation of slip systems

To create an orthonormal basis in the plane spanned by the two Schmid tensors, we introduce

$$\tilde{\mathbf{M}}_N = \frac{(T\tilde{\mathbf{M}}_1 + \tilde{\mathbf{M}}_2)}{\left\| T\tilde{\mathbf{M}}_1 + \tilde{\mathbf{M}}_2 \right\|} \qquad \tilde{\mathbf{M}}_T = b_1\tilde{\mathbf{M}}_1 + b_2\tilde{\mathbf{M}}_2 \tag{20}$$

including the ratio of the hardening parameters $T = \frac{x_1}{x_2}$, which is affected by the critical shear stress $\tau_\alpha^c = \tau_o x_\alpha$, where τ_o is reference shear stress. To accomplish the relations of orthogonality, namely $\tilde{\mathbf{M}}_N \cdot \tilde{\mathbf{M}}_N = \tilde{\mathbf{M}}_T \cdot \tilde{\mathbf{M}}_T = 1$ and $\tilde{\mathbf{M}}_T \cdot \tilde{\mathbf{M}}_N = 0$, we make use of the parameters b_1 and b_2. It follows that

$$b_1 = \left(\frac{1}{1 + p^2 - 2ap} \right)^{\frac{1}{2}} \quad \text{and} \quad b_2 = p b_1 \tag{21}$$

with $a := -(\sin^2(\psi) - \cos^2(\psi))^2 = \tilde{\mathbf{M}}_1 \cdot \tilde{\mathbf{M}}_2$ and $p = \dfrac{T-a}{Ta-1}$. The direction of the Piola-Kirchhoff stress tensor as well as of the rate of the right Cauchy-Green tensor in the plane can be fully characterized by a specified angle (Fig. 2, left).

$$\tilde{\mathbf{S}} = \cos(\varphi)\tilde{\mathbf{M}}_N + \sin(\varphi)\tilde{\mathbf{M}}_T \qquad \dot{\mathbf{C}} = \cos(\delta)\tilde{\mathbf{M}}_N + \sin(\delta)\tilde{\mathbf{M}}_T \qquad (22)$$

For isotropic elasticity, the first consistency parameter (10) can be summarized to

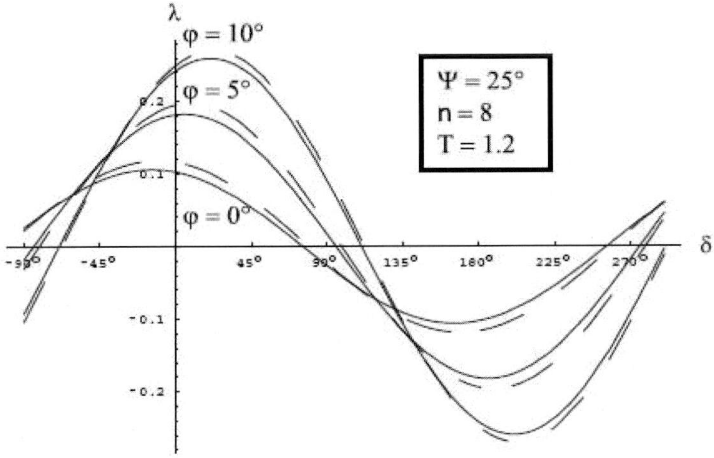

Fig. 3. Consistency parameter in dependence of the angle δ for different stress directions φ

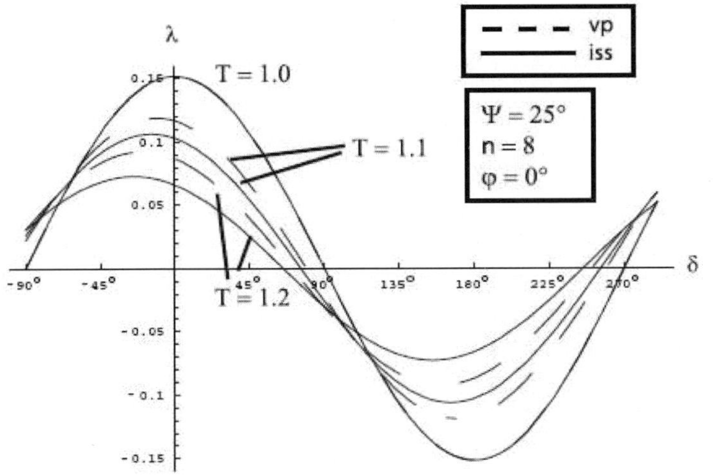

Fig. 4. Consistency parameter in dependence of the angle δ for different hardening ratios T

$$\lambda_1 = \frac{(\tilde{\mathbf{A}}_1 + \eta\tilde{\mathbf{A}}_2)\cdot\dot{\mathbf{C}}}{g_{11} + \eta(g_{12} + g_{21}) + \eta^2 g_{22}} \tag{23}$$

with

$$g_{ij} = \tilde{\mathbf{A}}_i \cdot \mathrm{sym}\tilde{\mathbf{M}}_j + h_{ij},$$
$$\tilde{\mathbf{A}}_i = 2G\,\mathrm{sym}\,\tilde{\mathbf{M}}_i$$

G as the shear modulus and $h_{\alpha\beta} = H(q - (q-1)\delta_{\alpha\beta})$ with the hardening modulus H and the latent hardening ratio q = 1.4. The differences caused by using the two distinct regularizations occur only in the parameter η:

$$\eta_{iss} = (T)^{n+1}\left(\frac{\tau_2}{\tau_1}\right)^n, \qquad \eta_{vp} = (T)^n\left(\frac{\tau_2}{\tau_1}\right)^n \tag{24}$$

Analysing (7), a dependence of the consistency parameter can be reduced to four variables. The angle Ψ is chosen with a constant value, reflecting the choice of a particular material. Accordingly, this parameter is dropped out of the following analysis. This procedure leaves three parameters (Tab.1)

$$\lambda_\beta = \lambda_\beta(\delta, \varphi, T) \tag{25}$$

that are investigated in the next section.

Table 1. Variables of the consistency parameter

T	ratio of the hardening parameters	$\frac{x_1}{x_2}$
δ	angle of the rate of the right Cauchy-Green tensor	$\dot{\mathbf{C}} = \cos(\delta)\tilde{\mathbf{M}}_N + \sin(\delta)\tilde{\mathbf{M}}_T$
φ	angle of the Mandel stress tensor	$\tilde{\mathbf{S}} = \cos(\varphi)\tilde{\mathbf{M}}_N + \sin(\varphi)\tilde{\mathbf{M}}_T$

3.2 Examination of the Parameters

The dependence of the consistency parameter λ_1 on the angle δ describing the direction of the right Cauchy-Green tensor $\dot{\mathbf{C}}$ is shown in Figure 3. The oscillating shape of the curve is due to equation (22). The major difference of the regularized approaches can be detected at the maxima and minima of the shown curves. The special case of identical hardening for both slip systems (x_1=x_2 $\Longrightarrow$ T=1, Fig. 4) leads to a coincidence of the regularizations, caused by generating an identical value for ηin (24). The analysed regularizations tend to coincide with increasing exponent n, contained in the approximations (15), (16) (Fig. 5, above).

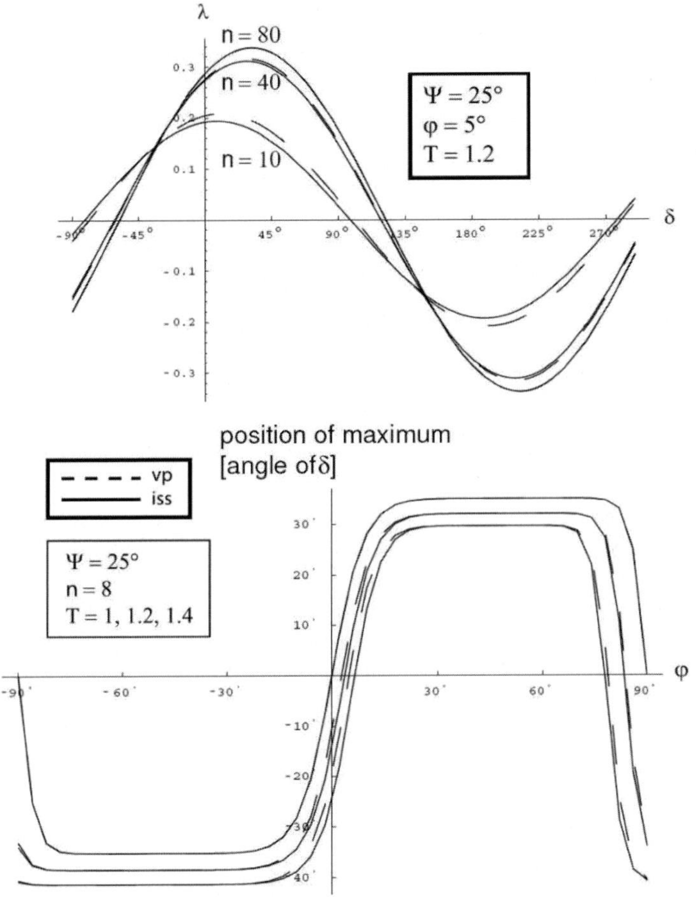

Fig. 5. Above: effect of increasing exponent; below: maximum location in dependence of the stress direction

The influence of the stress-direction φ on the maximum location, detected in Figs. 3 and 4 as a characteristic point, is shown in Figure 5 (below). The increasing/decreasing areas exhibit a difference of the investigated approaches, whereas a coincidence can be observed inside the plateau areas. A decoupling of the parameters φ and δover a range of about thirty degrees exists in the latter area. Allocating the decoupling to an unbalanced ratio of the shear stresses, a reduction of (7) to

$$\lambda_1 = \frac{\tilde{\mathbf{A}}_1 \cdot \dot{\mathbf{C}}}{\tilde{\mathbf{A}}_1 \cdot \mathrm{sym}\tilde{\mathbf{M}}_1 + h_{11}} \tag{26}$$

for the case $\tau_1 >> \tau_2 \Rightarrow \eta \sim 0$ is achieved.

Therefore, as an intermediate result achieved by interpreting Figure 5 (below), the following can be stated.

1. Deformation processes with shear stresses of similar amounts ($\tau_1 \approx \tau_2$) are assigned to stress directions pointing on intersection points of two slip systems (see Fig. 2 left), representing a vertex of the yield surface. Accordingly, the use of different regularizations has an effect in the surrounding area of a vertex point in the yield surface.
2. A prevailed activity of only one slip system, characterized by the dominance of one shear stress, is assigned to a stress direction pointing outside a surrounding area of a vertex point. Here the approaches coincide.

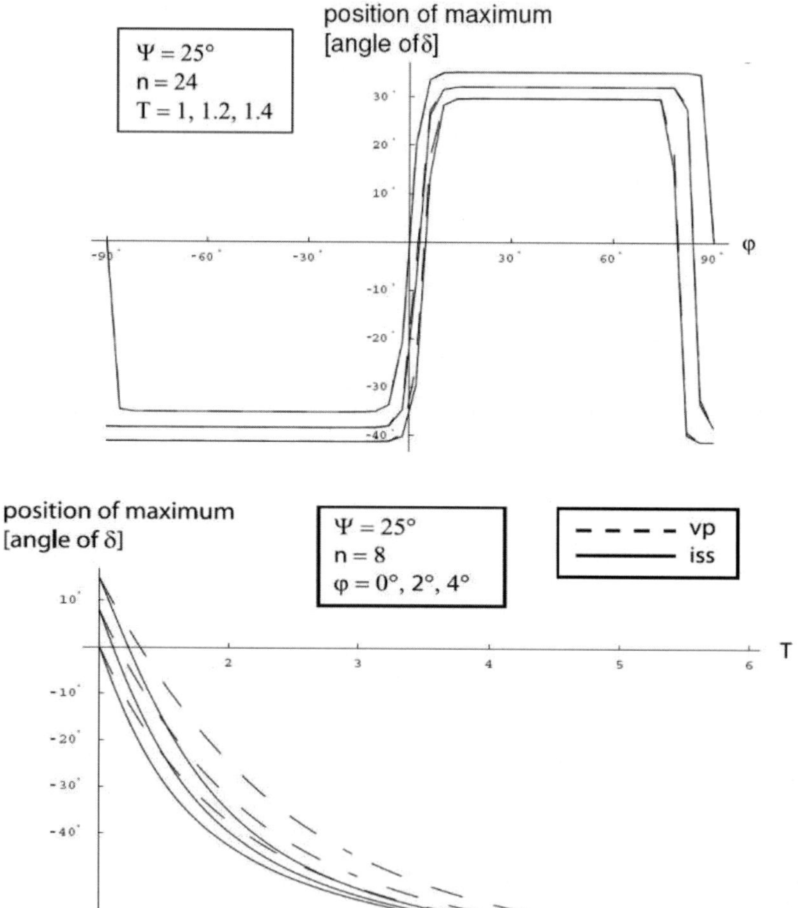

Fig. 6. Above: effect of increased exponent; below: maximum location in dependence of the hardening ratio

An increase of the exponent n contained in the regularizations (15) is generating a closer approximation of a vertex point in the yield surface, Fig. 1. This procedure generates a reduced surrounding area of a vertex point and therefore contracts the sloped range and increases the flat range in Fig. 6 (above).

The influence of the hardening ratio T on the maximum location (Fig. 4) is shown in the lower part of Figure 6. The coincidence of the approaches for identical hardening, here representing the starting point of the curve, has been mentioned before by consulting (24). Comparing the approaches for different stress directions, a restricted influence of the angle φ for an increasing hardening ratio can be noted. The mutual convergence of the curves for different regularizations, observed for a growing hardening ratio, is also caused by a dominated activity of only one slip system. Here the unbalanced activity is due to a massive difference in hardening.

As a conclusion of the analytical examination one can conclude the following.

1. During the dominant activity of one single slip system, perceivable differences of the two approaches can be found. Such a state is reached by a stress direction pointing out of the surrounding area of the vertex point. The size of this area in turn depends on the exponent n, for large values, it shrinks. Also an extreme outbalanced hardening ratio produces differences between the regularizations, even near a vertex. This can be established by the identification of pairs of the hardening ratio T and the stress direction φ. There always exists pairs that cause an effect by applying different regularizations, identifying the vertex point that is shifted by changing T.
2. In contrast to the first point, an effect due to different regularizations is eliminated by deformation processes causing a similar activity of both slip systems. Furthermore, the special case of identical hardening is always eliminating the effect of different regularizations, independent of the respective stress direction

3.3 Numerical Examination

For the numerical examination we consider an inelastic process with twelve octahedral slip systems and material properties as suitable for copper (Kocks 1970). The numerical approach of Miehe et al. (1999) is used,

$$\mathbf{Q} = -\frac{1}{\eta}\left[\left(\langle\Phi\rangle+1\right)^{m} - 1\right]\frac{\partial\psi}{\partial\tilde{\mathbf{C}}\tilde{\mathbf{S}}} \tag{27}$$

with the viscosity parameter $\eta = 1$ and the strain-rate-sensitivity exponent m = 5 for best numerical results. The flow rule Φ as well as the plastic potential Ψ are included and (27) can be subdivided in the consistency parameter $\lambda = -\frac{1}{\eta}\left[\left(\langle\Phi\rangle+1\right)^{m} - 1\right]$ and the flow normal $\frac{\partial\psi}{\partial\tilde{\mathbf{C}}\tilde{\mathbf{S}}}$.

Firstly, starting with the elastic process part, the ODE $\dot{\tilde{\mathbf{F}}} = \tilde{\mathbf{F}}\mathbf{L}_e$ with the ansatz

$$\mathbf{L}_e = \tfrac{1}{1000}\left(\cos(\beta_o)\mathbf{D}_1 + \sin(\beta_o)\mathbf{D}_2\right) \tag{28}$$

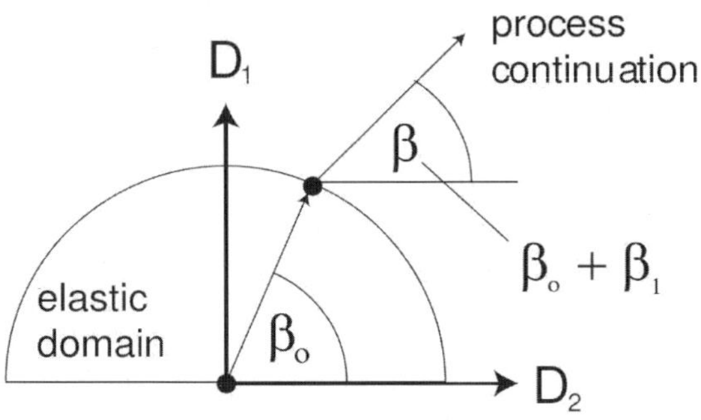

Fig. 7. Process directions during elastic and plastic process parts

for the velocity gradient is in use. Equation (28) includes the tensors

$$\mathbf{D}_1 = \frac{1}{\sqrt{1{,}5}}\begin{pmatrix} 1 & 0 & 0 \\ 0 & -\frac{1}{2} & 0 \\ 0 & 0 & -\frac{1}{2} \end{pmatrix} \qquad \mathbf{D}_2 = \frac{1}{\sqrt{2}}\begin{pmatrix} 0 & 1 & 0 \\ 1 & 0 & 0 \\ 0 & 0 & 0 \end{pmatrix}$$

as a symmetric shear, and a volume preserving tension, while the angle β_0 describes the process direction inside of the elastic domain (Fig. 7).

The plastic process part is described by the differential equations

$$\dot{\mathbf{F}} = \mathbf{F}\mathbf{L}_p, \quad \dot{\tilde{\mathbf{F}}} = \tilde{\mathbf{F}}\left(\mathbf{L}_p + \mathbf{Q}\right), \quad \dot{\mathbf{P}} = \mathbf{P}\mathbf{Q} \tag{29}$$

with the Maccauley bracket $\langle x \rangle = \frac{1}{2}\left(|x| + x\right)$, the velocity gradient $\mathbf{L}_p = \frac{1}{1000}\left(\cos(\beta)\mathbf{D}_1 + \sin(\beta)\mathbf{D}_2\right)$ analogously to the elastic process part and $\beta = \beta_0 + \beta_1$ with β_1 as the direction of the plastic process continuation (Fig. 7).

The relation $\mathbf{F} = \tilde{\mathbf{F}}\mathbf{P}^{-1}$ leads to the fact, that only two of the listed ODEs have to be solved. A combination of the Adams method and the BDF algorithm has been used in the solver LSODA, which is part of the ODEPACK (Hindmarsh 1983).

As a model process, the length of the plastic process part has been set equal to the elastic part.

The temporal evolution of the consistency parameter for the ISS and VP approach shows Figure 8 for different directions of the elastic process, where β_0 has been varied in 10-degree steps from 10 to 70 degrees. In the plastic range, the direction of strain is not changed, $\beta_1 = 0$. Besides the different scaling in amount and duration that is a consequence of the different approximation of the elastic domain, a qualitative match of the progressions can be noticed. A similar result can be found for processes that start straight into a corner of the yield locus and continue in different directions, $\beta_0 = 0$, β_1 is again varied in 10-degree steps from 10 to 70 degrees, see Figure 9. As expected, a deviation from the straight direction ($\beta_1 = 0$) shows a decreasing tendency of λ, i.e. towards unloading. This does not mean that the vertex

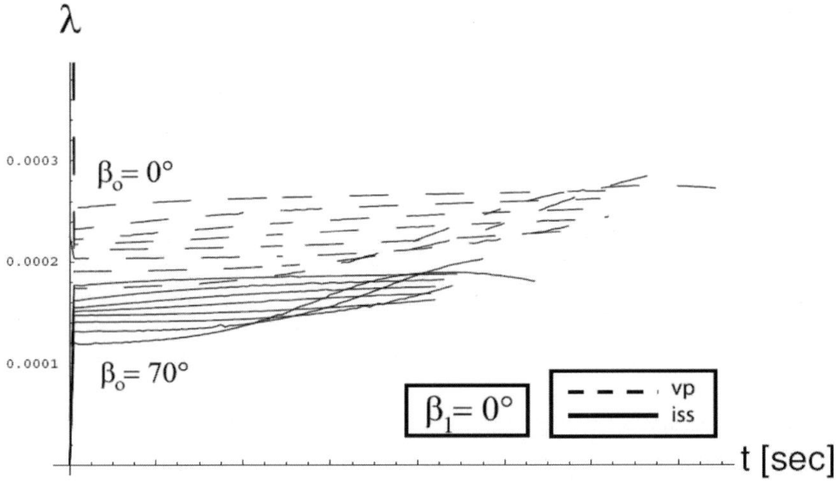

Fig. 8. Consistency parameter for different elastic process directions

effect is included in the equation. It is a consequence of the scalar product in (10). A further examination of selective components of the deformation gradient as well as the Piola-Kirchhoff stress tensor and the plastic transformation (not shown here) confirmed the result of an identical qualitative progression of the regularized approaches.

Here, the disclosed difference shown in section 3.2 could not be reproduced.

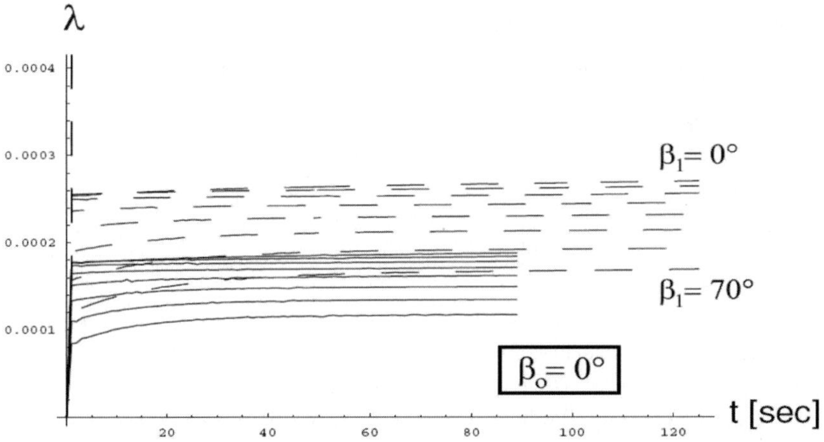

Fig. 9. Consistency parameter for different plastic process directions

4 Conclusion

Comparison of two regularizing models of crystal plasticity shows that noticeable differences can be found in situations where the slip systems exhibit strong differences in their activity. This can be in single slip situations or in polyslip at a vertex of the yield locus for extremely outbalanced hardening states of the slip systems. For similar hardening the differences are small.

Acknowledgments. Both authors gratefully acknowledge the interdisciplinary research environment they enjoyed while being associated students of the graduate school GK828.

References

Anand, L., Kothari, M.: A computational procedure for rate-independent crystal plasticity. J. Mech. Phys. Solids 44(4), 525–558 (1996)

Arminjon, M.: A regular form of the schmid law, application to the ambiguity problem. Textures and Microstructures 14(18), 1121–1128 (1991)

Asaro, R.J.: Micromechanics of crystals and polycrystals. Advances in Applied Mechanics 23 (1983)

Bertram, A.: An alternative approach to finite plasticity based on material isomorphisms. Int. J. Plast 15, 353–374 (1999)

Bertram, A.: Elasticity and Plasticity of Large Deformations – an Introduction, 2nd edn. Springer, Heidelberg (2008)

Bishop, J.F.W., Hill, R.: A theory of the plastic distortion of a polycrystal aggregate under combined stresses. Phil. Mag. 42, 414–427 (1951a)

Bishop, J.F.W., Hill, R.: A theoretical derivation of the plastic properties of a polycrystalline face-centred metal. Phil. Mag. 42, 1298–1307 (1951b)

Boyd, S., Vandenberghe, L.: Convex Optimization. Cambridge University Press, Cambridge (2004), http://www.stanford.edu/~boyd/cvxbook.html

Chin, G.Y., Mammel, W.L.: Generalization and equivalence of the minimum work (taylor) and maximum work (bishophill) principles for crystal plasticity. Trans. Metall. Soc. AIME, 1211–1214 (1969)

Gambin, W.: Plasticity of crystals with interacting slip systems. Enging. Trans. 39, 303–324 (1991)

Hindmarsh, A.C.: ODEPACK, a systematized collection of ode solvers. IMACS Transactions on Scientific Computation 1, 55–64 (1983)

Kiryk, R., Petryk, H.: A self-consistent model of rate-dependent plasticity of polycrystals. Arch. Mech. 50, 247–263 (1998)

Knockaert, R., Chastel, Y., Massoni, E.: Rate-independent crystalline and polycrystalline plasticity, application to fcc materials. Int. J. Plast 16, 179–198 (2000)

Kocks, U.F.: The relation between polcrystal deformation and single-crystal deformation. Metall. Trans. 1, 1121 (1970)

Kocks, U.F., Tomé, C.N., Wenk, H.R. (eds.): Texture and Anisotropy. Cambridge University Press, Cambridge (1998)

Koiter, W.T.: Stress-strain relations, uniqueness and variational theorems for elastic-plastic materials with a singular yield surface. Quarterly of Applied Mathematics 11, 350–354 (1953)

Lubliner, J.: Plasticity Theory. Macmillan Publishing Comp., New York (1990)

Miehe, C., Schröder, J., Schotte, J.: Computational homogenization analysis in finite plasticity simulation of texture development in polycrystalline materials. Comput. Methods Appl. Mech. Engrg. 171, 387–418 (1999)

Miehe, C., Schotte, J., Lambrecht, M.: Homogenization of inelastic solid materials at finite strains based on incremental minimization principles. application to the texture analysis of polycrystals. J. Mech. Phys. Solids 50, 2123–2167 (2002)

Mollica, F., Srinivasa, A.R.: A general framework for generating convex yield surfaces for anisotropic metals. Acta Mechanica 154, 61–84 (2002)

A Lower Bound Estimation of a Twinning Stress for Mg by a Stress Jump Analysis at the Twin-Parent Interface

R. Glüge and J. Kalisch

Institut für Mechanik, Otto-von-Guericke-Universität Magdeburg

Abstract. The stress jump at a twin-parent interface is analysed by using the geometrically linear strain measure. This is done by permitting the twin-parent interface to deviate from the ideally compatible orientation, and then asking for the elastic strains necessary to fulfil the kinematic compatibility. For linear elastic isotropic material behaviour an expression for the jump of the shear stress in the twin system in dependence of the interface alignment is given. In conjunction with a Schmid law this is used to estimate a lower bound for a critical shear stress of twinning for magnesium. The estimation is compared to experimental findings given in the literature.

1 Introduction

Mechanical twinning was probably first reported by Ewing and Rosenhain (1900). Since then, many authors have studied twinning experimentally. One possible appearance of deformation twins is a manifestation as lens-shaped regions within a crystal (the parent) that undergoes a simple shear deformations (see Fig. 1). In this region, the crystal lattice is oriented mirrored on the lens principal plane to the orientation of the surrounding parent crystal. The simple shear deformation and the accompanying orientation changes occur almost instantaneously by a sudden reposition of the atoms in the domain. Twinning occurs preferred in materials with a low stacking fault energy. The variety of materials that can undergo mechanical twinning is large. Prominent materials are hexagonal closest packed (hcp) single crystals, as well as intermetallic compounds (e.g., fcc TiAl, bct NiMn). For an introduction to twinning see (Hosford 1993), for a comprehensive review on twinning see, e.g., (Christian and Mahajan 1995; Pitteri and Zanzotto 2002; Boyko et al. 1994), where a dislocation-based approach to twinning is presented. Twinning has an important influence on the material behaviour on the macroscale. The reorientation of the crystal lattice leads to a strong texture evolution. Especially in hcp metals, the c-axis (which is the principal anisotropy axis in hcp metals) is reoriented about 85°. Moreover, the grain refinement by twinning influences the macroscopic yield stress. The newly formed grain boundaries are preferred sites for crack initiation (Yoo 1981). Another feature of twinning, caused by its polarity, is a strongly anisotropic yield limit (e.g., in extruded magnesium that is compressed or stretched along the extrusion axis, the yield stresses differ about a factor of 2 (Reed-Hill 1973)). Moreover, twinning represents an important deformation mode when slip systems can hardly be activated, e.g. at low temperatures. By twinning, high manganese alloyed steels can accommodate very fast high strains at low temperatures, which is interesting for crash energy absorption, deep

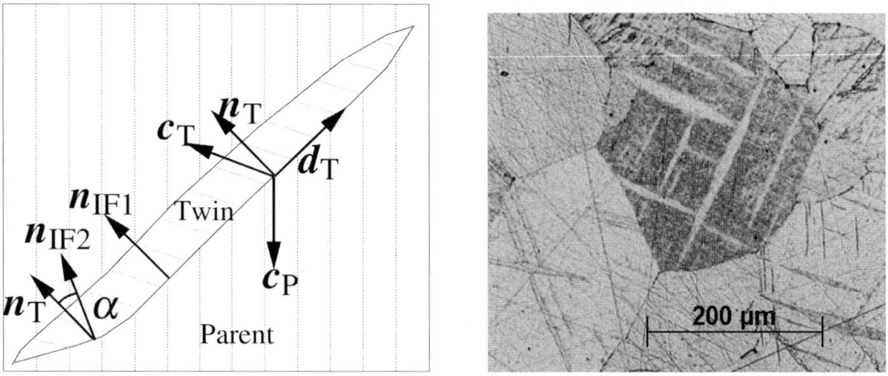

Fig. 1. (a) Sketch of a twin and geometrical variables, (b) twin network in magnesium

drawing, and forging processes (Schröder 2004). Due to the great variety of effects accompanying twinning, one is interested in its proper modelling. Some recent models describe a certain aspect of twinning on the macroscale (Tomé et al. 1991; Staroselsky and Anand 2003 and many more). But now, since the representative volume element technique (RVE) became a standard method, much of the aforementioned effects of twinning like the evolution of the grain morphology and the texture could be simulated by only one micromechanical model of twinning combined with the RVE-technique. There are some approaches to the micromechanical modelling of twinning (Orowan 1964; Forest and Parisot 2000; Idesman et al. 2000; Fischer et al. 2003). One problem when modelling twinning on the microscale is to deal with the discontinuities that emerge from the instantaneous strain increment and the instantaneous lattice reorientation. Another problem is that in many crystals that undergo twinning some of the potentially active twin systems form pairs of conjugate shear systems (Zanzotto 1992) that result in equal material stretching tensors. For this reason a purely elastic modelling like presented by Silling (1989) for a hypothetical material cannot be used without further efforts. Promising approaches to microscale twinning models are the phase field microelasticity theory (Wang et al. 2004) and the theory of multiple reference configurations (Rajagopal and Srinivasa 1995, 1997), as well as general approaches to phase transitions in solids (Pitteri and Zanzotto 2002; Abeyaratne and Knowles 2007).

In this article, we focus on the analytical approximation of the stress jump at the twin boundaries. An expression for the dependence of the stress jump on the misorientation of the twin-parent interface with respect to the shear plane is derived. The result is used to estimate a twinning stress in conjunction with a Schmid law. An example for the case of $\left\langle 101\,\bar{1}\right\rangle\left\{1012\right\}$ twinning $\left(\left\langle \boldsymbol{d}_T\right\rangle\left\{\boldsymbol{n}_T\right\}\right)$ in magnesium is given. This is done by determining the interface alignment in a Mg-crystal. From the misorientation from the ideally compatible interface alignment (see e.g. (Bhattacharya 2003; Pitteri and Zanzotto 2002)) the accommodating elastic strain jump and the stress jump are calculated. We employ a linear isotropic elasticity law, which is suitable for Mg. The fact that the interface lies static in the crystal permits in conjunction with a Schmid law an estimation of a lower bound for the critical Schmid stress.

1.1 Notation

Throughout the paper a direct tensor notation is preferred. If an expression cannot be represented in the direct notation without introducing new conventions, its components are given with respect to orthonormal base vectors $\boldsymbol{e}_i$, using the summation convention. Vectors are symbolised by lowercase bold letters $\boldsymbol{v} = v_i \boldsymbol{e}_i$, second order tensors by uppercase bold letters $\boldsymbol{T} = T_{ij} \boldsymbol{e}_i \otimes \boldsymbol{e}_j$ or bold Greek letters. The second order identity tensor is denoted by $\boldsymbol{I}$. Fourth-order tensors are symbolised like $\mathbb{C}$. A dot between two symbols represents one scalar contraction, e.g. $\boldsymbol{v} = \boldsymbol{A} \cdot \boldsymbol{w}$, $\alpha = \boldsymbol{A} \cdot\cdot \boldsymbol{B}$ and $\boldsymbol{\sigma} = \mathbb{C} \cdot\cdot \boldsymbol{\varepsilon}$. In order to shorten the expressions, $[\![X]\!] = X_{\mathrm{T}} \text{-} X_{\mathrm{P}}$ is introduced as the jump of X between twin and parent. The indices T and P denote a variable in the twin or the parent domain. Further, the abbreviation $(\boldsymbol{a} \otimes \boldsymbol{b})_{\mathrm{sym}} = 1/2(\boldsymbol{a} \otimes \boldsymbol{b} + \boldsymbol{b} \otimes \boldsymbol{a})$ is introduced.

1.2 Geometrical Description

Since twinning plays a major role in hcp metals, we introduce the vectors $\boldsymbol{c}_{\mathrm{P}}$ and $\boldsymbol{c}_{\mathrm{T}}$ that are normalised and parallel to the c-axis directions of the hcp unit cell in the parent and the twin, respectively. All normal vectors $\boldsymbol{n}$ are normalised to unit length, and point into the twin. The shear direction vector $\boldsymbol{d}_{\mathrm{T}}$ is normalised as well, and points into the shear direction. The vectors $\boldsymbol{n}_{\mathrm{T}}$ and $\boldsymbol{n}_{\mathrm{IF}}$ represent the shear plane normal and the interface normal, respectively. In hcp metals, the *c*/*a*-ratio is an important material constant. It relates the height of the crystallographic unit cell to the edge length of the base hexagon. For the most important twinning mode in hcp metals, namely the $\langle 10\bar{1}\bar{1} \rangle$ $\{1012\}$ twin system, the $\boldsymbol{c}$ as introduced here can be obtained by geometrical considerations as

$$\boldsymbol{c}_{\mathrm{P}} = -\left(\frac{3a^2}{c^2}+1\right)^{\frac{1}{2}} \boldsymbol{d}_{\mathrm{T}} - \left(\frac{c^2}{3a^2}+1\right)^{-\frac{1}{2}} \boldsymbol{n}_{\mathrm{T}} \tag{1}$$

$$\boldsymbol{c}_{\mathrm{T}} = -\left(\frac{3a^2}{c^2}+1\right)^{-\frac{1}{2}} \boldsymbol{d}_{\mathrm{T}} + \left(\frac{c^2}{3a^2}+1\right)^{-\frac{1}{2}} \boldsymbol{n}_{\mathrm{T}} \tag{2}$$

pointing away from the interface under consideration into its crystal.

2 A Schmid Law for Twinning

By means of a resolved shear stress criterion, a twin-parent interface moves towards the twin or the parent when a critical shear stress in the twin system is reached. The applicability of such a law depends strongly on the material. Two extreme examples are Zn and Mg. In Zn, the propagation stress of a twin is well below the nucleation stress (Bell and Cahn 1960), which induces a jerky yield behaviour. Further, the twin-parent interfaces are almost uncurved, and are aligned only in some specific orientations, the preference of which is temperature dependent (Straumal et al. 2001). Moreover, regarding that twinning in Zn is connected to kinking (Moore 1955), it is clear that a Schmid law for twinning in Zn would be too simplistic.

However, from a microscopic point of view, the application of a Schmid law as a twinning criterion seems to be reasonable due to the fact that the twin formation can be explained by the movement of partial dislocations, sometimes called misfit or transformation dislocations (Boyko et al. 1994; Vaidya and Mahajan 1980; Lavrentev and Bosin 1978; Matthews 1970), since a Schmid law works well for the dislocation movement underlying crystallographic slip. In fact, it is applied successfully to Mg (Barnett 2003), but seemed to be useless due to the large scattering of experimentally measured critical shear stresses (Christian and Mahajan 1995). One problem is that such measurements are difficult to perform. Twins nucleate at inhomogeneities in the crystal, e.g. at intersectioning points of slip lines, at grain boundaries or at crack tips. If one is interested in determining a critical nucleation stress, it would be necessary to determine the local stress where and when the twin emerges. Due to the inhomogeneities a reliable estimation of such a critical stress state is rather difficult. Another problem is that different experimenters take different criteria for twinning, e.g. one may look out for the movement of transformation dislocations, another for the appearance of the first twin, while another takes plastic deformations in a textured polycrystal as indication, or uses macroscopic stress-strain curves. Therefore, for $\langle 10\bar{1}\bar{1}\rangle\ \{10\bar{1}2\}$ twinning in Mg, one can find values ranging from 2.7 to 2.8MPa (Koike 2004) to 14 to 19MPa (Zhou et al. 2008) and 40 to 50MPa for a Mg alloy (Wang and Huang 2007).

However, in the literature an agreement of a Schmid stress for $\langle 10\bar{1}\bar{1}\rangle\ \{10\bar{1}2\}$ twinning in Mg seemed to converge to $\tau_{\mathrm{crit}} \approx 4\ \tau_{\mathrm{b}}$, i.e. the value given by Koike (2004). Another problem is the applicability of a Schmid law due to the characteristics of the transformation dislocation movement. In some cases, the critical stress needed to overcome the initial resistance to the dislocation movement is larger than the drag stress needed for the continuing movement, resulting in a jerky twin propagation (e.g. in Zn). Although the applicability of a Schmid law is to simplistic for many materials, we will restrict ourselves to such behaviour in the article and see how far we can get.

When applying a Schmid law to twinning, one has to bear in mind that twinning is unidirectional, and the inequalities have to be set up corresponding to the definition of the positive shear direction and shear plane normal. In the geometrical description the shear system has been introduced such that the shear strain γ_0 occurring when a twin develops is positive with respect to the shear system formed by $\boldsymbol{d}_{\mathrm{T}}$ and $\boldsymbol{n}_{\mathrm{T}}$. The twin boundary moves when the atoms sketched in Fig. 2 jump into the positions indicated with the arrows. By means of the Schmid law the twin would grow or shrink (i.e. the twin boundary would move towards the parent or the twin) if one of the inequalities

$$\tau_{\mathrm{TS,T}} \geq -\tau_{\mathrm{crit}}, \text{shrinking of the twin if violated (arrow 1 in Fig. 2)} \tag{3}$$

$$\tau_{\mathrm{TS,P}} \leq \tau_{\mathrm{crit}}, \text{growth of the twin if violated (arrow 2 in Fig. 2)} \tag{4}$$

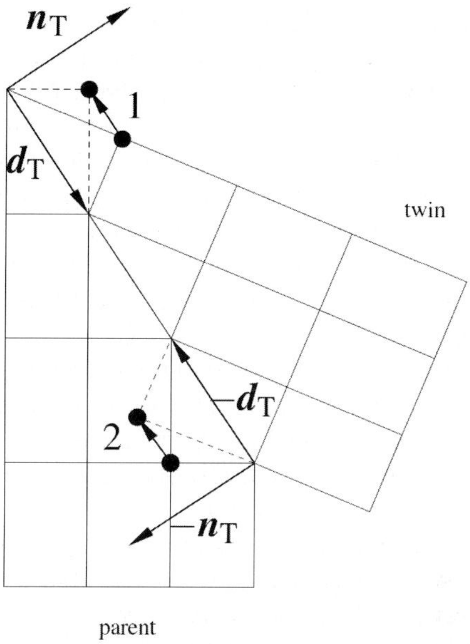

Fig. 2. Atom movement for twinning (arrow 2, growth of the twin, the interface moves towards the parent) and de-twinning (arrow 1, growth of the parent, the interface moves towards the twin). It is pointed out that the viewpoint which side of the interface is parent and which is twin is arbitrary. In this article, the definition is such that $\boldsymbol{n}_{\mathrm{T}}$ points into the twin.

is violated. $\tau_{\mathrm{TS,T}}$ and $\tau_{\mathrm{TS,P}}$ denote the shear stresses in the twin system on both sides of the interface, respectively. Multiplying Eq. (4) by -1 and adding to Eq. (3) yields

$$\tau_{\mathrm{TS,T}} - \tau_{\mathrm{TS,P}} \geq -2\ \tau_{\mathrm{crit}} \tag{5}$$

$$-\frac{1}{2}[\![\tau_{\mathrm{TS}}]\!] = \tau_{\mathrm{crit}} \tag{6}$$

If the stress jump $[\![\boldsymbol{\sigma}]\!]$ at the static interface is known, the jump of the shear stress in the twin system can be calculated by

$$[\![\tau_{\mathrm{TS}}]\!] = [\![\boldsymbol{\sigma}]\!] \cdot\cdot \left(\boldsymbol{d}_{\mathrm{TS}} \otimes \boldsymbol{n}_{\mathrm{TS}}\right) \tag{7}$$

and, inserted into Eq. (6), used to estimate lower bound of τ_{crit}.

3 Stress Jump at the Twin Interface

We neglect the body forces and the calculation is carried out at a static interface that divides the twin and the parent. To calculate the stress jump that emerges from the strain incompatibility at the twin-parent interface, one needs the jump balance of momentum and the kinematic compatibility of the deformation gradient,

$$[\![\boldsymbol{\sigma}]\!] \cdot \boldsymbol{n}_{\mathrm{IF}} = 0 \tag{8}$$

$$[\![\boldsymbol{F}]\!] - \boldsymbol{a} \otimes \boldsymbol{n}_{\mathrm{IF}} = 0 \tag{9}$$

which is known as the Hadamard condition. $\boldsymbol{a}$ is an unknown normal jump of $\boldsymbol{F}$ on the interface (see, e.g., (Liu 2002)). The stress-free state (subscript "sf") of the parent crystal is used as the reference placement. Therefore, the stress-free deformation gradients for the parent and the twin are given by

$$\boldsymbol{F}_{\mathrm{P,sf}} = \boldsymbol{I}\,, \tag{10}$$

$$\boldsymbol{F}_{\mathrm{T,sf}} = \boldsymbol{I} + \gamma_0 \boldsymbol{d}_{\mathrm{T}} \otimes \boldsymbol{n}_{\mathrm{T}}\,. \tag{11}$$

Throughout the paper we restrict ourselves to the geometrically linear strain measure given by

$$\boldsymbol{\varepsilon} = (\boldsymbol{F} - \boldsymbol{I})_{\mathrm{sym}} = \frac{1}{2}(\boldsymbol{F} + \boldsymbol{F}^T - 2\boldsymbol{I})\,, \tag{12}$$

which is reasonable for small strains. Typical shear strains which accompany twinning are in the range of 0 to 0▷2 ($\langle 10 1 \bar{1} \rangle$ $\{1012\}$ twinning mode, $\gamma_{\mathrm{Mg}} = 0.129$, $\gamma_{\mathrm{Cd}} = 0.171$, $\gamma_{\mathrm{Zn}} = 0.138$, $\gamma_{\mathrm{Ti}} = 0.174$). In the case of Mg, the polar decomposition of $\boldsymbol{F}_{\mathrm{T,sf}}$ yields a rotation of $\approx$ 3.7°and a length change of 0 and $\approx \pm 6\%$ in the principal stretching directions. This is considered to be sufficiently small to permit the use of the geometrically linear strain measure. The small strain assumption is also necessary for enabling us to calculate the stress jump, as indicated below. With (12) the stress-free strains

$$\boldsymbol{\varepsilon}_{\mathrm{T,sf}} = (\boldsymbol{d}_{\mathrm{T}} \otimes \boldsymbol{n}_{\mathrm{T}})_{\mathrm{sym}} \tag{13}$$

$$\boldsymbol{\varepsilon}_{\mathrm{P,sf}} = 0 \tag{14}$$

can be computed. The unknown strains on both sides of the interface are also needed. From the definition of the strain measure one can calculate

$$\begin{aligned} \boldsymbol{\varepsilon}_{\mathrm{T}} - \boldsymbol{\varepsilon}_{\mathrm{P}} &= (\boldsymbol{F}_{\mathrm{T}} + \boldsymbol{F}_{\mathrm{T}}^{\mathrm{T}} - 2\boldsymbol{I} - \boldsymbol{F}_{\mathrm{P}} - \boldsymbol{F}_{\mathrm{P}}^{\mathrm{T}} + 2\boldsymbol{I}) \\ &= (\boldsymbol{F}_{\mathrm{T}} - \boldsymbol{F}_{\mathrm{P}})_{\mathrm{sym}} \\ &= (\boldsymbol{a} \otimes \boldsymbol{n}_{\mathrm{IF}})_{\mathrm{sym}} \end{aligned} \tag{15}$$

and obtain one equation relating $\boldsymbol{\varepsilon}_{\mathrm{T}}$ and $\boldsymbol{\varepsilon}_{\mathrm{P}}$. Furthermore, we need an elasticity law. We assume a linear elastic material behaviour

$$\boldsymbol{\sigma} = \mathbb{C} \cdot\cdot (\boldsymbol{\varepsilon} - \boldsymbol{\varepsilon}_{\mathrm{sf}}) \tag{16}$$

The jump balance of momentum can now be rewritten as

$$\begin{aligned} 0 &= (\boldsymbol{\sigma}_{\mathrm{T}} - \boldsymbol{\sigma}_{\mathrm{P}}) \cdot \boldsymbol{n}_{\mathrm{IF}} \\ &= (\mathbb{C}_{\mathrm{T}} \cdot\cdot (\boldsymbol{\varepsilon}_{\mathrm{T}} - \boldsymbol{\varepsilon}_{\mathrm{T,sf}}) - \mathbb{C}_{\mathrm{P}} \cdot\cdot \boldsymbol{\varepsilon}_{\mathrm{P}}) \cdot \boldsymbol{n}_{\mathrm{IF}} \end{aligned} \tag{17}$$

If the deformation of the lattice is neglected, the stiffness tetrads $\mathbb{C}_{\mathrm{T}}$ and $\mathbb{C}_{\mathrm{P}}$ are related by

$$\mathbb{C}_{\mathrm{T}} = \boldsymbol{R} * \mathbb{C}_{\mathrm{P}} \tag{18}$$

$$\mathbb{C}_{\mathrm{T}} = R_{im} R_{jn} R_{ko} R_{lp} C_{\mathrm{P}mnop}\,, \tag{19}$$

where

$$\boldsymbol{R} = -\boldsymbol{I} + 2\boldsymbol{n}_{\mathrm{T}} \otimes \boldsymbol{n}_{\mathrm{T}} \tag{20}$$

is a proper orthogonal tensor that maps the parent lattice vectors into the twin lattice vectors, being a rotation of 180° around $\boldsymbol{n}_{\mathrm{T}}$. $\boldsymbol{R}$ has the property of being orthogonal and symmetric,

$$\boldsymbol{R} = \boldsymbol{R}^{-1} = \boldsymbol{R}^{\mathrm{T}}\,. \tag{21}$$

Inserting $\boldsymbol{\varepsilon}_{\mathrm{P}}$ from Eq. (15) into (17) and using Eq. (19) we are left with

$$0 = \left((\boldsymbol{R} * \mathbb{C}_{\mathrm{P}}) \cdot\cdot (\boldsymbol{\varepsilon}_{\mathrm{T}} - \boldsymbol{\varepsilon}_{\mathrm{T,sf}}) - \mathbb{C}_{\mathrm{P}} \cdot\cdot \left(\boldsymbol{\varepsilon}_{\mathrm{T}} - (\boldsymbol{a} \otimes \boldsymbol{n}_{\mathrm{IF}})_{\mathrm{sym}}\right)\right) \cdot \boldsymbol{n}_{\mathrm{IF}} \tag{22}$$

However, by looking at Eq. (17) one notes that the calculation is simplified considerably if the elastic material behaviour is assumed to be isotropic. In this case one can directly replace $\boldsymbol{\varepsilon}_{\mathrm{T}} - \boldsymbol{\varepsilon}_{\mathrm{P}}$ by $(\boldsymbol{a} \otimes \boldsymbol{n}_{\mathrm{IF}})_{\mathrm{sym}}$ from Eq. (15). Otherwise, further assumptions are needed, since one can only eliminate either ε_{T} or ε_{P}, but one of the two unknowns remains. The situation is the same when a geometrically non-linear strain measure is used, since Eq. (15) does not hold anymore. Due to the fact that the elastic anisotropy of magnesium is rather weak (see (Böhlke and Brüggemann 2001) for a graphical representation) we assume elastic isotropy. A quantitative estimation of the anisotropy of Mg is given in the appendix of the article. With the assumption of elastic isotropy Eq. (17) becomes

$$\left(\mathbb{C}_{\mathrm{iso}} \cdot\cdot (\boldsymbol{\varepsilon}_{\mathrm{T}} - \boldsymbol{\varepsilon}_{\mathrm{P}} - \boldsymbol{\varepsilon}_{\mathrm{T,sf}})\right) \cdot \boldsymbol{n}_{\mathrm{IF}} = 0 \tag{23}$$

with the isotropic elastic stiffness tetrad

$$\mathbb{C}_{\mathrm{iso}} = 3K\mathbb{P}_{I} + 2G\mathbb{P}_{II}\,, \qquad \mathbb{P}_{I} = \frac{1}{3}\boldsymbol{I} \otimes \boldsymbol{I}\,, \qquad \mathbb{P}_{II} = \mathbb{I} - \mathbb{P}_{I} \tag{24}$$

$\mathbb{I}$ maps every second order tensor into its symmetric part. K and G are the compression modulus and the bulk modulus, respectively $\mathbb{P}_{I}$ and $\mathbb{P}_{II}$ are called first and second isotropic projector (see, e.g., (Böhlke 2001)), which extract the dilatational and deviatoric parts of the strain, respectively. Inserting Eqs. (13) and (15) into Eq. (23) we are left with

$$0 = \left(\mathbb{C}_{\mathrm{iso}} \cdot\cdot \left((\boldsymbol{a} \otimes \boldsymbol{n}_{\mathrm{IF}})_{\mathrm{sym}} - \gamma_0 (\boldsymbol{d}_{\mathrm{T}} \otimes \boldsymbol{n}_{\mathrm{T}})_{\mathrm{sym}}\right)\right) \cdot \boldsymbol{n}_{\mathrm{IF}} \tag{25}$$

Using $\mathbb{P}_I \cdot\cdot (\boldsymbol{a} \otimes \boldsymbol{b}) = 1/3(\boldsymbol{a} \cdot \boldsymbol{b})\boldsymbol{I}$, $\mathbb{P}_{II} \cdot\cdot (\boldsymbol{a} \otimes \boldsymbol{b}) = (\boldsymbol{a} \otimes \boldsymbol{b})_{\mathrm{sym}} - 1/3(\boldsymbol{a} \cdot \boldsymbol{b})\boldsymbol{I}$ and recalling $(\boldsymbol{d}_{\mathrm{T}} \cdot \boldsymbol{n}_{\mathrm{T}}) = 0$, Eq. (25) can be rewritten as

$$\left(K(\boldsymbol{a} \cdot \boldsymbol{n}_{\mathrm{IF}})\boldsymbol{I} + 2G\left((\boldsymbol{a} \otimes \boldsymbol{n}_{\mathrm{IF}})_{\mathrm{sym}} - \frac{\boldsymbol{a} \cdot \boldsymbol{n}_{\mathrm{IF}}}{3}\boldsymbol{I} - \gamma_0 (\boldsymbol{d}_{\mathrm{T}} \otimes \boldsymbol{n}_{\mathrm{T}})_{\mathrm{sym}} \right) \right) \cdot \boldsymbol{n}_{\mathrm{IF}} = 0 \tag{26}$$

or, equivalently,

$$\boldsymbol{a} \cdot \left(G\boldsymbol{I} + \left(K + \frac{1}{3}G \right) \boldsymbol{n}_{\mathrm{IF}} \otimes \boldsymbol{n}_{\mathrm{IF}} \right) = 2G\gamma_0 (\boldsymbol{d}_{\mathrm{T}} \otimes \boldsymbol{n}_{\mathrm{T}})_{\mathrm{sym}} \cdot \boldsymbol{n}_{\mathrm{IF}} \,. \tag{27}$$

This inhomogeneous linear system of three equations can be solved explicitly for the components of $\boldsymbol{a}$. By means of the relation

$$\left(G\boldsymbol{I} + \left(K + \frac{1}{3}G \right) \boldsymbol{n}_{\mathrm{IF}} \otimes \boldsymbol{n}_{\mathrm{IF}} \right)^{-1} = \frac{1}{G}\boldsymbol{I} - \frac{K + \frac{1}{3}G}{G\left(G + \left(K + \frac{1}{3}G \right) \right)} \boldsymbol{n}_{\mathrm{IF}} \otimes \boldsymbol{n}_{\mathrm{IF}} \tag{28}$$

we finally find

$$\boldsymbol{a} = \gamma_0 \left((\boldsymbol{n}_{\mathrm{IF}} \cdot \boldsymbol{d}_{\mathrm{T}})\boldsymbol{n}_{\mathrm{T}} + (\boldsymbol{n}_{\mathrm{IF}} \cdot \boldsymbol{n}_{\mathrm{T}})\boldsymbol{d}_{\mathrm{T}} - \frac{2K + \frac{2}{3}G}{K + \frac{4}{3}G} (\boldsymbol{n}_{\mathrm{IF}} \cdot \boldsymbol{d}_{\mathrm{T}})(\boldsymbol{n}_{\mathrm{IF}} \cdot \boldsymbol{n}_{\mathrm{T}})\boldsymbol{n}_{\mathrm{IF}} \right). \tag{29}$$

Now we may rewrite the stress jump in terms of the input variables,

$$[\![\boldsymbol{\sigma}]\!] = 2G\gamma_0 \left((\boldsymbol{n}_{\mathrm{IF}} \cdot \boldsymbol{d}_{\mathrm{T}})(\boldsymbol{n}_{\mathrm{IF}} \cdot \boldsymbol{n}_{\mathrm{T}}) \left(\frac{3K - 2G}{3K + 4G}\boldsymbol{I} - \frac{6K + 2G}{3K + 4G}\boldsymbol{n}_{\mathrm{IF}} \otimes \boldsymbol{n}_{\mathrm{IF}} \right) + \right. \tag{30}$$

$$\left. + (\boldsymbol{n}_{\mathrm{IF}} \cdot \boldsymbol{d}_{\mathrm{T}})(\boldsymbol{n}_{\mathrm{T}} \otimes \boldsymbol{n}_{\mathrm{IF}})_{\mathrm{sym}} + (\boldsymbol{n}_{\mathrm{IF}} \cdot \boldsymbol{n}_{\mathrm{T}})(\boldsymbol{d}_{\mathrm{T}} \otimes \boldsymbol{n}_{\mathrm{IF}})_{\mathrm{sym}} - (\boldsymbol{d}_{\mathrm{T}} \otimes \boldsymbol{n}_{\mathrm{T}})_{\mathrm{sym}} \right). \tag{31}$$

3.1 Special Cases: $\mathbf{n}_{\mathrm{IF}}$ Equals $\mathbf{n}_{\mathrm{T}}$ or $\mathbf{d}_{\mathrm{T}}$

$[\![\tau]\!] = 0$ must hold for $\boldsymbol{n}_{\mathrm{IF}} = \boldsymbol{n}_{\mathrm{T}}$ or $\boldsymbol{n}_{\mathrm{IF}} = \boldsymbol{d}_{\mathrm{T}}$ independent of the elasticity law. This is due to $[\![\boldsymbol{\sigma}]\!] \cdot \boldsymbol{n}_{\mathrm{IF}} = 0$ and the symmetry of $[\![\boldsymbol{\sigma}]\!]$. The latter case is not observed in practice, but interesting from a theoretical point of view. Far from the tips of the twins, the twin-parent interface is almost parallel to the shear plane, i.e. $\boldsymbol{n}_{\mathrm{IF}} = \boldsymbol{n}_{\mathrm{T}}$. After a short calculation we obtain $[\![\boldsymbol{\sigma}]\!] = 0$, i.e. the stresses are continuous. The same has been found by a finite element analysis of Fischer et al. (2003) (see Fig. 4a and 4b), where a parallelogram-shaped twin is embedded into a crystal. Far from the nucleation sites the strain energy is zero in both the twin and the parent, which indicates vanishing stresses, even though the strain jump at the interface is quite remarkable.

In the case $\boldsymbol{n}_{\mathrm{IF}} = \boldsymbol{d}_{\mathrm{T}}$ the corresponding twin would propagate along the shear plane normal. Again, $[\![\boldsymbol{\sigma}]\!] = 0$ in the case of elastic isotropy. This type of twin is not observed in practice, but predicted by the finite element simulations done by Forest and Parisot (2000), where a Schmid law is used. Forest and Parisot (2000) suggest the use of higher order strain gradients to distinguish such kink-twins from admissible twins.

3.2 Arbitrary Alignment of the Interface and the Effect on the Schmid Stress

Now let us examine $[\![\sigma]\!]$ near the tips of the twin, where $\boldsymbol{n}_{\mathrm{IF}}$ is no longer parallel to $\boldsymbol{n}_{\mathrm{T}}$ (see Fig. 1a). The jump of the resolved shear stress in the twin system is the most interesting stress component, since it would directly enter the Schmid law. Inserting Eq. (31) into Eq. (7), we find the jump of the resolved shear stress in the twin system to be

$$[\![\tau]\!] = G\gamma_0\left(-1+(\boldsymbol{n}_{\mathrm{IF}}\cdot\boldsymbol{d}_{\mathrm{T}})^2+(\boldsymbol{n}_{\mathrm{IF}}\cdot\boldsymbol{n}_{\mathrm{T}})^2-C(\boldsymbol{n}_{\mathrm{IF}}\cdot\boldsymbol{d}_{\mathrm{T}})^2(\boldsymbol{n}_{\mathrm{IF}}\cdot\boldsymbol{n}_{\mathrm{T}})^2\right), \tag{32}$$

$$C=\frac{12K+4G}{3K+4G}. \tag{33}$$

Now we can easily investigate the influence of a deviation between $\boldsymbol{n}_{\mathrm{IF}}$ and $\boldsymbol{n}_{\mathrm{T}}$ on $[\![\tau]\!]$. Let us assume that $\boldsymbol{n}_{\mathrm{IF}}$ is rotated from $\boldsymbol{n}_{\mathrm{T}}$ towards $\boldsymbol{d}_{\mathrm{T}}$ by the angle α, what describes the situation when one approaches the tip of the twin coming from the nearly parallel faces bounding the twin. The scalar products in Eq. (33) are given by $\boldsymbol{n}_{\mathrm{IF}}\cdot\boldsymbol{n}_{\mathrm{T}}=\cos(\alpha)$ and $\boldsymbol{n}_{\mathrm{IF}}\cdot\boldsymbol{d}_{\mathrm{T}}=\sin(\alpha)$, which yields

$$[\![\tau]\!] = -G\gamma_0\frac{C}{4}\sin(2\alpha)^2. \tag{34}$$

One can confirm that $[\![\tau]\!]=0$ for $\alpha=0$ and $\alpha=\pi/2$, which corresponds to $\boldsymbol{n}_{\mathrm{IF}}=\boldsymbol{n}_{\mathrm{T}}$ and $\boldsymbol{n}_{\mathrm{IF}}=\boldsymbol{d}_{\mathrm{T}}$, respectively. Further, Eq. (34) is symmetric in α, which reflects the geometrical symmetry found at the tips of the twins. One can see that the propagation plane of the twins is approximately bisectioning the opening angle (see Fig. 1b).

4 Lower Bound Estimation of a Twinning Stress for Mg

Inserting (34) into Eq. (6) yields a lower bound estimation for τ_{crit}

$$\tau_{\mathrm{crit}} \geq G\gamma_0\frac{C}{8}\sin(2\alpha)^2. \tag{35}$$

If one is able to measure $\boldsymbol{n}_{\mathrm{IF}}$ with respect to the crystal lattice and further knows γ_0, K and G, one can estimate a lower bound for the critical resolved shear stress for twinning. As an example, for magnesium we have approximately γ_0 = 0.129, K = 50GPa and G = 17GPa, which yields

$$\tau_{\mathrm{crit}} \geq 840\sin(2\alpha)^2\,\mathrm{MPa}. \tag{36}$$

Such estimation possibly proves to be useful in view of the difficulties associated with the measurements of such a twinning stress. The derived lower bound estimation provides an non-direct access to local stresses where direct measurements can hardly be applied. However, it is difficult to find the required geometrical information in the literature, because it is not sufficient to measure angles and distances on a micrograph. One needs the full three-dimensional geometric information on how the interface is oriented with respect to the twin system. However, Eq. (36) is already restricted to $\boldsymbol{n}_{\mathrm{IF}}$ in the plane spanned by $\boldsymbol{n}_{\mathrm{T}}$ and $\boldsymbol{d}_{\mathrm{T}}$. In Hartt and Reed-Hill (1968), Fig. 7, (here Fig. 3) one can find a micrograph of a $\langle 10\bar{1}\bar{1}\rangle\{10\bar{1}2\}$ twin, where we are able to measure the angle between the opposing interfaces of the twin lamella to be approximately 3°. Generally, one can find any opening angle at the tip of the twin, because the micrograph plane can cut the $\boldsymbol{n}_{\mathrm{T}}$-plane in a sharp angle, which artificially scales up the measured opening angle. In the case discussed here the micrograph plane contains $\boldsymbol{n}_{\mathrm{T}}$ and $\boldsymbol{d}_{\mathrm{T}}$, i.e. we do not need to recalculate the angle. Assuming that $\boldsymbol{n}_{\mathrm{IF}}$ lies also in this plane, we can apply Eq. (36) (with $2\alpha = 3°$), and yield $\tau_{\mathrm{crit}} \geq 2.3\mathrm{MPa}$. Earlier estimates of the twinning stress cited by Barnett (2003) came up with $\tau_{\mathrm{crit}} \approx 4\tau_{\mathrm{b}}$, where $\tau_{\mathrm{b}} \approx 0.6\mathrm{MPa}$ is the critical Schmid stress for basal slip (Barnett 2003). This yields a τ_{crit} of approximately 2.4MPa. It seems that the presented lower bound estimation gives reasonable results, although one has to bear in mind that the twinning stress reported by Barnett (2003) is a mean value for Mg and Mg-Al alloys that have been reported by different authors.

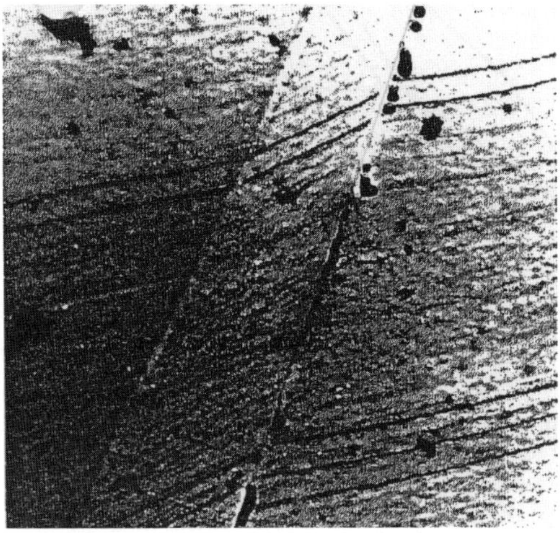

Fig. 3. Micrograph of a $\langle 10\bar{1}\bar{1}\rangle\{10\bar{1}2\}$ twin in Mg, taken from Hartt and Reed-Hill (1968). $\boldsymbol{n}_{\mathrm{T}}$ and $\boldsymbol{d}_{\mathrm{T}}$ lie in the plane of the micrograph.

5 Pole Figures of Some Interesting Quantities Regarding the Stress Jump

For the sake of completeness, some important quantities regarding the stress jump should be visualised. In Fig. 4, the jump of the pressure $[\![p]\!] = -1/3\, \boldsymbol{I} \cdot [\![\boldsymbol{\sigma}]\!]$, the norm of the deviatoric part of $[\![\sigma]\!]$, namely $\| [\![\sigma]\!]' \| = \| [\![\sigma]\!] + [\![p]\!] \boldsymbol{I} \|$ and the jump of the resolved shear stress $[\![\tau]\!]$. The layout of the figures is such that every point in the fourth of the circles represents a certain $\boldsymbol{n}_{\mathrm{IF}}$, namely the parallel projection of $\boldsymbol{n}_{\mathrm{IF}}$ into the shear plane, with the shear direction aligned horizontal. For example, the centre point represents $\boldsymbol{n}_{\mathrm{IF}} = \boldsymbol{n}_{\mathrm{T}}$, or the outermost point at the right represents $\boldsymbol{n}_{\mathrm{IF}} = \boldsymbol{d}_{\mathrm{T}}$. Due to the symmetry only one fourth of the domain is plotted. The values have been evaluated for Mg, with G =17GPa, K = 50GPa and $\gamma_0 = 0.129$. It is interesting to see that $[\![\tau]\!]$ grows stronger when $\boldsymbol{n}_{\mathrm{IF}}$ is rotated from $\boldsymbol{n}_{\mathrm{T}}$ towards $\boldsymbol{d}_{\mathrm{T}}$ compared to a rotating $\boldsymbol{n}_{\mathrm{IF}}$ around $\boldsymbol{d}_{\mathrm{T}}$, which can be seen when looking at the isolines in Fig. 4c. This suggests that the latter deviation of $\boldsymbol{n}_{\mathrm{IF}}$ from $\boldsymbol{n}_{\mathrm{T}}$ should be favoured. Another interesting observation is that the jump of the pressure is negative for all $\boldsymbol{n}_{\mathrm{IF}}$, i.e. inside the twin the pressure must always be lower than in the parent.

(a) $0 > [\![p]\!] > \dfrac{-3K}{3K+4G} G\gamma_0$ (b) $0 < \| [\![\sigma]\!]' \| < \sqrt{2G\gamma_0}$ (c) $0 > [\![\tau_{\mathrm{TS}}]\!] > -G\gamma_0$

Fig. 4. Jumps of pressure, norm of the deviatoric part of [[s]] and jump of the resolved shear stress in the twin system for different $\boldsymbol{n}_{\mathrm{IF}}$. $\boldsymbol{n}_{\mathrm{IF}}$ projected parallel into the shear plane. The intensity represents the absolute value of the quantity that is plotted from zero (white) to the maximum absolute value (black) given in each figures caption.

6 Conclusions

Within the framework of geometrically linear strains and isotropic linear elastic material behaviour the stress jump at a twin-parent interface has been analysed. Its dependence on the alignment of the interface with respect to the twin system has been given. The results have been used to estimate a lower bound for a twinning stress for the weakly anisotropic Mg. It has been found that the stress jump is zero if the interface normal is parallel to the shear plane normal or the shear direction. The former represents twins which propagate along the shear direction and, therefore, have the interface aligned approximately parallel to the shear plane. The latter is not observed in practice, but predicted by finite element calculations which employ a Schmid law for twinning (Forest and Parisot 2000).

Moreover, any deviation from the two alignments mentioned above is penalized by pushing $\tau_{\mathrm{TS,T}}$ and $\tau_{\mathrm{TS,P}}$ towards their critical values of twinning or de-twinning, respectively. Presuming that a Schmid law holds, a lower bound for the twinning stress can be estimated by $-1/2\, [\![\tau_{\mathrm{TS}}]\!] < \tau_{\mathrm{crit}}$ when the interface alignment is known. The lower bound estimation given in the article yields for Mg a $\tau_{\mathrm{crit}} \approx 2.3$MPa, which corresponds quite well to a $\tau_{\text{crit}} \approx 4\tau_{\text{b}}$ found in the literature (Wonsiewicz and Backofen 1967; Barnett 2003; Koike 2004). Regarding the jump of the pressure, it is found that $[\![p]\!] \leq 0$ for any inclination of the interface to the twin plane normal. This means that at least near the tips of the twins the pressure inside the twin must be below the pressure that is found in the vicinity of the twin tip in the parent.

One should note that if one drops the assumption of elastic isotropy, or the geometrically linear strain measure, the problem of calculating the normal jump $\boldsymbol{a}$ of the deformation gradient becomes non-linear and though more complicated, and further assumptions are needed. Further, one has to bear in mind that the applicability of a Schmid law has to be carefully checked individually for every material and twinning mode.

Acknowledgments. We would like to thank Prof. F.D. Fischer for helpfully commenting and discussing this work.

References

Abeyaratne, R., Knowles, J.: Evolution of Phase Transitions - A Continuum Theory, Cambridge (2007)

Barnett, M.: A Taylor Model Based Description of the Proof Stress of Magnesium AZ31 during Hot Working. Metall. Mater. Trans. A 34, 1799–1806 (2003)

Bell, R., Cahn, R.: The Dynamics of Twinning and the Interrelation of Slip and Twinning in Zinc Crystals. Proceedings of the Royal Society of London. Series A, Mathematical and Physical Sciences 239, 494–521 (1960)

Bhattacharya, K.: microstructure of martensite- why it forms and how it gives rise to the shape-memory effect. Oxford Series on Materials Modelling (2003)

Böhlke, T.: Crystallographic Texture Evolution and Elastic Anisotropy. Shaker Verlag (2001)

Böhlke, T., Brüggemann, C.: Graphical representation of the generalized Hooke's law. Technische Mechanik 21(2), 145–158 (2001)

Boyko, V., Garber, R., Kossevich, A.: Reversible Crystal Plasticity. AIP Press, New York (1994)

Christian, J., Mahajan, S.: Deformation Twinning. Progress in Materials Science 39, 157 (1995)

Ewing, J., Rosenhain, W.: Crystalline Structure of Metals. Phil. Trans. Royal Soc. 193A, 353 (1900)

Fischer, F., Schaden, T., Appel, F., Clemens, H.: Mechanical Twinning, their development and growth. European Journal of Mechanics A/Solids 22, 709–726 (2003)

Forest, S., Parisot, R.: Material Crystal elasticity and Deformation Twinning. Rend. Sem. Mat. Univ. Torino 58, 99–111 (2000)

Hartt, W., Reed-Hill, R.: Internal Deformation and Fracture of Second-Order {10-11}-<10-12>-Twins in Magnesium. Trans. Metal. Soc. AIME 242, 1127–1133 (1968)

Hosford, F.: The Behaviour of Crystals and Textured Polycrystals. Oxford University Press, Oxford (1993)

Idesman, A., Levitas, V., Stein, E.: Structural changes in elastoplastic material: a unified approach to phase transformation, twinning and fracture. International Journal of Plasticity 16, 893–949 (2000)

Koike, J.: Enhanced Deformation Mechanisms by Anisotropic Plasticity in Polycrystalline Mg Alloys at Room Temperature. Metall. Mater. Trans. A 36, 1689–1696 (2004)

Lavrentev, F., Bosin, M.: The Peculiar Effect of Forest Dislocations on Single Twin Layer Development in Zinc and Beryllium Single Crystals. Mater. Sci. Eng. 33, 243–248 (1978)

Liu, I.-S.: Continuum Mechanics. Springer, Berlin (2002)

Matthews, J.: Role of deformation twinning in the fracture of single-crystal films. Acta Metall. 18, 175–181 (1970)

Moore, A.: Twinning and accommodation kinking in zinc. Acta Metall. 1, 163–169 (1955)

Orowan, E.: Dislocations in metals. AIME, New York (1964)

Pitteri, M., Zanzotto, G.: Continuum Models for Phase Transitions and Twinning in Crystals. Chapman and Hall/CRC (2002)

Rajagopal, K., Srinivasa, A.: On the inelastic behaviour of solids. Part I: twinning. International Journal of Plasticity 11, 653–778 (1995)

Rajagopal, K., Srinivasa, A.: Inelastic behaviour of materials. Part II. Energetics associated with discontinuous deformation twinning. International Journal of Plasticity 13, 1–35 (1997)

Reed-Hill, R.: Inhomogeneity of Plastic Deformation. A.S.M (1973)

Schröder, T.: Ausgekochter Stahl für das Auto von Morgen. MaxPlanck Forschung 3, 36–41 (2004)

Silling, S.: Phase changes induced by deformation in isothermal elastic crystals. J. Mech. Phys. Solids 37, 293–316 (1989)

Simmons, G., Wang, H.: Single Crystal Elastic Constants and Calculated Aggregate Properties: A Handbook. MIT Press, Cambridge (1971)

Staroselsky, A., Anand, L.: A constitutive model for hcp materials deforming by slip and twinning: application to magnesium alloy AZ31B. International Journal of Plasticity 19, 1843–1864 (2003)

Straumal, B., Sursaeva, V., Polyakov, S.: Faceting and Roughening of the Asymmetric Twin Grain Boundaries in Zinc. Interface science 9, 275–279 (2001)

Tomé, C., Lebensohn, R., Kocks, U.: A model for texture development dominated by deformation twinning: application to zirconium alloys. Acta metall. mater 39, 2667–2680 (1991)

Vaidya, S., Mahajan, S.: Accommodation and formation of {11-21} twins in Co single crystals. Acta Metall. 28, 1123–1131 (1980)

Wang, Y., Huang, J.: The role of twinning and untwinning in yielding behaviour in hot-extruded Mg-Al-Zn alloy. Acta Mater. 55, 897–905 (2007)

Wang, Y., Jin, Y., Khachaturyan, A.: The Effects of Free Surfaces on Martensite Microstructures: 3D Phase Field Microelasticity Simulation Study. Acta Mater. 52(3), 1039–1050 (2004)

Wonsiewicz, B., Backofen, W.: Plasticity of Magnesium Crystals. Trans. Metal. Soc. of AIME 239, 1422–1431 (1967)

Yoo, M.: Slip, Twinning, and Fracture in Hexagonal Close-Packed Metals. Metall. Trans. 12A, 409–418 (1981)

Zanzotto, G.: On the Material Symmetry Group of elastic crystals and the Born Rule, Arch. Rational Mech. Anal. 121, 1–36 (1992)

Zhou, A., Basu, S., Barsoum, M.: Kinking nonlinear elasticity, damping and microyielding of hexagonal close-packed metals. Acta Mater. 56, 60–67 (2008)

Appendix

1: It follows from the representation theory that the transversal isotropic and the hexagonal elastic stiffness tetrads coincide and have the components

$$
\begin{aligned}
C_{ijkl} = {} & a_1 \delta_{ij} \delta_{kl} + \\
& + a_2 (\delta_{ij} c_k c_l + c_i c_j \delta_{kl}) + \\
& + a_3 (\delta_{ik} \delta_{jl} + \delta_{il} \delta_{jk}) + \\
& + a_4 c_i c_j c_k c_l + \\
& + a_5 (\delta_{il} c_j c_k + \delta_{ik} c_j c_l + \delta_{jl} c_i c_k + \delta_{jk} c_i c_l)
\end{aligned} \tag{37}
$$

involving 5 independent elastic constants $a_{1..5}$ and the components c_i of $\boldsymbol{c}_{\mathrm{P}}$ or $\boldsymbol{c}_{\mathrm{T}}$, with respect to orthonormal base vectors.

2: The degree of anisotropy of a stiffness tetrad can be quantified by (Böhlke 2001)

$$
a = \frac{\left\| \mathbb{C} - 3K_{\mathrm{eq}} \mathbb{P}_I - 2G_{\mathrm{eq}} \mathbb{P}_{II} \right\|}{\left\| \mathbb{C} \right\|} \tag{38}
$$

$$
3K_{\mathrm{eq}} = \mathbb{C} \cdots\cdot \mathbb{P}_I \tag{39}
$$

$$
2K_{\mathrm{eq}} = \frac{1}{5} \mathbb{C} \cdots\cdot \mathbb{P}_{II}, \tag{40}
$$

where $\mathbb{P}_I$ and $\mathbb{P}_{II}$ are the isotropic projectors used in this article. In the case of an isotropic stiffness tetrad $\mathbb{C}$, it turns out that $a = 0$. The more anisotropic is $\mathbb{C}$, the larger becomes a. For magnesium, the elastic constants found in Simmons and Wang (1971) ($C_{11} = 56.49$; $C_{33} = 58.73$; $C_{44} = 16.81$; $C_{12} = 23.16$; $C_{13} = 18.1$ [GPa] with respect to the normalised Voigt notation) yield $a_{\mathrm{Mg}} \approx 0.06$. In comparison, the much more anisotropic titanium ($C_{11} = 123.1$; $C_{33} = 152.9$; $C_{44} = 30.7$; $C_{12} = 99.6$; $C_{13} = 68.8$ [GPa] Simmons and Wang (1971)) gives $a_{\mathrm{Ti}} \approx 0.17$. An example for a very anisotropic material is Li ($C_{11} = 13.4$; $C_{12} = 11.3$; $C_{44} = 9.6$ [GPa]), which yields an a of almost $a_{\mathrm{Li}} \approx 0.38$.

Part II

Fibre and Particle Reinforced Solids

Numerical Evaluation of Effective Material Properties of Piezoelectric Fibre Composites

S. Kari, H. Berger, and U. Gabbert

Institut für Mechanik, Otto-von-Guericke-Universität Magdeburg

Abstract. This paper presents a method for the evaluation of effective material properties of piezoelectric fibre composites using homogenisation techniques based on the finite element method (FEM) with representative volume element (RVE) method. Numerical studies are performed to estimate the influence of diameter and arrangement of fibres on effective material properties. All effective material properties of transversely randomly distributed uni-directional piezoelectric fibre composites are evaluated and comparisons are made with regular packing like square and hexagonal arrangements.

1 Introduction

Piezoelectric materials have the property of converting electrical energy into mechanical energy, and vice versa. This reciprocity in the energy conversion makes piezoelectric ceramics such as PZT (lead, zirconium, titanate) very attractive materials toward sensors and actuators applications. In the last years, composite piezoelectric materials have been developed by combining piezo-ceramic fibres with passive non-piezoelectric polymers. Such fibrous composites exhibit a good behaviour, and they are typically applied within the linear regime for such applications. It is of interest to know the overall coupled electro-mechanical properties and the local fields in the constituent phases. The micromechanical methods provide the homogenized properties of the piezoelectric fibre composites from the known properties of their constituents (fibre and matrix) through an analysis of a periodic RVE model. A number of methods have been developed to predict and to simulate the coupled piezoelectric and mechanical behaviour of composites. Basic analytical approaches have been reported (e.g., [10,20], which are not capable of predicting the response to general loading, i.e., they do not give the full set of overall material parameters. Semi analytical and Hashin/Shtrikman-type bounds for describing the complete overall behaviour (i.e., all elements of the material tensors) have been developed ([6,7]) which are useful tools for theoretical considerations. However, the range between the bounds can be very wide for certain effective moduli. Mechanical mean field type methods have been extended to include electro-elastic effects ([4,13,22,11]) based on an Eshelby-type solution for a single inclusion in an infinite matrix ([3,14]). Such mean field type methods are capable of predicting the entire behaviour under arbitrary loads. However, they use averaged representations of the electric and mechanical field within the constituents of the composite. This restriction can be overcome by employing periodic micro field approaches (commonly referred to as unit cell models) where the fields are typically solved numerically with high resolution, e.g., by the finite element method ([15]). In such models the representative unit cell and the boundary conditions are designed to capture a few special load cases, which are connected to specific deformation patterns (e.g. [8,9,12,16]). This allows the prediction of only a few key material parameters; for example, only normal loads can be applied consistently using the symmetry boundary conditions.

A different method, which can handle arbitrary loading scenarios, is the asymptotic homogenisation approach ([1,21]). The local problems are considered and the effective elastic, piezoelectric, and dielectric moduli are explicitly determined analytically.

Fig. 1. Cross section of a fibre composite

Most of these methods are restricted to quite regular packings of fibres like a rectangular arrangement and a hexagonal arrangement. This offers great simplicity to the analysis of the problem. However, in most practical situations of uni-directional piezoelectric fibre composites, the fibres are aligned in their longitudinal direction, while their arrangement in the matrix in transverse cross-section is usually distributed randomly as shown in Fig. 1. To the knowledge of the authors, there is not much development to handle the problem of transversely randomly distributed piezoelectric fibre composites. The aim of the present paper is to predict the full set of piezoelectric, dielectric, and mechanical effective material coefficients of these composites. Also different investigations are made to analyse the influence of the diameter of the fibres on effective material properties. In the following, the FEM based numerical homogenisation approach is applied to transversely random and unidirectional piezoelectric fibre composites subjected to different loading conditions with periodic boundary conditions to predict their effective coefficients. The developed RVE method is directed to produce a general procedure for the evaluation of effective properties of composites with complex geometrical reinforcements.

2 Piezoelectricity and Piezoelectric Composites

Coupled piezoelectric problems are those in which an electric potential gradient causes deformations (converse piezoelectric effect), while strains cause an electric potential gradient in the material (direct piezoelectric effect). The coupling between mechanical and electrical fields is characterized by piezoelectric coefficients. This paper considers piezoelectric materials that respond linearly to changes in the electric field, electric displacement, or mechanical stress and strain. These assumptions are compatible with the piezoelectric ceramics, polymers, and composites applied with a low electric field as in most cases of application [19]. Therefore, the behaviour of the piezoelectric medium can be described by the following piezoelectric constitutive equations, which correlate stresses T, strains S, electric fields E, and electrical displacements D as follows (the superscript t for transpose)

$$\begin{bmatrix} T \\ D \end{bmatrix} = \begin{bmatrix} C & -e^t \\ e & \varepsilon \end{bmatrix} \begin{bmatrix} S \\ E \end{bmatrix}, \tag{1}$$

where C is the elasticity matrix, ε is the permittivity matrix, and e is the piezoelectric strain coupling matrix.

For a transversely isotropic piezoelectric solid, there are 11 independent material properties in the stiffness matrix, the piezoelectric matrix, and the dielectric matrix. In the case of aligned fibres made of a transversely isotropic piezoelectric solid (PZT), embedded in an isotropic polymer matrix, the resulting composite is a transversely isotropic piezoelectric material (6mm) for a hexagonal array, and tetragonal (4mm) for a square array. Consequently, the constitutive Eq. (1) for the composite can be written as

$$\begin{bmatrix} \overline{T}_{11} \\ \overline{T}_{22} \\ \overline{T}_{33} \\ \overline{T}_{23} \\ \overline{T}_{31} \\ \overline{T}_{12} \\ \overline{D}_1 \\ \overline{D}_2 \\ \overline{D}_3 \end{bmatrix} = \begin{bmatrix} C_{11}^{eff} & C_{12}^{eff} & C_{13}^{eff} & 0 & 0 & 0 & 0 & 0 & -e_{13}^{eff} \\ C_{12}^{eff} & C_{11}^{eff} & C_{13}^{eff} & 0 & 0 & 0 & 0 & 0 & -e_{13}^{eff} \\ C_{13}^{eff} & C_{13}^{eff} & C_{33}^{eff} & 0 & 0 & 0 & 0 & 0 & -e_{33}^{eff} \\ 0 & 0 & 0 & C_{44}^{eff} & 0 & 0 & 0 & -e_{15}^{eff} & 0 \\ 0 & 0 & 0 & 0 & C_{44}^{eff} & 0 & -e_{15}^{eff} & 0 & 0 \\ 0 & 0 & 0 & 0 & 0 & C_{66}^{eff} & 0 & 0 & 0 \\ 0 & 0 & 0 & 0 & e_{15}^{eff} & 0 & \varepsilon_{11}^{eff} & 0 & 0 \\ 0 & 0 & 0 & e_{15}^{eff} & 0 & 0 & 0 & \varepsilon_{11}^{eff} & 0 \\ e_{13}^{eff} & e_{13}^{eff} & e_{33}^{eff} & 0 & 0 & 0 & 0 & 0 & \varepsilon_{33}^{eff} \end{bmatrix} \begin{bmatrix} \overline{S}_{11} \\ \overline{S}_{22} \\ \overline{S}_{33} \\ \overline{S}_{23} \\ \overline{S}_{31} \\ \overline{S}_{12} \\ \overline{E}_1 \\ \overline{E}_2 \\ \overline{E}_3 \end{bmatrix} \tag{2}$$

in both cases except that $C_{11}^{eff} - C_{22}^{eff} = 2C_{66}^{eff}$ for 6mm symmetry.

3 Numerical Homogenisation Technique

The object of study is regarded as a large-scale/macroscopic structure. The common approach to model the macroscopic properties of 3D piezoelectric fibre composites is to create an RVE, which captures the major features of the underlying microstructure. The fibres are regarded as a small-scale micro structure.

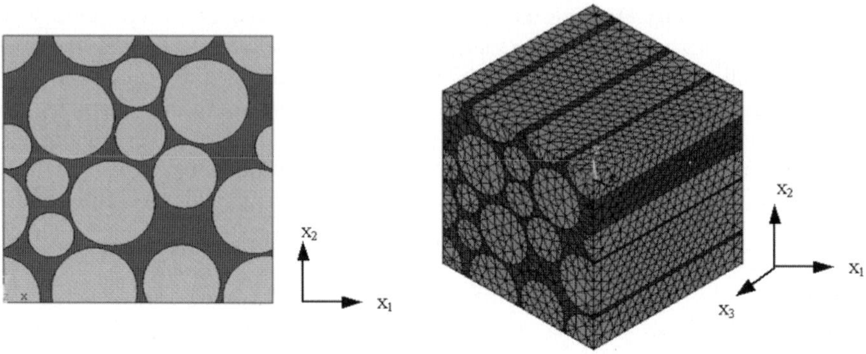

Fig. 2. Representative unit cell (left) and corresponding finite element mesh (right)

One of the most powerful tools to speed up the modelling process with both the composite discretisation and the computer simulation of composites in realistic conditions, is the homogenisation method. The main idea of the method is to find a globally homogeneous medium equivalent to the original composite, where the strain energy stored in both systems is approximately the same. The first step in the homogenisation technique is to create a representative volume element, which captures the overall behaviour of a composite structure. In the second step the effective material properties are calculated by applying appropriate load cases and periodic boundary conditions to the unit cell. The effective material properties should represent the effective material properties of the original composite structure.

3.1 Generation of RVE Models

The RVE was generated using the RSA algorithm [23] modified to provide a user specified minimum distance between any two fibres and for periodicity between opposite boundary surfaces.

First, the random coordinates are generated for the centre of each circular disc with a prescribed radius, where the circular disc is placed on the x_1-x_2 plane. The next generated coordinates of centres of circles are checked for non-overlapping condition with previously placed circles. If there is no overlap and the periodicity is satisfied, then the current circle will be placed on the plane. If a circle cuts the boundary of the unit cell, then on the opposite site also a circle has to be placed to grantee the periodicity. This process will be terminated when the desired volume fraction is achieved or when no more discs can be added because of the jamming limit, which can occur at a volume fraction higher than 55%. For higher volume fractions, different diameters of circles are used, and these are placed on the x_1-x_2 plane in a descending manner. With this approach the volume fraction achieved is about 80% with the adequate finite element meshing. Figure 2 shows the examples of generated RVEs with variable diameter of fibres, and their corresponding 3D finite element meshes.

3.2 Periodic Boundary Conditions

Composite materials can be represented as a periodical array of the RVEs. Therefore, periodic boundary conditions must be applied to the RVE models. This implies that each RVE in the composite has the same deformation mode, and there is no separation or overlap between the neighbouring RVEs after deformation. These periodic boundary conditions described in Cartesian coordinates are given by [21,24],

$$u_i = \bar{S}_{ij} x_j + v_i \tag{3}$$

In the above Eq. (3) $\bar{S}_{ij}$ are the average strains, v_i is the periodic part of the displacement components (local fluctuation) on the boundary surfaces, which is generally unknown and depends on the applied global loads. A more explicit form of periodic boundary conditions suitable for RVE models can be derived from the above general expression. The displacements on a pair of opposite boundary surfaces (with their normal along the x_j axis) are

$$u_i^{K^+} = \bar{S}_{ij} x_j^{K^+} + v_i^{K^+} \tag{4}$$

$$u_i^{K^-} = \bar{S}_{ij} x_j^{K^-} + v_i^{K^-} \tag{5}$$

where the index ' K^+ ' means along the positive x_j direction, and ' K^- ' means along the negative x_j direction on the corresponding surfaces of the 3D RVE. The local fluctuations $v_i^{K^+}$ and $v_i^{K^-}$ around the average macroscopic value are identical on two opposing faces due to the periodicity conditions on the RVE. So, the difference between the above two equations is the applied macroscopic strain condition

$$u_i^{K^+} - u_i^{K^-} = \bar{S}_{ij}(x_j^{K^+} - x_j^{K^-}) \tag{6}$$

Similarly, the periodic boundary conditions for electrical potential are given by applied macroscopic electric field condition

$$\Phi^{K^+} - \Phi^{K^-} = \bar{E}_i(x_i^{K^+} - x_i^{K^-}) \tag{7}$$

where Φ represents the voltage and $\bar{E}_i$ represents the average electric field.

It is assumed that the average mechanical and electrical properties of a unit cell are equal to the average properties of the particular composite. The average stresses and strains in the RVE are defined by

$$\bar{S}_{ij} = \frac{1}{V}\int_V S_{ij}\,dV \tag{8}$$

$$\bar{T}_{ij} = \frac{1}{V}\int_V T_{ij}\,dV \tag{9}$$

where V is the volume of the periodic RVE. Analogously, the average electric fields and electrical displacements are defined by

$$\overline{E}_i = \frac{1}{V}\int_V E_i dV \tag{10}$$

$$\overline{D}_i = \frac{1}{V}\int_V D_i dV \tag{11}$$

3.3 Finite Element Modelling of RVE

All finite element calculations are made with the commercial FE package ANSYS. The matrix and the fibres are meshed with 10 node tetrahedron elements (SOLID 98) with displacement degrees of freedom (DOF) and additional electric potential (voltage) degrees of freedom. These allow for a fully coupled electromechanical analyses. To obtain the homogenized effective material properties, periodic boundary conditions are applied to the RVE by coupling opposite nodes on the opposite boundary surfaces. In order to apply these periodic boundary conditions on the FE model of the RVE, the meshes on the opposite boundary surfaces must be the same. For each pair of displacement components at the two corresponding nodes with identical in-plane coordinates on two opposite boundary surfaces, a constraint equation (periodic boundary condition Eq. (6) and (7) in the previous section) is imposed. For this purpose any three faces of the RVE along the three coordinate directions are meshed with triangular area elements, and these area meshes are copied to the opposite faces of the RVE. Then, the meshing of the volumes of the fibres and of the matrix is performed. Finally, the area mesh is deleted. In this way identical meshes are achieved on the opposite boundary surfaces of the RVE, which are required in order to apply periodic boundary conditions. The mesh size used is fine enough to represent accurately the geometry of the fibres and the matrix. Applying the constraint equations on all opposite nodes at opposite boundary surfaces interactively in ANSYS is a very time consuming task due to a great number of coupled nodes. So, we automated this process by using the ANSYS Parametric Design Language (APDL) to generate all required constraint equations. Furthermore, we used the APDL for the evaluation of needed average strains and stresses, and evaluated the effective material properties in the end. The developed APDL-scripts in combination with the ANSYS batch processing provide a powerful tool for a fast calculation of homogenized material properties for composites with a great variety of inclusion geometries.

3.4 Boundary Conditions for Evaluation of the Different Effective Coefficients

For the evaluation of the effective coefficients, the boundary conditions have to be applied to the RVE in such a way that, except of one component of the strain/electrical field vector in Eq. (2), all other components are made equal to zero. Then each effective coefficient can be easily determined by multiplying the corresponding row of material matrix by the strain/electrical field vector. This can be achieved by applying the appropriate boundary conditions and constraint equations to the different surfaces of the RVE [5].

As an example of the application of the algorithm, the one leading to the calculation of the effective coefficients C_{13}^{eff} and C_{33}^{eff} is explained. The boundary conditions and constraint equations have to be applied to the RVE in such a way that, except of the strain in the x_3 direction ($\overline{S}_{33}$), all other mechanical strains and electrical field ($\overline{E}_i$) become zero. Due to the zero strains and electric fields, except of $\overline{S}_{33}$, the first row becomes $\overline{T}_{11} = C_{13}^{eff} \overline{S}_{33}$. Then C_{13}^{eff} can be calculated as the ratio of $\overline{T}_{11} / \overline{S}_{33}$. Similarly, C_{33}^{eff} can be evaluated as the ratio of $\overline{T}_{33} / \overline{S}_{33}$ from the third row of the matrix Eq. (2). For the calculation of the total average values $\overline{S}_{33}$, $\overline{T}_{11}$ and $\overline{T}_{33}$ according to Eqs. (8) and (9), the integral is replaced by a sum over averaged element values multiplied by the respective element volume. Analogously all other coefficients, the formulae of which are based on the average normal strain, can be evaluated.

4 Results and Discussion

Transversely randomly distributed uni-directional piezoelectric fibre composite RVE models are created using modified RSA algorithm. With the same diameter of the fibre using this algorithm, it is possible to create RVE models within our finite element meshing tool up to 50% volume fraction. It is not possible to generate higher volume fraction models because the jamming limit can be reached. A different approach is used to generate higher volume fraction RVE models, i.e., with different diameters of fibres. These will be placed in the RVE in a descending order. At first the fibres with the largest diameter are placed in the RVE, and then, the next possible largest diameter fibres are inserted, and so forth. This procedure has to assure that a prescribed minimum distance between any two fibres and the periodicity condition on the opposite boundary surfaces is guaranteed. With this approach it is possible to generate up to 80% volume fraction RVE models with an adequate FE mesh. Since in this approach, different diameters of fibres are used for higher volume fraction RVE models, different investigations are made to analyse the influence of the diameter of fibres on effective material properties in the linear micromechanical analysis.

For the calculation of the effective coefficients we consider a composite with the piezoelectric (PZT-5) fibres embedded randomly in a soft non-piezoelectric material (polymer) in the transverse cross section. Here we are assuming that the fibres and the matrix are ideally bonded, and that the fibres are straight and parallel to the x_3 axis. The fibre section is circular, and the RVE is having a square cross section. The piezoelectric fibres are uniformly polarized along the x_3 direction.

Table 1. Material properties of the composite constituents fibre (PZT-5) and matrix (Polymer)

	C_{11}	C_{12}	C_{13}	C_{33}	C_{44}	C_{66}	e_{15}	e_{13}	e_{33}	ε_{11}	ε_{33}
PZT-5	12.1	7.54	7.52	11.1	2.11	2.28	12.3	5.4	15.8	8.11	7.35
Polymer	0.386	0.257	0.257	0.386	0.064	0.064	-	-	-	0.07965	0.07965

The material properties of the polymer and PZT-5 are listed in Table 1, where elastic properties, piezoelectric constants and permittivities are given in GPa, C/m^2 and nF/m, respectively.

4.1 Effect of the Fibre Diameter on Effective Material Properties

Many investigations are performed to study the influence of the diameter of piezoelectric fibres on the effective material properties of these composites. Here the length of the RVE (*L*) remains constant, and by varying the diameter of fibres (*D*), using homogenisation techniques, different effective material properties are obtained at 45 % volume fraction. It can be observed that by increasing the *L/D,* the variations in the effective material properties are negligible. Also the fluctuation (error) of effective material properties around the mean value, which are obtained from the ensemble averages of the effective material properties of five RVE samples, are negligible.

The numerical homogenisation techniques are also applied to two different types of the RVE models, one with the identical diameter of the piezoelectric fibres and another with random diameter (0.32mm-0.12mm) of the fibres at 50% volume fraction. In both cases, five different RVE samples are considered, and the effective material properties are obtained from the ensemble average of the effective material properties of the five different samples. It can also be observed that the differences in the effective material properties are again negligible, and the differences between the random and identical diameter of piezoelectric fibres are at most 2% (see Table 2).

Table 2. Comparison of piezoelectric fibres, C[GPa], E [C/m^2], Eps [F/m]

50% volume fraction	C11	C12	C22	C23	C33	C13	C66	C44
Same diameter	9.40	4.93	9.07	5.47	31.96	5.58	1.87	1.99
Random diameter	9.23	4.93	9.13	5.49	31.91	5.52	1.87	2.00

50% volume fraction	E15	E13	E33	Eps11*10^{-09}	Eps33*10^{-09}
Same diameter	0.0021	-0.224	9.81	0.273	3.86
Random diameter	0.0021	-0.219	9.80	0.266	3.85

4.2 Comparison of Effective Material Properties for Different Arrangement of Fibres

The effective electrical and mechanical properties of transversely randomly distributed uni-directional piezoelectric fibre composites are evaluated for different volume fractions up to 80%. The effective material properties, which are obtained for these cases, are compared with square arrangement and hexagonal arrangement of the piezoelectric fibres. For the square arrangement the maximum theoretically achievable volume fraction is 78.54%. Due to the meshing limits with our finite element approach a maximum volume fraction of 70% can be generated only.

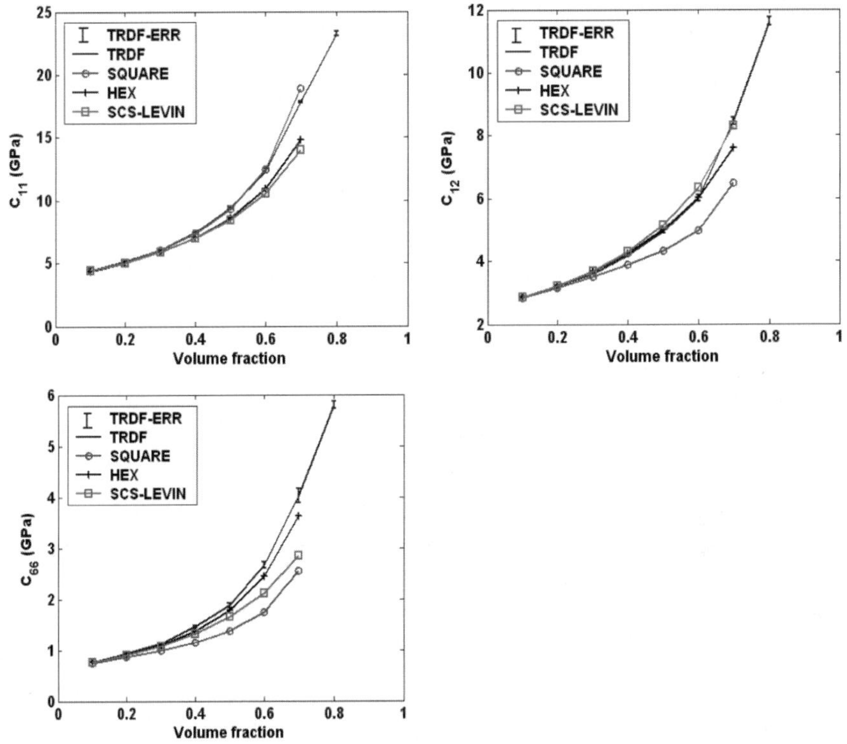

Fig. 3. Comparison of effective mechanical properties of transversely randomly distributed piezoelectric fibre composites (TRDF) with square (SQUARE) and hexagonal (HEX) array with the analytical self-consistent scheme by Levin [17] (SCS-Levin)

Figs. 3 and 4 represent the effective mechanical properties and the piezoelectric properties, respectively, calculated for different fibre arrangements in the composite. The figures compare the numerically calculated results based on different fibre arrangements, such as the transversely randomly distributed arrangement (TRDF), the square arrangement (SQUARE) and the hexagonal arrangement (HEX) with the results calculated by the self consistent schema (SCS-LEVIN) [17]. From Figure 3 it can be observed that the transverse mechanical properties are tending to increase for transversely randomly distributed composites when compared with regular array composites, especially for hexagonal array, but not in all other cases. As a comparison between the square array and the hexagonal array, the hexagonal array has a 6-fold axis of symmetry along fibre direction, and results in a transverse isotropic behaviour, i.e., $C_{11}^{eff} - C_{22}^{eff} = 2C_{66}^{eff}$, whereas for the case of the square array, it has only 4-fold axis of symmetry, and it will give rise to a tetragonal behaviour resulting in a higher transverse stiffness. For the transverse shear modulus C_{66}^{eff}, it is observed that the square array composite has a lower transverse shear modulus, and for hexagonal array composite it is has a higher value and satisfies the transverse isotropy.

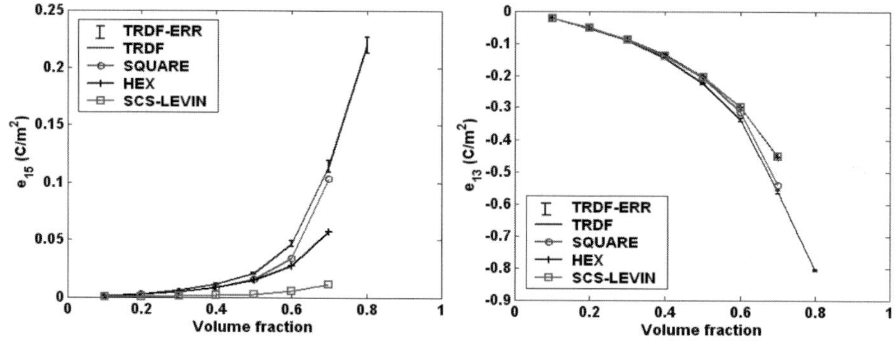

Fig. 4. Comparison of effective piezoelectric properties of transversely randomly distributed fibre composites (TRDF) with square (SQUARE) and hexagonal (HEX) array with the analytical self-consistent scheme by Levin [17] (SCS-Levin)

For the effective coefficient C_{33}^{eff}, all three cases yield approximately the same result, and this is due to the fact that in all three cases fibres are aligned in their longitudinal direction (x_3-direction) and uniformly polarized in x_3 direction.

For the piezoelectric material coefficient e_{33}^{eff}, and the dielectric material coefficient ε_{33}^{eff}, a similar behaviour can be observed like C_{33}^{eff} as stated above. Therefore, these diagrams are not presented in the paper. In general, from our analysis it can be observed that the assumption of a transversely randomly distributed fibre composites results in higher transverse material properties when compared with a regular array of fibre arrangement . The longitudinal material properties are almost the same like for a regular array of composites. Also the numerical results of the effective material coefficients like C_{11}^{eff}, ε_{11}^{eff} along the transverse direction of a transversely randomly distributed fibre composites match well with the results of SCS at lower volume fractions in the considered fraction range between 10% and 40%. Beyond this volume fraction range, the effective coefficients of SCS are underestimated. Along the longitudinal direction, the numerical results fit very well with the SCS for all volume fractions. From the overall observation, the results of SCS are closer to the numerical results for hexagonal arrangement. The fibres in the transverse cross section are placed with the uniform random distribution. So the resulting structure in the transverse cross section is isotropic, and, consequently, only 11 independent effective material coefficients are required to describe the behaviour of these composites.

The transverse isotropy was checked for all generated RVE samples and for all effective material coefficients.

5 Conclusions

The numerical homogenisation techniques are applied to evaluate the effective material properties of piezoelectric composites for different volume fractions. Several investigations are made to analyse the influence of the fibre diameter on the effective material properties, and it was found that the influences are very small. The results show that the effective material properties depend mainly on the volume fraction. This statement is valid only for the linear case for the evaluation of the effective material properties. There may be some influence in the non-linear case, debonding, and damage predictions. Further investigations have to be carried out to determine the influence of the size of the fibres on the behaviour of the composites at the macro level. By considering this fact using the RSA algorithm with different diameters, higher volume fraction RVE models are generated up to 80%. A comparison of the effective material properties are made between the regular packing and the transversely randomly distributed composites, and from this it can be concluded that the transverse effective material coefficients have higher values in case of transversely randomly distributed composites when compared to the regular packing composites like square and hexagonal array of fibres. Also the comparison of the effective coefficients obtained by using a numerical method is made with analytical methods like the self-consistent method and the Schulgasser universal relations [18], and a good agreement is achieved. A generalized procedure has been developed to calculate all effective coefficients automatically for all volume fractions based on the ANSYS Parametric Design Language. It reduces the manual work and time and can be used as a template to evaluate the effective coefficients of the piezoelectric fibre composites with arbitrary arrangement of fibres.

Acknowledgments. This work has been supported by DFG Germany, Graduiertenkolleg 828 "Micro-Macro Interactions in Structured Media and Particle Systems". This support is gratefully acknowledged.

References

[1] Bakhvalov, N., Panasenko, G.: Homogenization: Averaging Processes in Periodic Media - Mathematical Problems in the Mechanics of Composite Materials. Kluwer, Dordrecht (1989)
[2] Bensoussan, A., Lions, J.L., Papanicolaou, G.: Asymptotic analysis for periodic structures. North-Holland, Amsterdam (1978)
[3] Benveniste, Y.: The determination of elastic and electric fields in a piezoelectric inhomogeneity. J. Appl. Phys 72(3), 1086–1095 (1992)
[4] Benveniste, Y.: Universal relations in piezoelectric composites with eigenstress and polarization fields. Part I: Binary media –local fields and effective behavior. J. Appl. Mech. 60(2), 265–269 (1993)
[5] Berger, H., Kari, S., Gabbert, U., Rodriguez-Ramos, R., Guinovart-Diaz, R., Otero, J.A., Bravo-Castillero, J.: An analytical and numerical approach for calculating effective material coefficients of piezoelectric fibre composites. Int. J. Sol. Struct. 42, 5692–5714 (2005)
[6] Bisegna, P., Luciano, R.: Variational bounds for the overall properties of piezoelectric composites. J. Mech. Phys. Solids 44(4), 583–602 (1996)
[7] Bisegna, P., Luciano, R.: On methods for bounding the overall properties of periodic piezoelectric fibrous composites. J. Mech. Phys. Solids 45(8), 1329–1356 (1997)
[8] Böhm, H.J.: Numerical investigation of microplasticity effects in unidirectional long fibre reinforced metal matrix composites. Modell. Simul. Mater. Sci. Engng. 1(5), 649–671 (1993)

[9] Brockenbrough, J.R., Suresh, S.: Plastic deformation of continuous fibre-reinforced metal-matrix composites: effects of fibre shape and distribution. Scr. Metall. Mater. 24, 325–330 (1990)

[10] Chan, H.L., Unsworth, J.: Simple model for piezoelectric polymere 1-3 composites used in ultrasonic transducer applications. IEEE Trans. Ultrason. Ferroelectrics Frequency Control 36(4), 434–441 (1989)

[11] Chen, T.: Piezoelectric properties of multiphase fibrous composites: some theoretical results. J. Mech. Phys. Sol. 41(11), 1781–1794 (1993)

[12] Cleveringa, H.H.M., van der Giessen, E., Needleman, A.: Comparison of discrete dislocations and continuum plasticity predictions for a composite material. Acta Metall. Mater. 45(8), 3163–3179 (1997)

[13] Dunn, M.L., Taya, M.: Micromechanics predictions of the effective electroelastic moduli of piezoelectric composites. Int. J. Sol. Struct. 30(2), 161–175 (1993)

[14] Dunn, M.L., Wienecke, H.A.: Inclusions and inhomogeneities in transversely isotropic piezoelectric solids. Int. J. Sol. Struct. 34(27), 3571–3582 (1997)

[15] Gaudenzi, P.: On the electromechanical response of active composite materials with piezoelectric inclusions. Comput. Struct. 65(2), 157–168 (1997)

[16] Gunawardena, S.R., Jansson, S., Leckie, A.: Modeling of anisotropic behavior of weakly bonded fibre reinforced MMC's. Acta Metall. Mater 41(11), 3147–3156 (1993)

[17] Levin, V.M., Rakovskaja, M.I., Kreher, W.S.: The effective thermoelectroelastic properties of microinhomogeneous materials. Int. J. Solids Struct. 36, 2683–2705 (1999)

[18] Schulgasser, K.: Relationships between the effective properties of transversely isotropic piezoelectric composites. J. Mech. Phys. Solids 40, 473–479 (1992)

[19] Silva, E.C.N., Fonseca, J.S.O., Kikuchi, N.: Optimal design of periodic piezocomposites. Comput. Methods Appl. Engng. 159, 49–77 (1998)

[20] Smith, W.A., Auld, B.A.: Modeling 1-3 composite piezoelectrics: thickness-mode oscillations. IEEE Trans. Ultrason. Ferroelectrics Frequency Control 38, 40–47 (1991)

[21] Suquet, P.: Elements of homogenization theory for inelastic solid mechanics. In: Sanchez-Palencia, E., Zaoui, A. (eds.) Homogenization Techniques for Composite Media, pp. 194–275. Springer, Berlin (1987)

[22] Wang, B.: Three-dimensional analysis of an ellipsoidal inclusion in a piezoelectric material. Int. J. Sol. Struct. 29, 293–308 (1992)

[23] Wang, J.S.: Random sequential adsorption, series expansion and monte carlo simulation. Physics A 254, 179–184 (1998)

[24] Xia, Z., Zhang, Y., Ellyin, F.: A Unified periodical boundary conditions for representative volume elements of composites and applications. Int. J. Sol. Struct. 40, 1907–1921 (2003)

Evolutionary Optimisation of Composite Structures

N. Bohn and U. Gabbert

Institut für Mechanik, Otto-von-Guericke-Universität Magdeburg

Abstract. The optimisation of composite structures is one of the major demands of the aerospace industry. Such optimisation tasks lead often to problems where the objective functions are non-smooth, non-differentiable and multimodal. For these reasons, analytical optimisation methods seem to be an inappropriate choice. In the present study, the use of evolution strategies, which exhibit excellent global search properties and require no gradient information, is proposed. An optimisation algorithm is developed on the basis of this method. The software can be coupled with finite element tools to obtain the information necessary for the evaluation of the objective function. Such a general optimisation approach allows for the treatment of various problem classes. In this paper three applications from the fields of topology optimisation, material design and laminate optimisation are presented. The results demonstrate that evolution strategies are a suitable method for the optimisation of composite structures.

1 Introduction

Composite materials play a major role in meeting the increasing demand of the aerospace industry for lightweight and low-cost structures. Compared to classical monolithic engineering materials, composites offer higher specific strength and specific stiffness values. Therefore, they can be efficiently used to minimize the overall mass while the structure is still meeting certain requirements. Typically fibrous particles are applied to reinforce a matrix material where the non-symmetric fibre geometry leads to macroscopically orthotropic composite properties. Optimal structures can be obtained by tailoring the material to the specific technical application. For a cost effective optimisation, numerical methods should be applied. Structural models on the basis of the finite element method (FEM), where the mechanical response of the models depends on a N-dimensional vector of design variables $\mathbf{x}$, are usually employed. The optimisation goal is defined by an appropriate objective function $f(\mathbf{x})$, which returns a scalar value to estimate the quality of the structural behaviour. The optimisation procedure does not require the objective function as analytical function. It is sufficient to be able to evaluate the objective function pointwise, e.g., by a finite element analysis.

The general formulation of the optimisation problem reads in standard mathematical terms as follows

$$\min_{\mathbf{x}\in S} f(\mathbf{x}), \quad S = \left\{ \mathbf{x} \in \Re^N \;\middle|\; h_i(\mathbf{x}) = 0,\; g_j(\mathbf{x}) \le 0 \right\}, \tag{1}$$

where $h_i(\mathbf{x})$, $i = 1,\ldots,n$ are the equality constraints, and $g_j(\mathbf{x})$, $j = 1,\ldots,m$ denote the inequality constraints. Since equality constraints can be implemented by using two inequality restrictions, consequently, only inequality constraints are considered. The restricted optimisation problem (1) can be transformed into a free one via the penalty-method [1]. In this case the basic idea is to use a modified objective function

$$\tilde{f}(\mathbf{x}) = f(\mathbf{x}) + \gamma \cdot \sum_{j=1}^{m} \left(\max \left\{ g_j(\mathbf{x}), 0 \right\} \right)^2 , \tag{2}$$

where the newly introduced *penalty-terms* ensure that the constraints are not violated. The penalty parameter γ has to be adjusted according to the optimisation task. By applying this method, we restrict ourselves to free optimisation problems of the type

$$\min_{\mathbf{x} \in S} \tilde{f}(\mathbf{x}), \quad S = \Re^N . \tag{3}$$

A solution of (3) can be obtained iteratively by means of a descent direction $\mathbf{d}^{(g)}$ and a step-size $t^{(g)}$

$$\mathbf{x}^{(g+1)} = \mathbf{x}^{(g)} + t^{(g)} \cdot \mathbf{d}^{(g)} . \tag{4}$$

Therefore, in each step g a suitable search direction has to be found, and then a step-size along this direction must be computed. Two main approaches are available to solve this problem: the analytical methods and the direct methods. For analytical methods, derivatives of the objective function are employed to determine the search direction $\mathbf{d}^{(g)}$.

Various methods using different approaches, have been developed. Among the most important are the steepest descent method, the conjugate gradient method, Newton's method, and the quasi-Newton methods [1, 2]. Since all these techniques require the computation of gradients, their operability can only be guaranteed if the objective function is continuously differentiable and unimodal.

Direct methods use a different approach. The search direction as well as the step-sizes are computed from information about the objective function values only. Thus, these methods can be applied more or less directly to noisy, non-smooth or multimodal optimisation problems. Since a wide variety of algorithms exists, a general concept for the construction of the search directions and the step-sizes cannot be given here. Further details can be found in [3].

When the optimisation of composite materials is considered, the objective functions are more likely to be non-smooth, non-differentiable and multimodal [4, 5]. Furthermore, if the objective function has to be evaluated by means of a finite element analysis, it is not possible to compute the derivatives analytically, and numerical difference methods must be used. The design variables are then perturbed individually, and a new function evaluation has to be performed for each perturbed variable. Such a sensitivity analysis is very time consuming even for rather small problem dimensions. For these reasons, the analytical methods are generally not suitable for the optimisation of composite structures, and we suggest applying direct methods. The present study focuses particularly on evolution strategies, which operate on a set of feasible solutions and provide an efficient method with excellent global search properties [6].

The algorithm can be easily implemented, and various external software tools can be integrated without a great effort for the evaluation of the objective function. More details about evolution strategies can be found in the second section. The third section shows some applications.

2 Evolution Strategies

This section provides a short introduction to evolution strategies (ES), which, belonging to the class of direct optimisation methods, require no gradient information. The optimum is sought for by the application of the principles of the natural evolution, i.e., recombination, mutation and selection to a set μ of feasible solutions, the individuals

$$\mathbf{w} = \lfloor x_1, x_2, \ldots, x_N, \ \sigma_1, \sigma_2, \ldots, \sigma_{N_\sigma} \rfloor, \quad k = 1, \ldots, \mu . \tag{5}$$

In this context, N stands for the problem dimension, x_i represent the design variables of the optimisation problem, and σ_i are the step-sizes for the mutation process. For well-conditioned optimisation problems, it is sufficient to set $N_\sigma = 1$, while more flexibility on badly scaled objective functions can be obtained by choosing $N_\sigma = N$ [7]. Thus, one individual represents one complete design and contains additional information to control the optimisation algorithm. The generation of new solutions starts with recombination, where a number of $\lambda \geq \mu$ offspring individuals are created by exchanging or averaging the properties of randomly selected parents

$$w'_{ki} = \begin{cases} w_{pi} \text{ or } w_{qi} \\ \frac{1}{2}(w_{pi} + w_{qi}), \end{cases} \quad k = 1, \ldots, \lambda, \ \ i = 1, \ldots, N, \ \ p, q \sim U(1, \mu). \tag{6}$$

For more recombination variants see [7]. Note that recombination is applied to the object variables as well as to the mutation step-sizes. The parameters $p, q \sim U(1, \mu)$ represent realizations of random variants underlying the uniform distribution $U(1, \mu)$. After recombination an intermediate population of λ individuals

$$\mathbf{w}' = \lfloor x'_1, x'_2, \ldots, x'_N, \ \sigma'_1, \sigma'_2, \ldots, \sigma'_{N_\sigma} \rfloor, \quad k = 1, \ldots, \lambda , \tag{7}$$

exists. A mutation is carried out by the application of small random changes in each component of an individual. Normal distributed random numbers with a zero mean value are used, and the standard deviation of this distribution function is represented by the mutation step-sizes which are part of the individuals. The process starts with the variation of the mutation step-sizes

$$\sigma''_{ki} = \sigma'_{ki} \cdot h(\bar{z}_k, \tau_1, \ldots, \tau_r) . \tag{8}$$

The function h depends on the standard normal distributed random variants $\bar{z} \sim N(0,1)$ as well as on γ heuristic factors τ_j, $j = 1, \ldots, \gamma$. Detailed information can again be found in [6, 11, 12]. The object variables are then mutated according to

$$x''_{ki} = x'_{ki} + z_{ki}, \quad k = 1, \ldots, \lambda , \tag{9}$$

where $z_{ki} \sim N(0, \sigma''_{ki})$ is a normal distributed random variant which depends on the individual mutation step-sizes. The population consists finally of λ offsprings

$$\mathbf{w}'' = \lfloor x''_1, x''_2, ..., x''_N, \ \sigma''_1, \sigma''_2, ..., \sigma''_{N_\sigma} \rfloor, \quad k = 1, ..., \lambda . \tag{10}$$

After the evaluation of the objective function is performed for each individual, the μ best individuals are selected to become the parents for the next iteration. In the case of a (μ + λ)-selection, the best μ individuals are chosen from the set of the union of parents and offspring, and in the case of a (μ, λ)-selection the best μ individuals are chosen only from the set of the offsprings. For multimodal objective functions, the (μ + λ) scheme is likely to get trapped in local minima, while a (μ, λ) strategy is able to locate even the global optimum for sufficiently large populations. The generation loop, consisting of recombination, mutation and selection, is then repeated until a termination criterion, such as a lower limit for the mutation step-sizes or a maximal number of generations, is fulfilled. Due to the fact that the mutation step-sizes are part of the individuals and undergo recombination and mutation as well, the selection process leads to their adaption during the search: the best individuals not only contain good object parameter settings, but they also exhibit well adjusted mutation step-sizes, which guarantee further improvement with respect to the objective function. Evolution strategies can also be used to solve discrete optimisation problems if the mutation operators are appropriately converted [8, 9]. For detailed information about the method it is referred to [6, 7, 10-12].

3 Applications

The optimisation of composite structures involves two major steps. First, a suitable software tool is needed for the development of a structural model and the computation of the mechanical response. In this case the finite element method is frequently applied. Furthermore, it is necessary to couple this model representation with a reliable and robust optimisation algorithm. A general software tool is developed on the basis of evolution strategies with an interface to the commercial finite element code Ansys.

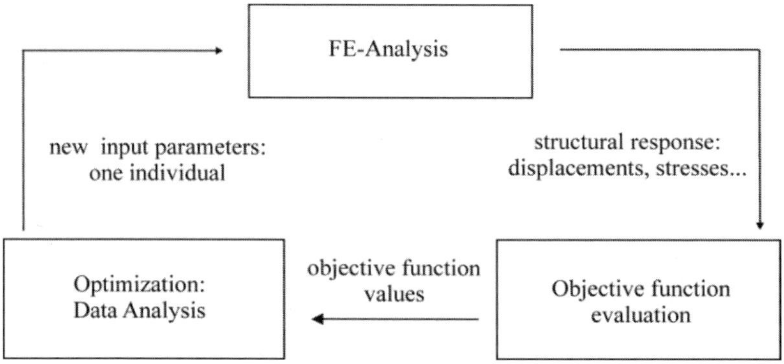

Fig. 1. Main steps during the optimisation

Our software possesses a modular structure, allowing for the implementation of various intermediate steps in the optimisation process. A flow chart of the main steps in the optimisation tool is shown in Fig. 1. In the following three examples are presented.

3.1 First Example: Topology Optimisation

The first application is related to the field of topology optimisation. We consider a plate consisting of two materials having the Young's moduli $E_1 = 1 \cdot 10^{10}\,\mathrm{Nm}^2$, $E_2 = 1 \cdot 10^{8}\,\mathrm{Nm}^2$, the Poisson's ratios $\nu_1 = \nu_2 = 0.3$, and the mass densities $\rho_1 = 2.7 \cdot 10^3\,\mathrm{kg/m}^3$, $\rho_2 = 4.0 \cdot 10^2\,\mathrm{kg/m}^3$. The plate has the dimensions 0.8m × 0.15m, and the applied load has an intensity of $q = 10\,\mathrm{kN/m}$. Fig. 2 shows the model and the boundary conditions.

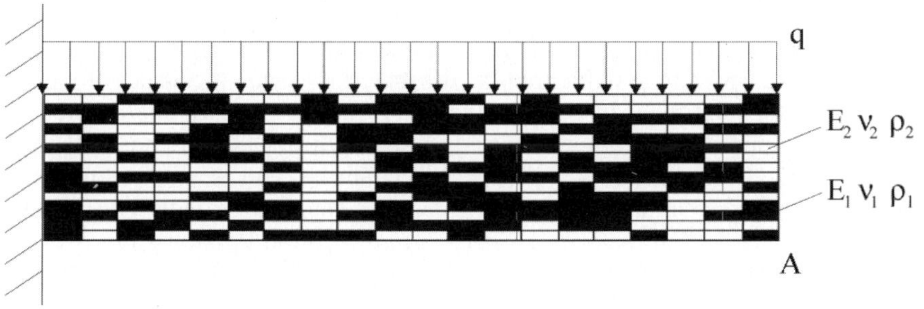

Fig. 2. Composite plate, geometry and boundary conditions

The objective is to find the optimal distribution of material one, that provides a minimal weight M and a deflection at point A lower than a prescribed value $\hat{v}_A = 6 \cdot 10^{-3}\,\mathrm{m}$. Thus, the object variables are represented by the material properties of each finite element. We apply a (10+20)-ES with a discrete recombination scheme according to equation (6). Since only two material types are present, the mutation process needs no step-sizes, and it is sufficient to alter the material properties for three randomly selected finite elements. The displacement restriction is implemented in the objective function by using a penalty term, and the problem statement is then

$$\min\ f(M, v_A) = M + \gamma \cdot \begin{cases} (v_A - \hat{v}_A)^2 & \text{for } |v_A| \geq |\hat{v}_A| \\ 0 & \text{else} \end{cases} \tag{11}$$

A penalty parameter $\gamma = 10^{20}$ is used. The evolution of the material distribution can be seen in Fig. 3, where an intermediate design and the optimal material layout are shown. Concerning manufacturing considerations, the optimal structure should be of course constructed by a continuous reinforcement. The mass of the plate is reduced from 90.52kg to 47.12kg.

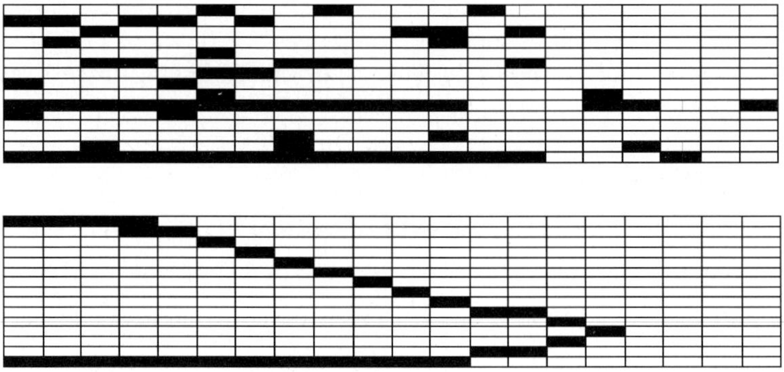

Fig. 3. Composite plate, intermediate and optimal design

3.2 Second Example: Optimisation of Short Fibre Reinforced Composites

Evolution strategies can be applied to tailor a material for a given structure. Consider a quadratic plate with a hole in the middle, the material of which consists of a polysulfon matrix being reinforced by short aramid fibres. Fig. 4 shows one quarter of the model with the initial fibre orientations. A side length of 200mm and a 30mm hole radius are used. The load intensity is p = 300N/mm, and an isotropic, linear elastic material behaviour is assumed for the fibres and the matrix with given Young's moduli $E_f = 8.3 \cdot 10^4$N/mm^2 and $E_m = 2.6 \cdot 10^3$N/mm^2 and the Poisson's ratios $\nu_f = 0.22$ and $\nu_m = 0.35$ (m refers to the matrix, and f indicates the fibre properties). The amount of fibres in the plate is limited to a prescribed average fibre density of Φ^0_{av}. The purpose of the optimisation is to distribute a fixed amount of fibres in the matrix material in such a way, that a maximal structural stiffness is obtained under restrictions concerning the maximal principal stress σ_{h1} and the average fibre density in the plate. The constraints are $\sigma_{h1} \le \sigma^0_{h1} = 900\text{MPa}$ and $\Phi_{av} = \Phi^0_{av} = 0.3$. We further restrict the local element fibre densities Φ_i to lie within the interval $0 \le \Phi_i \le 0.6$. The fibre orientation α_i in each finite element of the model is subjected to the condition $-90° \le \alpha_i \le 90°$. Both the fibre density and fibre orientation are chosen as parameters for the optimisation.

Therefore, element-wise orthotropic material properties are assumed, where the material axes are defined by the variable α_i, and the effective material parameters per element are functions of the local fibre densities Φ_i (i is the element number).

Eq. (12) is used to calculate the effective material properties [13], where E_1 represents the Young's modulus in the fibre direction, E_2 the Young's modulus perpendicular to the fibre direction, G_{12} the shear modulus, and ν_{12} the Poisson's ratio,

$$E_1^i = \Phi_i \cdot E_f + (1-\Phi_i) \cdot E_m, \quad \nu_{12}^i = \Phi_i \cdot \nu_f + (1-\Phi_i) \cdot \nu_m,$$
$$E_2^i = \frac{E_m}{1-\sqrt{\Phi_i} \cdot \left(1-\frac{E_m}{E_f}\right)}, \quad G_{12}^i = \frac{G_m}{1-\sqrt{\Phi_i} \cdot \left(1-\frac{G_m}{G_f}\right)}. \tag{12}$$

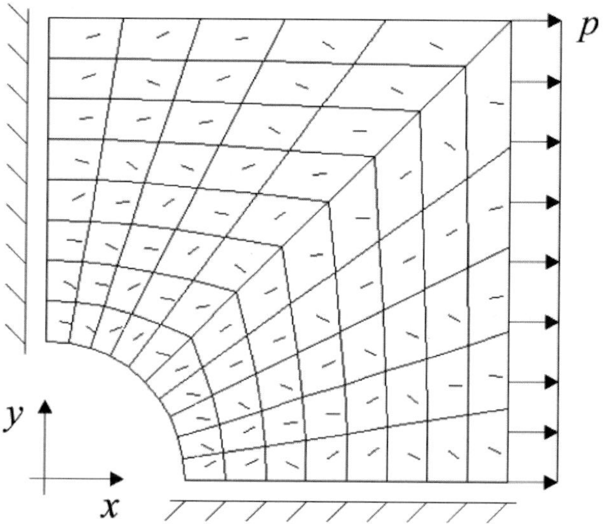

Fig. 4. Plate with hole, model with initial fibre orientations

With the plate volume V_p and the volume of the i-th finite element V_i, the average fibre density in the plate can be computed by

$$\Phi_{av} = \frac{\sum V_i \cdot \Phi_i}{V_p} \quad . \tag{13}$$

The objective function to be minimized is finally

$$f(\boldsymbol{\alpha},\boldsymbol{\Phi}) = \int_{(V)} \boldsymbol{\sigma}^{\mathrm{T}}\boldsymbol{\varepsilon}\,\mathrm{d}V + \gamma_1 \cdot \begin{cases} \left(\Phi_{av} - \Phi_{av}^0\right)^2 & \text{if} \quad \Phi_{av} \neq \Phi_{av}^0 \\ 0 & \text{else} \end{cases} + \gamma_2 \cdot \begin{cases} \left(\sigma_{h1} - \sigma_{h1}^0\right)^2 & \text{if} \quad \sigma_{h1} > \sigma_{h1}^0 \\ 0 & \text{else.} \end{cases} \tag{14}$$

$\boldsymbol{\sigma}$ and $\boldsymbol{\varepsilon}$ represents the stresses and the strains, respectively. The first term in Eq. (14) is therefore equivalent to the elastic energy in the structure. The second term ensures[1] that the average fibre density in the plate remains equal to Φ_{av}^0, and the third term penalizes violations of the stress restriction. The penalty parameters $\gamma_1 = \gamma_2 = 10^{20}$ are used. A (10,70)-ES with discrete recombination and Schwefel's mutation type is applied [12]. In this case, Eq. (8) has to be specified by using the formula $h = \exp(\tau_1 \cdot \bar{z}_k)$, where the relation $\tau \sim 1/\sqrt{N}$ holds (for detailed information see [7, 10, 11]).

[1] Simply two inequality constraints are implemented [1].

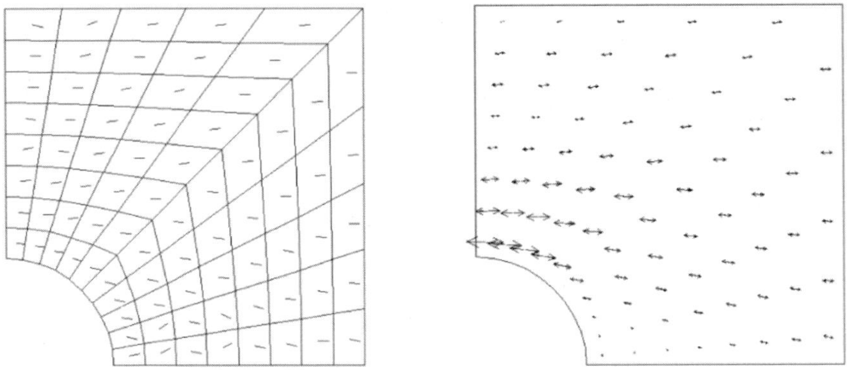

(a) fiber orientations (b) principal stress directions

Fig. 5. Optimal short fibre orientations and principal stress directions

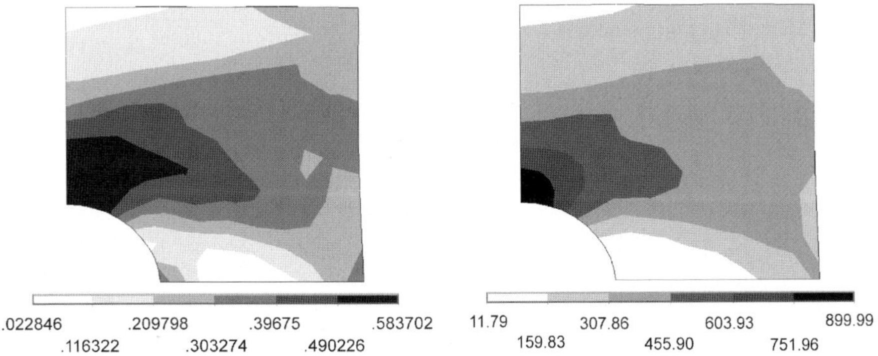

(a) fiber density (b) maximal principal stress

Fig. 6. Optimal plate design: fibre density distribution and maximal principal stress σ_{h1}

Fig. 5 shows the optimal fibre orientations and the directions of the principal stresses. Obviously the fibre orientations correspond to the main stress axes.

The optimal fibre density distribution can be obtained from Fig. 6 (a). The fibres tend to concentrate in the regions close to the upper side of the hole, where the maximal stresses occur. Such a result agrees with our expectations, since, due to the high stresses, these regions contribute to a great extent to the overall plate stiffness and can thus be considered as very efficient zones for the placement of the reinforcing phase. Figure 6 (b) shows the contours of the maximal principal stress σ_{h1} for the optimal design. The stress restriction $\sigma_{h1} \leq 900$MPa is obviously fulfilled.

The optimisation algorithm converges after about 80 generations leading to a decrease of 49% in the objective function when compared to the random initial design. A similar fibre distribution is obtained on the basis of a finite element homogenisation model (see [14]).

3.3 Third Example: Laminate Ply Angle Optimisation

A simply supported 6-layer laminate plate under a pressure load of $p = 1\text{N/mm}^2$ is considered. The plate thickness is $t = 5\text{mm}$ and the material is described by the Young's moduli $E_x = 82550\text{N/mm}^2$, $E_y = 6289\text{N/mm}^2$, the shear modulus $G_{xy} = 2909\text{N/mm}^2$ and the Poisson's ratio $v = 0.3$. The numbering of the layers is from bottom to top. The model is shown in Fig. 7.

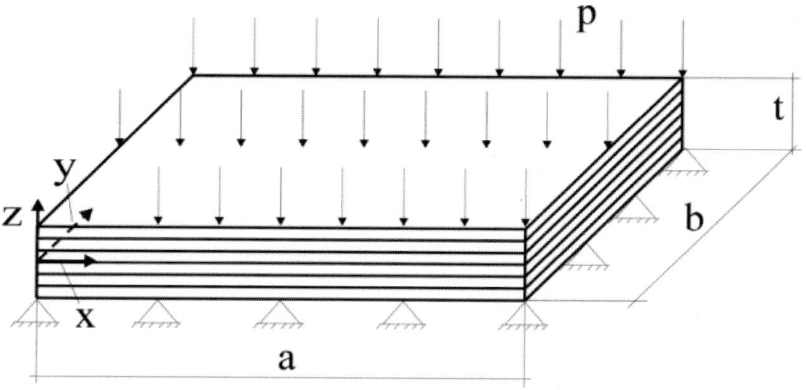

Fig. 7. Laminate plate, model and boundary conditions

An orientation angle of α_i, $i = 1,\ldots,6$, is assigned to each of the six layers of the composite, with the restrictions $-90° \leq \alpha_i \leq 90°$. The reference axis for the angles is the global x-axis. Positive orientations are measured in the mathematical positive sense from the x-axis towards the y-axis, and negative orientations indicate that the angle is measured in the mathematical negative sense from the x-axis towards the y-axis. We aim to optimise the ply orientations α_i to obtain a structure of maximal stiffness. Thus, we minimize the objective function

$$f(\boldsymbol{\alpha}) = \int_{(V)} \boldsymbol{\sigma}^{\mathbf{T}} \boldsymbol{\varepsilon} \, dV. \tag{15}$$

Table 1 shows the resulting ply orientations for various aspect ratios of the side lengths a and b.

Table 1. Optimal ply orientations for various aspect ratios of the side lengths a and b

Side lengths a:b [mm]	Optimal ply orientation angle [degree]					
	α_1	α_2	α_3	α_4	α_5	α_6
600:600	45.0 /	-44.9 /	-45.5 /	-45.3 /	-45.0 /	44.9
600:400	40.2 /	-37.7 /	-37.2 /	-37.4 /	-37.8 /	40.1
600:300	35.8 /	-29.0 /	-29.5 /	-28.7 /	-29.4 /	35.6
600:200	2.5 /	-89.9 /	-88.6 /	-89.9 /	-89.9 /	3.1

The results show that the optimal designs tend to exhibit one orientation for the outer layers and another common ply orientation for all inner layers. This phenomenon can be observed for all plate geometries. For the quadratic plate, orientations of 45° in the top and the bottom layers go together with orientations of -45° in the inner layers. The optimal angles in the outer and the inner layers decrease with decreasing side length b, and approximately 0° at the outer sides and -90° for the inner layers are obtained for the last geometry variant. Such results can be expected, since the bending stiffness along the direction of b increases with decreasing values of b. Thus, a non-equal bending stiffness in both directions results for different side lengths a and b. This stiffness discrepancy is then compensated by the material orientations.

4 Conclusions

In the present paper evolution strategies are introduced as a reliable and efficient method for the optimisation of composite structures. The algorithm operates on a set of feasible solutions and requires no derivatives of the objective function. Thus it has superior global search qualities and can be used even for non-differentiable, non-smooth and multimodal problems, which arise frequently in the context of the optimisation of composite structures. The test examples underlined the good search properties of evolution strategies.

Acknowledgments. The support of the DFG-Graduiertenkolleg 828 *Micro-Macro-Interactions in Structured Media and Particle Systems* is gratefully acknowledged.

References

[1] Fletcher, R.: Practical Methods of Optimization. Wiley & Sons, Chichester (1996)
[2] Haftka, R., Gürdal, Z.: Elements of Structural Optimization. Kluwer, Dordrecht (1992)
[3] Powell, M.J.D.: Direct search algorithms for optimization calculations. Acta Numerica 7, 287–336 (1998)
[4] Zohdi, T.: Genetic design of solids possessing a random-particulate microstructure. Phil. Trans. R. Soc. Lond. A 361, 1021–1043 (2003)
[5] Zohdi, T., Wriggers, P.: Introduction to Computational Micromechanics. Springer, Berlin (2005)
[6] Bäck, T., Schwefel, H.P.: An Overview of Evolutionary Algorithms for Parameter Optimization. Evolutionary Computation 1(1), 1–23 (1993)
[7] Bäck, T.: Evolutionary algorithms in theory and practice: evolution strategies, evolutionary programming, genetic algorithms. Oxford University Press, Oxford (1996)
[8] Herdy, M.: Application of the Evolutionsstrategie to Discrete Optimization Problems. In: Schwefel, H.-P., Männer, R. (eds.) PPSN 1990. LNCS, vol. 496, pp. 188–192. Springer, Heidelberg (1991)
[9] Cai, J., Thierauf, G.: A parallel evolution strategy for solving discrete structural optimization. Adv. Eng. Softw. 27(1-2), 91–96 (1996)

[10] Bäck, T., Hoffmeister, F., Schwefel, H.P.: A survey of evolution strategies. In: Belew, R., Booker, L. (eds.) Proceedings of the 4th International Conference on Genetic Algorithms. Oxford University Press, Oxford (1991)
[11] Bäck, T., Schwefel, H.P.: Evolution Strategies I: Variants and their Computational Implementation. In: Winter, G., Periaux, J., Galan, M., Cuesta, P. (eds.) Genetic Algorithms in Engineering and Computer Science, ch. 6, pp. 111–126. Wiley & Sons Ltd, Chichester (1995)
[12] Beyer, H.G., Schwefel, H.P.: Evolution strategies - a comprehensive introduction. Natural Computing 1(1), 3–52 (2002)
[13] Chamis, C.: Simplified composite micromechanics equations for hygral, thermal and mechanical properties. SAMPE Quarterly 15, 14–23 (1984)
[14] Brighenti, R.: Fibre distribution optimisation in fibre-reinforced composites by a genetic algorithm. Composite Structures 71, 1–15 (2005)

Fibre Rotation Motion in Homogeneous Flows

H. Altenbach[1], K. Naumenko[1], S. Pylypenko[2], and B. Renner[1]

[1] Lehrstuhl Technische Mechanik, Martin-Luther-Universität Halle-Wittenberg
[2] GKMM, Otto-von-Guericke-Universität Magdeburg

Abstract. The rotation of an inertialess ellipsoidal particle in a Newtonian fluid has been firstly analysed by Jeffery [15]. He found that in the case of the shear flow the particle rotates such that the end of its axis of symmetry describes a closed periodic orbit. Below the effect of particle inertia on the rotary motion of a slender fibre in uniform flow fields is discussed. The equations of motion and the constitutive equation for the hydrodynamic moment are introduced. These equations are solved for various flow fields. For the plane flow fields with dominant vorticity (elliptic and rotational flows) the effect of inertia is a slow particle drift toward the flow plane.

1 Introduction

The analysis of motion of a rigid particle in flow fields is important for understanding the dynamics of fibre suspensions in various technological applications. An example is the injection molding of polymer composites, where the flow of the fibre suspension during the filling of a cavity leads to the formation of fibre orientation microstructure. Several models have been developed with the aim to simulate the flow of fibre suspensions and in particular to predict the fibre orientation states. An essential feature of all approaches is the behaviour of a single fibre in viscous flow fields. For example, in the theory of dilute suspensions proposed in [1, 6] the evolution equation for the fibre microstructure is derived from the angular velocity of a single particle. The micropolar theories of concentrated suspensions (e.g., [4, 11]) treat the angular velocity, the tensor of inertia and the rotation tensor of particles as field quantities. Furthermore, to account for the rotary interactions between the fluid and suspended particles, the additional skew symmetric part of the stress tensor is introduced [4]. To verify the corresponding constitutive equation, studies of particle motion in homogeneous flow fields are helpful.

The rotation of a neutrally buoyant ellipsoid of revolution in a Newtonian fluid has been firstly analysed by Jeffery [15]. He found that in the case of the shear flow the ellipsoid rotates such that the end of its symmetry axis describes a closed periodic orbit. Bretherton [8] generalized the Jeffery's formula for the angular velocity to axisymmetrical particles of various shapes. In these pioneering works both the particle inertia and the fluid inertia are ignored. Problems of particle dynamics in flow fields under the consideration of the fluid inertia but neglecting the particle inertia are discussed in [17, 19].

Here the problem of the particle rotation will be analysed by taking into account its rotary inertia. We recall the governing equations of rigid body dynamics for the slender undeformable particle and apply the viscous type constitutive equation for the hydrodynamic moment. For several plane flows we present approximate analytical and numerical solutions to the equations of motion. The results are compared with solutions, where the rotary inertia is ignored. We show that the principal effect associated with the particle inertia is the slow drift of the rotating fibre towards the flow plane. The drift from the Jeffery orbit has been documented in several experimental studies (see, for example, the introduction section in [8]).

2 Basic Assumptions

Let us consider a slender fibre with the length l and the line mass density ρ (mass per unit length) as shown in Fig. 1.

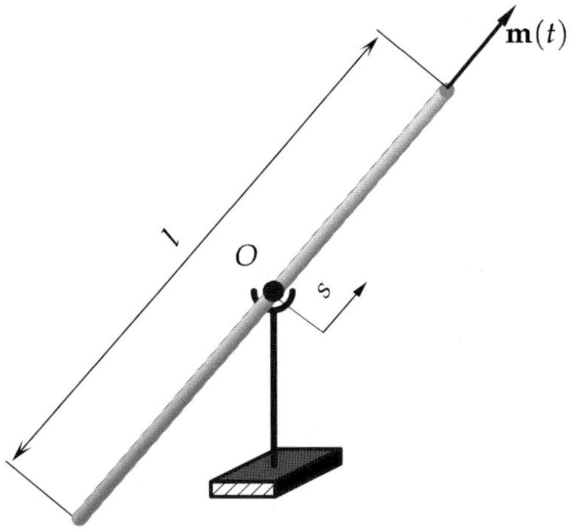

Fig. 1. Slender fibre: geometry and assumptions

The unit vector **m** defines the actual direction of the fibre axis and s is the axial coordinate ($-l/2 \leq s \leq l/2$). The angular momentum with respect to the centre of mass O can be calculated as follows (e.g. [13])

$$\mathbf{K}_O = \mathbf{J} \cdot \boldsymbol{\omega}, \quad \mathbf{J} = -\int_{-l/2}^{l/2} (\mathbf{r}_p - \mathbf{r}_O) \times \mathbf{E} \times (\mathbf{r}_p - \mathbf{r}_O) \rho \mathrm{d}s, \tag{1}$$

where $\mathbf{J}$ is the tensor of inertia, $\boldsymbol{\omega}$ is the angular velocity vector of the fibre, $\mathbf{r}_p$ defines the position of fibre points with respect to the reference frame, $\mathbf{r}_O$ is the position vector of the centre of mass and $\mathbf{E}$ is the second rank unit tensor. Here and in the following derivations the direct tensor calculus in the sense of Gibbs and Lagally [16, 20] is applied. The symbols $\cdot$, $\times$, and $\otimes$ denote the scalar (dot), the vector (cross), and the dyadic products, respectively.

With $\mathbf{r}_p = \mathbf{r}_O + s\mathbf{m}$ the tensor of inertia takes the form

$$\mathbf{J} = \rho \frac{l^3}{12}(\mathbf{E} - \mathbf{m} \otimes \mathbf{m}) . \tag{2}$$

The equations of motion can be formulated as follows

$$\frac{\mathrm{d}}{\mathrm{d}t}(m\mathbf{v}_O) = \mathbf{F}, \quad \frac{\mathrm{d}}{\mathrm{d}t}(\mathbf{J} \cdot \boldsymbol{\omega}) = \mathbf{M}_O . \tag{3}$$

m = ρl is the particle mass, $\mathbf{v}_O = \dot{\mathbf{r}}_O$ is the translation velocity of the centre of mass, $\mathbf{F}$ is the resultant force, and $\mathbf{M}_O$ is the resultant moment with respect to the centre of mass.

The orientation of the fibre can be described by the rotation tensor $\mathbf{Q}(t)$, $\mathbf{Q} \cdot \mathbf{Q}^T = \mathbf{E}$, $\det \mathbf{Q} = 1$. Specifying the reference direction of the fibre axis by $\mathbf{m}_0$, the actual direction $\mathbf{m}$ is defined by

$$\mathbf{m}(t) = \mathbf{Q}(t) \cdot \mathbf{m}_0 . \tag{4}$$

For the given angular velocity the rotation tensor can be calculated by solving the left Darboux problem [21]

$$\dot{\mathbf{Q}} = \boldsymbol{\omega} \times \mathbf{Q}, \quad \mathbf{Q}(0) = \mathbf{E} . \tag{5}$$

The right dot product of the first equation in (5) with $\mathbf{m}_0$ results in

$$\dot{\mathbf{m}} = \boldsymbol{\omega} \times \mathbf{m} . \tag{6}$$

Let $\mathbf{v}_f(\mathbf{r})$ be the velocity of the undisturbed fluid flow (flow without particle) at a given point $\mathbf{r}$ of a reference system. Following [15] let us assume

$$\mathbf{L} = \nabla \mathbf{v}_f = \text{const}, \quad \operatorname{tr} \mathbf{L} = \nabla \cdot \mathbf{v}_f = 0, \quad \mathbf{v}_f(\mathbf{r}) = \mathbf{r} \cdot \mathbf{L} . \tag{7}$$

The velocity gradient can be decomposed as follows

$$\mathbf{L} = \mathbf{D} - \boldsymbol{\Phi} \times \mathbf{E}, \quad \mathbf{D} = \frac{1}{2}\left[\nabla \mathbf{v}_f + \left(\nabla \mathbf{v}_f\right)^T\right], \quad \boldsymbol{\Phi} = \frac{1}{2} \nabla \times \mathbf{v}_f , \tag{8}$$

where the tensor $\mathbf{D}$ is the symmetric part of the velocity gradient (deformation part) and the vector $\boldsymbol{\Phi}$ is the vorticity part.

In the case of plane flows these parts can be represented as follows

$$\mathbf{D} = d_0(\mathbf{i} \otimes \mathbf{j} + \mathbf{j} \otimes \mathbf{i}), \quad \boldsymbol{\Phi} = \phi_0 \mathbf{j} \times \mathbf{i},$$

where the orthogonal unit vectors $\mathbf{i}$ and $\mathbf{j}$ belong to the flow plane. The scalars d_0 and ϕ_0 are the rates of the deformation and the vorticity, respectively. In what follows we assume $d_0 \geq 0, \phi_0 \geq 0$.

In [12] Giesekus has classified the plane flows as the rotational flow ($d_0 - 0, \phi_0 > 0$), the elliptic flow ($\phi_0 > d_0 > 0$), the shear flow ($d_0 = \phi_0 > 0$), the stretching flow ($\phi_0 = 0, d_0 > 0$) and the hyperbolic flow $d_0 > \phi_0 > 0$. To analyse the behaviour of the fibre placed in an arbitrary position of the flow field, the equations of translation and rotary motion (3) should be solved. Here we limit our discussion to the rotary motion. To this end we assume that the centre of mass of the fibre is placed in the origin of the flow field, i.e. in the position $\mathbf{r} = \mathbf{0}$. Furthermore, the origin of the reference frame is selected in the centre of mass of the fibre.

3 Constitutive Equation for the Hydrodynamic Moment

If the fibre is placed in the homogeneous flow field, a local disturbance of the velocity field occurs, which leads to pressure gradients and a viscous stress tensor. From the distributions of stresses on the particle surface the resultant force and the resultant moment exerted on the particle can be computed. Such an analysis in the case of the Stokes flows is presented in [7, 15], for example. Here we are primarily interested in the behaviour of the fibre rather than the disturbance field of the surrounding fluid. Therefore we prefer the phenomenological approach to derive the constitutive equation for the hydrodynamic moment exerted on the fibre.

Following Brenner [7] in the case of the laminar flow, the moment is a linear function of the particle angular velocity and the velocity gradient of the undisturbed flow

$$\mathbf{M}_O = \mathbf{G} \cdot (\boldsymbol{\Phi} - \boldsymbol{\omega}) + {}^{(3)}\mathbf{H} \cdot\cdot \mathbf{D}, \tag{9}$$

where the second rank tensor $\mathbf{G}$ and the third rank tensor ${}^{(3)}\mathbf{H}$ are called the resistance tensors. For particles with the symmetry axis $\mathbf{m}$ they have the form

$$\begin{aligned} \mathbf{G} &= \eta_L \mathbf{m} \otimes \mathbf{m} + \eta_T (\mathbf{E} - \mathbf{m} \otimes \mathbf{m}), \\ {}^{(3)}\mathbf{H} &= \frac{1}{2}\eta(\mathbf{m} \times \mathbf{E} \otimes \mathbf{m} + \mathbf{m} \times \mathbf{e}^i \otimes \mathbf{m} \otimes \mathbf{e}_i), \end{aligned} \tag{10}$$

where the positive scalars ν_L, ν_T and η are resistance coefficients. The vectors $\mathbf{e}_i$, $i = 1,2,3$ denote the arbitrary basis, and the vectors $\mathbf{e}^i$ stand for the reciprocal basis, i.e. $\mathbf{e}_i \cdot \mathbf{e}^k = \delta_i^k$ with δ_i^k as the Kronecker symbol. To derive a non-linear expression for the hydrodynamic moment one may use the theory of invariants presented in [5]. Equation (9) by taking into account (10) can be also formulated as follows

$$\mathbf{M}_O = [\nu_L \mathbf{m} \otimes \mathbf{m} + \nu_T (\mathbf{E} - \mathbf{m} \otimes \mathbf{m})] \cdot (\boldsymbol{\Phi} - \boldsymbol{\omega}) + \eta \mathbf{m} \times \mathbf{D} \cdot \mathbf{m}\,. \tag{11}$$

In the case of a slender particle one may neglect the resistance ν_L since this property is related to the rotation about the axis $\mathbf{m}$. Furthermore, one may set $\nu_L = 0$, $\eta = \nu_T = \nu_0 l^3 / 12$ as shown in [3].

4 Solutions to Equations of Motion

With the tensor of inertia (2) and the hydrodynamic moment (11) the second equation in (3) and Eq. (6) take the form

$$\lambda\dot{\boldsymbol{\omega}}_T = \mathbf{m}\times\mathbf{L}^T\cdot\mathbf{m} - \boldsymbol{\omega}_T, \dot{\mathbf{m}} = \boldsymbol{\omega}_T\times\mathbf{m}, \lambda \equiv \frac{\rho}{\nu_0}, \boldsymbol{\omega}_T \equiv \boldsymbol{\omega}\cdot(\mathbf{E}-\mathbf{m}\otimes\mathbf{m}). \qquad (12)$$

Taking into account that $\dot{\boldsymbol{\omega}}_T = -\ddot{\mathbf{m}}\times\mathbf{m}$, the first Eq. in (12) can be written down as follows

$$\lambda(\ddot{\mathbf{m}} + \dot{\mathbf{m}}\cdot\dot{\mathbf{m}}\mathbf{m}) + \dot{\mathbf{m}} = \mathbf{L}^T\cdot\mathbf{m} - \mathbf{m}\cdot\mathbf{L}^T\cdot\mathbf{m}\mathbf{m}. \qquad (13)$$

Note that λ represents a small parameter in the case of fibres in viscous flows. Therefore, in many works the left hand side in the first Eq. in (13) is neglected. This approach originates from Jeffery [15]. By neglecting $\lambda\dot{\boldsymbol{\omega}}_T$ we get

$$\boldsymbol{\omega}_T = \mathbf{m}\times\mathbf{L}^T\cdot\mathbf{m} \quad\Rightarrow\quad \dot{\mathbf{m}} = \mathbf{L}^T\cdot\mathbf{m} - \mathbf{m}\mathbf{m}\cdot\mathbf{L}\cdot\mathbf{m}. \qquad (14)$$

The second equation in (14) is widely used in statistical models of fibre suspensions and injection molding simulations, e.g., [2, 9].

The solutions to the inertialess Eqs. (14) are presented in [9, 10, 15] among others. The general closed form solution to Eqs. (12) is not available. In what follows we analyse the influence of the inertia on the particle rotation for different cases of plane flows. For the rotational flow we derive an approximate analytical solution to Eqs. (12). For other types of plane flows, we present the results of numerical solutions to Eqs. (13) by use of the fourth order Runge-Kutta method. They will be compared with those based on Eqs. (14).

4.1 Resting Medium

First let us consider the free particle rotation in a resting medium, i.e. in the case that the undisturbed velocity is a zero vector. This problem has been analysed in [14] for the case of an arbitrary axisymmetrical rigid body, where a series solution for the angular velocity is derived. The numerical solution in terms of the right angular velocity and the rotation vector is presented in [18]. In the case of the slender particle the solution is rather simple. Indeed, by setting $\mathbf{L} = \mathbf{0}$ Eq. (12) becomes

$$\lambda\dot{\boldsymbol{\omega}}_T + \boldsymbol{\omega}_T = \mathbf{0}. \qquad (15)$$

With the initial condition $\boldsymbol{\omega}_T(0) = \boldsymbol{\omega}_0$, $\boldsymbol{\omega}_0\cdot\mathbf{m}_0 = 0$ the general solution of (15) is

$$\boldsymbol{\omega}_T = \boldsymbol{\omega}_0 \exp\left(-\frac{t}{\lambda}\right)$$

Let $\boldsymbol{\omega}_0 = \omega_0\mathbf{k}$, where $\mathbf{k}$ is a unit vector. Then the actual orientation of the particle is determined by

$$\mathbf{m}(t) = \mathbf{Q}(\varphi(t)\mathbf{k})\cdot\mathbf{m}_0, \quad \varphi(t) = \omega_0\lambda\left[1-\exp\left(-\frac{t}{\lambda}\right)\right]. \qquad (16)$$

Here and in the following derivations $\mathbf{Q}(\varphi\mathbf{k})$ denotes the rotation about the fixed axis $\mathbf{k}$ on the angle φ [21]

$$\mathbf{Q}(\varphi\mathbf{k}) \equiv \mathbf{k}\otimes\mathbf{k} + \cos\varphi(\mathbf{E}-\mathbf{k}\otimes\mathbf{k}) + \sin\varphi\mathbf{k}\times\mathbf{E}\,.$$

From (16) follows that $\mathbf{k}\cdot\mathbf{m} = \mathbf{k}\cdot\mathbf{m}_0 = 0$. The particle rotates about the vector $\mathbf{k}$ with a decaying angular velocity. At $t\to\infty$ the angle of particle orientation is $\varphi_\infty = \omega_0\lambda$.

4.2 Rotational Flow

In this case Eqs. (12) take the form

$$\lambda\dot{\boldsymbol{\omega}}_T = \boldsymbol{\Phi}_T - \boldsymbol{\omega}_T, \quad \boldsymbol{\Phi}_T \equiv \boldsymbol{\Phi}\cdot(\mathbf{E}-\mathbf{m}\otimes\mathbf{m}), \quad \dot{\mathbf{m}} = \boldsymbol{\omega}_T\times\mathbf{m}\,. \tag{17}$$

If we drop the inertia term, i.e. the left-hand side of Eq. (17) then the solution is

$$\boldsymbol{\omega}_T = \boldsymbol{\Phi}_T \quad\Rightarrow\quad \mathbf{m}(t) = \mathbf{Q}(\varphi(t)\mathbf{n})\cdot\mathbf{m}_0, \quad \varphi(t) = \phi_0 t, \quad \mathbf{n} = \frac{\boldsymbol{\Phi}}{\phi_0} \tag{18}$$

The meaning of Eqs. (18) is obvious. The fibre axis rotates with the fluid about the vorticity axis $\mathbf{n}$ with the constant angular velocity. The end of the particle axis describes the circular periodic orbit and the radius of the orbit depends on the reference orientation $\mathbf{m}_0$. Similar conclusions have been drawn by Jeffery [15] in the case of an ellipsoid of revolution in the shear flow. The more complex fibre orbits in the elliptic flows will be discussed later. Here, in the rather simple case of rotational flow, one may verify that the solution (18) does not satisfy (17) even in the case of small but finite λ. Furthermore, one may verify that the steady state solution to Eq. (17) is

$$\dot{\boldsymbol{\omega}}_T = \mathbf{0}, \quad \mathbf{m}(t)\cdot\mathbf{n} = 0 \quad\Rightarrow\quad \boldsymbol{\omega}_T = \boldsymbol{\Phi}, \quad \mathbf{m}(t) = \mathbf{Q}(\varphi(t)\mathbf{n})\cdot\mathbf{m}_0, \tag{19}$$

i.e. the fibre rotates in the flow plane about the vorticity axis with the constant angular velocity ϕ_0. In [3] an approximate asymptotic solution characterizing the slow transient fibre rotation from the arbitrary initial orientation $\mathbf{m}_0\cdot\mathbf{n}\neq 0$ to the stable rotation in the flow plane is presented. The result can be formulated as follows

$$\mathbf{m} = \mathbf{Q}[(\varphi-\varphi_0)\mathbf{n}]\cdot\mathbf{Q}\left[(\psi-\psi_0)\frac{\mathbf{m}_0\times\mathbf{n}}{|\mathbf{m}_0\times\mathbf{n}|}\right]\cdot\mathbf{m}_0, \tag{20}$$

where

$$\varphi(t) = \phi_0 t + \varphi_0, \quad \tan\psi(t) = \tan\psi_0\exp(-\phi_0^2\lambda t)\,. \tag{21}$$

Figure 2 shows the time variation of the angle ψ obtained by the numerical solution of Eq. (13). For the comparison the result according to Eqs. (20) and (21) is presented. A slight disagreement between the solutions is only observed at the beginning of the motion. The slow drift of the particle towards the flow plane is well described by Eqs. (20) and (21).

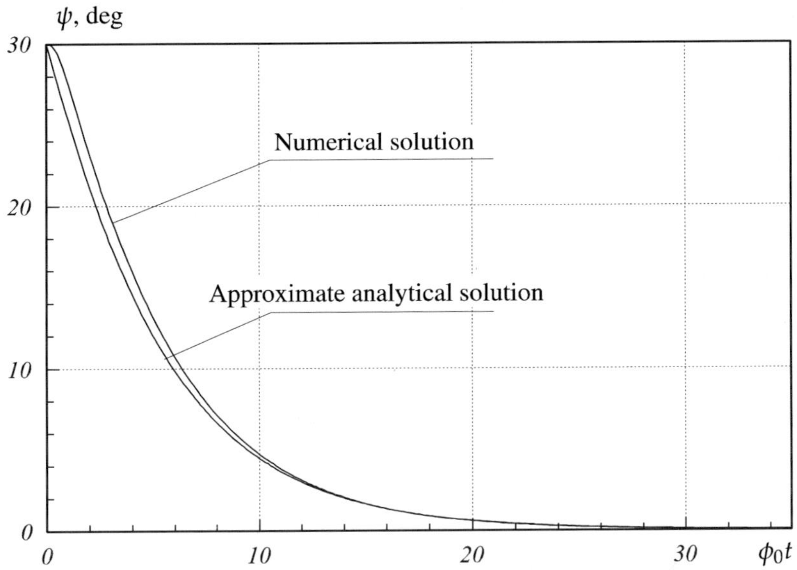

Fig. 2. Angle between the particle axis and the flow plane vs. time - rotational flow

4.3 Elliptic Flow

For the elliptic flow Eqs. (12) take the form

$$\lambda\dot{\boldsymbol{\omega}}_T = \phi_0 \mathbf{m}\times[(1+\kappa)\mathbf{i}\otimes\mathbf{j}-(1-\kappa)\mathbf{j}\otimes\mathbf{i}]\cdot\mathbf{m}-\boldsymbol{\omega}_T, \quad \kappa = \frac{d_0}{\phi_0}, \quad \dot{\mathbf{m}} = \boldsymbol{\omega}_T \times \mathbf{m}\,. \tag{22}$$

By neglecting the inertia term the particle angular velocity is

$$\boldsymbol{\omega}_T = \phi_0 \mathbf{m}\times[(1+\kappa)\mathbf{i}\otimes\mathbf{j}-(1-\kappa)\mathbf{j}\otimes\mathbf{i}]\cdot\mathbf{m}\,. \tag{23}$$

It can be shown that Eq. (23) has the same form as that derived by Jeffery [15] for the angular velocity of an ellipsoid of revolution in a shear flow. Jeffery found the particle orientation in terms of Euler angles. According to his solution the particle rotates such that the end of its axis of symmetry describes a closed periodic orbit. The ordinary differential equation for the particle axis can be obtained from (23) as follows

$$\dot{\mathbf{m}} = \phi_0(\mathbf{E}-\mathbf{m}\otimes\mathbf{m})\cdot[(1+\kappa)\mathbf{i}\otimes\mathbf{j}-(1-\kappa)\mathbf{j}\otimes\mathbf{i}]\cdot\mathbf{m}\,. \tag{24}$$

The integral of (24) represents the orbit as shown by Ericksen [10].

The influence of the particle inertia can be investigated with the help of numerical solutions. Figure 4 illustrates the results for the particle rotation in the elliptic flow with $\phi_0 = 1\,\mathrm{s}^{-1}$, $\kappa = 0.66$ and $\lambda = 0.2\mathrm{s}$.

The broken lines are the solutions to the inertialess Eq. (24). They illustrate the stable periodic variation of the angle between the particle and the flow plane, Fig. 3a as well as the Jeffery-type orbit, Fig. 3b. The solid lines in Fig. 3 correspond to the

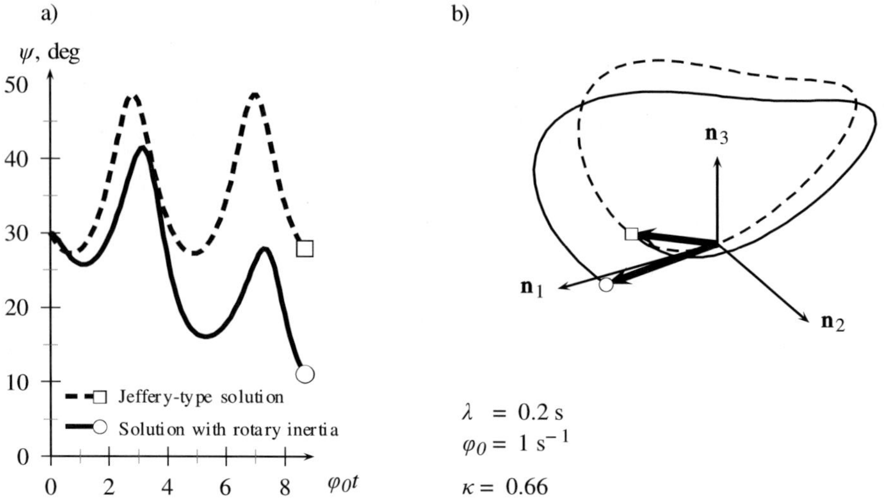

Fig. 3. Particle motion in elliptic flow – comparison with the Jeffrey-type solution. a) Angle between the particle axis and the flow plane vs. dimensionless time, b) trajectory of the particle axis.

solutions of Eq. (13) by taking into account the rotary inertia. We observe that similarly to the case of rotational flow, the orbit is non-stable and the particle drifts towards the flow plane.

4.4 Shear Flow

In this case Eqs (12) can be written as

$$\lambda \dot{\boldsymbol{\omega}}_T = 2\phi_0 \mathbf{m} \times \mathbf{i} \otimes \mathbf{j} \cdot \mathbf{m} - \boldsymbol{\omega}_T, \qquad \dot{\mathbf{m}} = \boldsymbol{\omega}_T \times \mathbf{m}, \tag{25}$$

Equations (25) possess an infinite number of equilibrium solutions $\mathbf{m} \cdot \mathbf{j} = 0$, $\boldsymbol{\omega}_T = \mathbf{0}$. In the case that $\lambda = 0$, two solutions $\mathbf{m} = \pm \mathbf{i}$ corresponding to the alignment of the vector **m** parallel to streamlines are found to be stable [10]. Furthermore, as shown in [10] the vector **m** approaches to the steady state alignment for any initial orientation $\mathbf{m}_0$. For $\lambda \neq 0$ Eq. (25) is analysed in [3] considering the special initial condition $\boldsymbol{\omega}_T(0) \cdot \mathbf{i} = 0$. The result can be formulated as follows

$$\mathbf{m} = \mathbf{Q}(\alpha \mathbf{k}') \cdot \mathbf{i}, \quad \mathbf{k}' = \mathbf{Q}(\beta_0 \mathbf{i}), \tag{26}$$

where β_0 is the initial condition and α is defined by the following differential equation

$$\lambda \ddot{\alpha} + \dot{\alpha} = -2\phi_0 \sin^2 \alpha \cos \beta_0 \,. \tag{27}$$

By setting $\lambda = 0$ the integral of Eq. (27) is

$$\cot \alpha(t) = 2\phi_0 t \cos \beta_0 + \cot \alpha_0 \,. \tag{28}$$

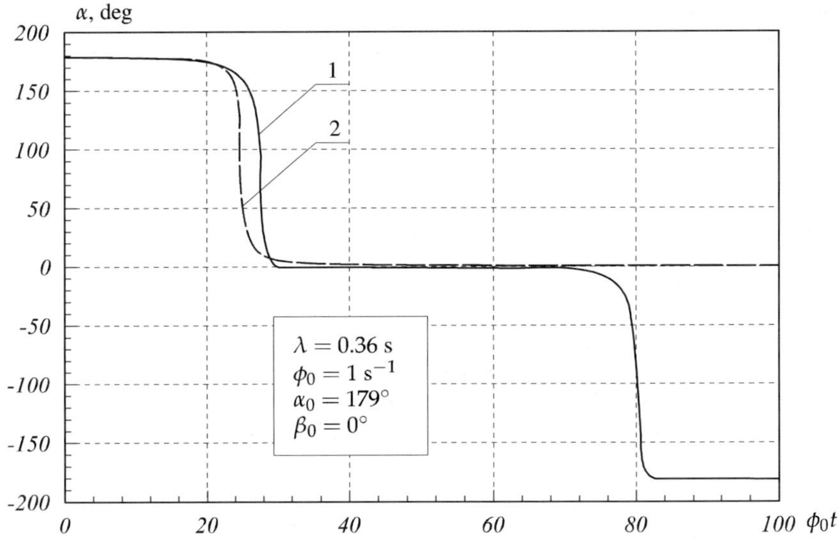

Fig. 4. Angle between the particle axis and the stream lines vs. time - shear flow: 1 – solution of Eq. (31), 2 – solution of Eq. (32)

Equation (28) shows that for all initial conditions except $\mathbf{m}_0 \cdot \mathbf{j} = 0$ the fibre aligns parallel to the streamlines. Our numerical tests with Eq. (27) suggest that for $\lambda \ll 1$ and $\dot{\alpha}(0) = 0$ the solution agrees well with the inertialess solution (28). The fibre rotates in the clockwise direction about the vector **k**´ and tends to the steady state orientation parallel to streamlines. After any small counterclockwise disturbance from the steady state, the fibre will return to the steady state orientation. If slightly rotated from the steady state in the clockwise direction the fibre will rotate on almost 180° and align parallel to the streamlines.

However, starting from a certain value of $\lambda < 1$ the result does not agree with the inertialess solution. As an example let us consider the behaviour of the fibre, initially placed in the flow plane, i.e. for β = 0°. Figure 4 presents the results for λ=0.36s, $\phi_0 = 1\,\mathrm{s}^{-1}$ and the initial condition α_0 = 179°. In this case the fibre alignment is not observable. The fibre jumps over the steady state position and continues to rotate in the clockwise direction about the vector **k**´.

5 Conclusions

The aim of this paper is to analyse the influence of the rotary inertia on the behaviour of the slender particle in stationary homogeneous flows. For several types of plane flows we obtained solutions illustrating the trajectories of the particle axis. From the results we may conclude that the inertialess Eqs (14), despite their simplicity, have a limited range of application. In the case of rotational and elliptic flows they do not reflect the slow drift of the fibre and lead to the incorrect conclusion about the stable orbits. In the case of shear flow the inertialess approach predicts the steady state fibre alignment along the streamlines, which is only possible up to certain values of λ.

References

[1] Advani, S.G., Tucker, C.L.: The use of tensors to describe and predict fibre orientation in short fibres composites. J. Rheol. 31(48), 751–784 (1987)
[2] Altan, M.C., Rao, B.N.: Closed-form solution for the orientation field in a center-gated disk. J. Rheol. 39(3), 581–599 (1995)
[3] Altenbach, H., Naumenko, K., Pylypenko, S., Renner, B.: Influence of rotary inertia on the fibre dynamics in homogeneous creeping flows. ZAMM 87, 81–93 (2007)
[4] Altenbach, H., Naumenko, K., Zhilin, P.: A micro-polar theory for binary media with application to phase-transitional flow of fibre suspensions. Continuum Mechanics and Thermodynamics 15, 539–570 (2003)
[5] Altenbach, H., Naumenko, K., Zhilin, P.: A note on transversely isotropic invariants. ZAMM 86, 162–168 (2006)
[6] Bay, R.S., Tucker, C.L.: Fibre orientation in simple injection moldings. Part 1: theory and numerical methods. Polym. Comp. 13, 317–331 (1992)
[7] Brenner, H.: The Stokes resistance of an arbitrary particle. 3. Shear fields. Chem. Eng. Sci. 19(9), 631–651 (1964)
[8] Bretherton, F.P.: The motion of rigid particles in a shear flow at low Reynolds number. J. Fluid Mech. 14, 284–304 (1962)
[9] Dinh, S.M., Armstrong, R.C.: A rheological equation of state for semiconcentrated fibre suspensions. J. Rheol. 28(3), 207–227 (1984)
[10] Ericksen, J.L.: Transversely isotropic fluids. Kolloid-Zeitschrift 173, 117–122 (1960)
[11] Eringen, A.C.: Continuum theory of dense rigid suspensions. Rheol. Acta 30, 23–32 (1991)
[12] Giesekus, H.: Strömungen mit konstantem Geschwindigkeitsgradienten und die Bewegung von darin suspendierten Teilchen. Teil II: Ebene Strömungen und eine experimentelle Anordnung zu ihrer Realisierung. Rheol. Acta 2(2), 112–122 (1962)
[13] Gummert, P., Reckling, K.A.: Mechanik. Vieweg, Braunschweig (1994)
[14] Ivanova, E.A.: A new approach to the solution of some problems of rigid body dynamics. ZAMM 81, 613–622 (2001)
[15] Jeffery, G.B.: The motion of ellipsoidal particles immersed in a viscous fluid. Proc. R. Soc. London A 102, 161–179 (1922)
[16] Lagally, M.: Vorlesungen über Vektorrechnung. Geest & Portig, Leipzig (1962)
[17] Leal, L.G.: Particle motions in a viscous fluid. Ann. Rev. Fluid Mech. 12, 435–476 (1980)
[18] Renner, B., Altenbach, H., Naumenko, K.: Numerical treatment of finite rotation for a cylindrical particle. Technische Mechanik 25, 151–161 (2005)
[19] Subramanian, G., Koch, D.L.: Inertial effects on fibre motion in simple shear flow. J. Fluid Mech. 535, 383–414 (2005)
[20] Wilson, E.B.: Vector Analysis, Founded upon the Lectures of G. W. Gibbs. Yale University Press, New Haven (1901)
[21] Zhilin, P.A.: A new approach to the analysis of free rotations of rigid bodies. ZAMM 76, 187–204 (1996)

Part III

Solids under Thermal Stressing

Distortion and Residual Stresses during Metal Quenching Process

A.K. Nallathambi[1], Y. Kaymak[1], E. Specht[1], and A. Bertram[2]

[1] Institut für Strömungstechnik und Thermodynamik,
Otto-von-Guericke-Universität Magdeburg
[2] Institut für Mechanik, Otto-von-Guericke-Universität Magdeburg

Abstract. Quenching is a complex thermo-mechano-metallurgical problem. During the quenching process, transient heat conduction, metallic phase transformations, and plastic behaviour of the metals introduce high residual stresses and distortions. This article presents the mathematical formulation of the physics behind the quenching process, numerical techniques and optimum of cooling strategies for the selected geometries. The Finite Element Method (FEM) is used to solve the coupled partial differential equations in the framework of an isothermal-staggered approach. Coupling effects such as phase transformation enthalpy, transformation-induced plasticity and dissipation are considered. Numerical examples are presented for an L profile made up of 100Cr6 steel.

1 Introduction

Quenching can be defined as *cooling of metals at a rate faster than cooling in still air.* Quenching is physically one of the most complex processes in engineering and very difficult to understand. Quenching used to be called *black hole* of heat treatment processes [1]. Most of the metallic parts have to be quenched after the thermal treatment processes to obtain the required properties such as hardness, micro-structure, etc. Quenching induces high residual stresses due to several mechanisms like phase transformation, thermal shrinkage, and transformation induced plasticity.

The distortion of the L profile can be better understood from Fig. 1, where the distorted shape of the profile is shown at different stages of cooling. Initially due to higher thermal shrinkage at the ends of the legs, the profile bends toward the legs (until 1). However, the ends of the legs soon undergo the phase transition which is also accompanied by a volume increase. Hence, the distortion changes its direction (1-2). As the phase transition penetrates through the legs, the distortion again changes its direction and the profile bends toward the legs one more time (2-3-4). Finally, the phase transition is completed throughout the profile and the distortion gradually decreases as the temperature becomes uniform (4-5). However, a permanent deformation remains due to the mechanical yielding and transformation induced plasticity.

The computer simulation of the quenching process includes three different analyses: (a) Thermal analysis for the computation of cooling curves, (b) Metallurgical analysis for the computation of micro structure composition, and (c) Mechanical analysis for computation of stresses and strains.

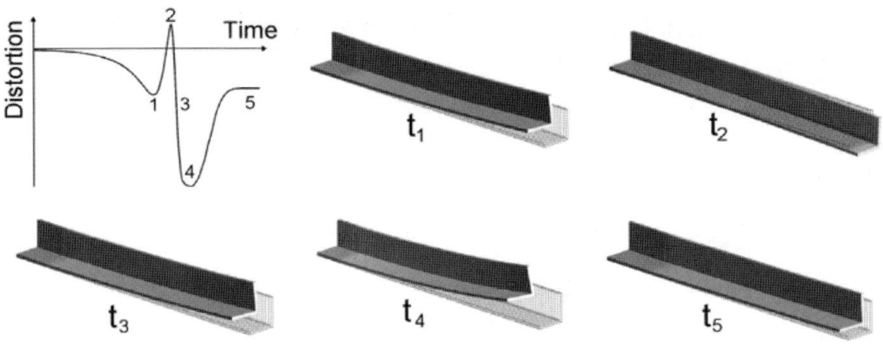

Fig. 1. Distortion of L profile at different stages of quenching

The latent heat released during the phase transformation increases the non-linearity of the problem. The heat flow method [2] is used to model the thermal field, and the FEM is employed for solving the thermal equilibrium equation. In steel like alloys, diffusive and displacive solid-solid phase transformations occur. The diffusive transformation is time-dependent and occurs in the high temperature zone. During the diffusive phase transformation, the parent austenite phase transforms into product phases such as pearlite and bainite. Unlike diffusive, the displacive transformation occurs in the lower temperature zone which is independent of the time. Martensite is the only product of the phase displacive transformation.

The shape change during the quenching process occurs due to the elastic, plastic, thermal phase changes, and transformation induced plastic strains. The complex mechanism behind the residual stress evolution during the quenching process is well explained by Todinov [3]. References [4-8] give more information about the distortion and residual stress calculation during metal quenching. This article is arranged in the following manner: Section 2 presents the mathematical formulation of the three physical fields. The FEM implementation with the isothermal staggered algorithm is described in Section 3. The simulation results are presented with numerical examples in Section 4.

2 Mathematical Formulation

During the quenching process, the temperature, micro-structure, and stresses at every material point change with respect to time. The thermal, metallurgical, and mechanical fields are modelled separately and discussed in this section.

2.1 Thermal Field

Let an open bounded domain $\Omega \subset \mathbb{R}^{nd}$ (n_d =1, 2, 3) be the configuration of a non-linear thermo-plastic body $\mathscr{B}$ with particles defined by $\vec{X} \in \tilde{\Omega}_0$, $\Gamma = \partial\Omega$ its smooth boundary and the time interval of analysis $t \in \Upsilon$ ($\Upsilon \subset \mathbb{R}^{+}$). As usual, $\tilde{\Omega} = \Omega \cup \Gamma$ and

$\Gamma = \Gamma_\theta \cup \Gamma_q$. The metal quenching problem consists of finding the absolute temperature field $\theta : \tilde{\Omega} \times \Upsilon \to \mathbb{R}^+$ such that [9]

$$\rho\, c_p \dot{\theta} = -\nabla \cdot \vec{q} + q_v \qquad \text{in } \tilde{\Omega} \times \Upsilon \tag{1}$$

subject to the boundary conditions

$$\theta = \theta_s \qquad \text{in } \Gamma_\theta \times \Upsilon \tag{2}$$

$$\vec{q} \cdot \vec{n} = -q_s \qquad \text{in } \Gamma_q \times \Upsilon \tag{3}$$

and the initial condition

$$\theta(\vec{X}, t)\big|_{t=0} = \theta_0(\vec{X}) \qquad \text{in } \Omega \tag{4}$$

Eq. (1) represents the energy balance obtained from the first law of thermodynamics. The density ρ and the specific heat capacity c_p, are both functions of the temperature and the phase fraction f_j. The heat generation per unit volume is denoted by q_v, and $\vec{q}$ is the heat flux vector. The internal heat generation accounts for both the phase transformation enthalpies and mechanical energy dissipation

$$q_v = \chi\, \sigma_y\, \dot{\varepsilon}^p + \sum_{j=1}^{n_p} L_j \dot{f}_j , \tag{5}$$

where χ is the fraction of mechanical energy converted into thermal energy, σ_y is the yield strength, $\dot{\varepsilon}^p$ is the rate of effective plastic strain, L_j is the latent heat of the individual phase transformation, $\dot{f}_j$ is the phase transition rate and n_p is the number of product phases. In Eq. (0.2), θ_s is the prescribed surface temperature on Γ_θ. On the heat flux boundary Γ_q, q_s is the normal heat flux due to convection-radiation phenomenon. Using the temperature-dependent overall Heat Transfer Coefficient (HTC) α, q_s can be stated according to Newton's law of convection as

$$q_s = -\alpha(\theta)(\theta - \theta_\infty) , \tag{6}$$

where θ is the surface temperature, and θ_∞ is the ambient temperature. Fourier's law of heat conduction states that the heat flux vector $\vec{q}$ is proportional to the temperature gradient

$$\vec{q} = -\mathbf{k}(\theta, f_j) \cdot \nabla\theta , \tag{7}$$

where $\mathbf{k}$ is the temperature and phase fraction-dependent second-order thermal conductivity tensor.

2.2 Phase Transformation Field

Phase transformations in solids can be classified as diffusive and displacive transformations. During the transformations in steel, the parent phase austenite may transform into product phases such as pearlite (diffusive) and martensite (displacive).

2.2.1 Diffusive Transformation

The evolution of the diffusive phase transitions is best described by Time-Temperature-Transformation (TTT) diagrams, which are constructed using the isothermal phase change data. The IT (also named TTT) diagrams can be obtained from the Johnson-Mehl-Avrami-Kolmogorov (JMAK) law [10]. In the IT diagram, the double *C*-curves are plotted for 1% (the transformation start time, t_s) and 99% (the transformation end time, t_e) of the product phase fraction at every temperature θ using the JMAK law. The isothermal formation of the new phase is described by a simple linear iso-kinetic rule [8]

$$\dot{f} = \frac{1}{t_e - t_s}, \tag{8}$$

which states that the rate of phase transformation is constant in the isothermal case. The t_s and t_e can be obtained form the IT diagram. In the non-isothermal case, the cooling curve is considered to be composed of small isothermal steps. The transformation begins at the incubation time t_{inc}, and it ends when the phase fraction reaches unity or the temperature is out of the transformation range. Using Scheil's additivity rule, the incubation time is given as

$$\int_0^{t_{inc}} \frac{1}{t_s(\theta(t))}\, dt = 1. \tag{9}$$

2.2.2 Displacive Transformation

Shear-dominant, diffusionless, martensitic transformations occur when the temperature of the steel drops rapidly below a critical temperature M_s. Martensite, which is hard and brittle, is a solid solution of carbon in tetragonally, distorted BCC iron. In this work, the displacive transformation is modelled using Koistinen-Marburger's law [8]

$$f_M = f_A\{1 - \exp[k_M(\theta - M_s)]\}, \qquad \text{when } \theta < \mathrm{M_s}, \tag{10}$$

where f_M and f_A are martensite and austenite phase fractions, M_s is the martensite start temperature and k_M (≈ 0.011) is the stress-dependent transformation constant.

2.3 Displacement Field

A thermo-plastic body $\mathscr{B}$ with interior $\Omega \subset \mathbb{R}^{n_d}$ (n_d =1, 2, 3) and displacement boundary Γ_u, traction boundary Γ_t, $\Gamma = \Gamma_u \cup \Gamma_t$ together as $\tilde{\Omega} = \Omega \cup \Gamma$ and the time interval of analysis $t \in \Upsilon$ ($\Upsilon \subset \mathbb{R}^+$), has to satisfy the equilibrium equation

$$\nabla \cdot \mathbf{T} + \vec{b} = \vec{0} \qquad \text{in } \tilde{\Omega} \times \Upsilon \tag{11}$$

subject to the boundary conditions

$$\vec{u} = \vec{u}_s \qquad \text{in } \Gamma_u \times \Upsilon \tag{12}$$

$$\mathbf{T}\cdot\vec{n}=\vec{t} \qquad \text{in } \Gamma_t\times\Upsilon, \tag{13}$$

where **T** is the stress tensor, $\vec{b}$ is the body force vector, $\vec{u}_s$ is the prescribed displacement vector, and $\vec{u}_t$ is the prescribed traction vector with unit outward normal $\vec{n}$. The total deformation observed in the quenching process is less than 4%. Therefore, using the advantage of small deformation theory, the total strain E can be additively decomposed into four components as in [6]

$$\mathbf{E}=\frac{1}{2}[\nabla\vec{u}+(\nabla\vec{u})^T]=\mathbf{E}^e+\mathbf{E}^p+\mathbf{E}^{tp}+\mathbf{E}^{trip}, \tag{14}$$

where $\mathbf{E}^e$ is the elastic strain tensor, $\mathbf{E}^p$ is the plastic part of strain tensor, $\mathbf{E}^{trip}$ is the Transformation Induced Plastic (TRIP) strain tensor, and $\mathbf{E}^{tp}$ is the volumetric strain tensor due to temperature and phase changes. Once estimating the plastic, the thermal phase change, and the TRIP strain tensors, the elastic strain tensor can be obtained from the total strain tensor, and its methods of estimation are discussed in the subsequent subsections in detail. Using the elastic part of the strain tensor $\mathbf{E}^e$, the stress tensor **T** can be determined from the constitutive law of the material.

2.3.1 Volumetric Thermal Phase Change Strain

During the quenching process, the temperature and phase fractions of the material change drastically. The density of the material undergoing a phase change is a function of the temperature and the phase fractions. For non phase-changing materials like aluminium, copper, nickel, etc, the density depends only on temperature. This nature of varying density produces a reversible strain $\mathbf{E}^{tp}$. Instead of using the coefficient of thermal expansion, $\mathbf{E}^{tp}$ is expressed in terms of the reference density ρ_R and the current density $\rho(\theta,f_j)$ of the mixture

$$\mathbf{E}^{tp}=\left(\sqrt[3]{\frac{\rho_R}{\rho(\theta,f_j)}}-1\right)\mathbf{I}. \tag{15}$$

2.3.2 Transformation Induced Plastic Strain (TRIP)

During the phase change period, austenite may transform into any combination of the following micro-structures: pearlite, bainite and martensite. There is an irreversible strain always associated with the phase change phenomena which is known as TRIP strain, and it is proportional to the stress deviator **T**' and the rates of phase transformation f_j. Even though the induced stress lies below the yield limit, the TRIP strain occurs in phase changing materials like steel. The TRIP strain rate can be calculated from the macroscopic material behaviour based on the micro-mechanical approach, and it is given as in [8]

$$\dot{\mathbf{E}}^{trip}=-\tfrac{3}{2}\mathbf{T}'\sum_{j=1}^{n_p}\{\Lambda_j\ln(f_j)\dot{f}_j\}, \tag{16}$$

where Λ_j is called the Greenwood-Johnson (GJ) coefficient which must be determined experimentally.

2.3.3 Plastic Strain

When the equivalent stress exceeds the yield stress, plastic deformations occur. Using a classical rate-independent, isotropic, thermo-plastic material model with a temperature- and phase fraction-dependent constitutive law, and by systematically employing the yield criterion, loading criterion, flow rule, hardening rule, and consistency condition which are discussed separately in detail, the plastic strain can be estimated. The isotropic constitutive law of the material can be written as in [11]

$$\mathbf{T} = \mathbb{C}^e \cdot \mathbf{E}^e = \kappa \operatorname{tr}\left(\mathbf{E} - \mathbf{E}^{tp}\right)\mathbf{I} + 2\mu\left(\mathbf{E}' - \mathbf{E}^{trip} - \mathbf{E}^p\right) \tag{17}$$

where $\mathbb{C}^e$ is the fourth order elasticity tensor, κ is the bulk modulus and μ is the shear modulus, together functions of temperature and phase fractions.

1. ***Yield criterion.*** The von-Mises yield criterion has the special feature of the smooth surface with convexity which is suitable for pressure-independent ductile materials and given as

$$\phi(\mathbf{T}', \varepsilon^p, T, f_j) = \| \mathbf{T}' \| - \sqrt{\tfrac{2}{3}}\sigma_y(\varepsilon^p, \theta, f_j), \tag{18}$$

where ε^P is the effective plastic strain which is used as a strain hardening internal variable, and σ_y is the temperature- and phase fractions-dependent yield strength.

2. ***Loading criterion.*** The loading criterion can be stated as

$$\phi = 0 \text{ and } \dot{\phi}\big|_{\varepsilon^p = const} > 0 \text{ loading}$$
$$\phi = 0 \text{ and } \dot{\phi}\big|_{\varepsilon^p = const} = 0 \text{ neutral loading}$$
$$\phi = 0 \text{ and } \dot{\phi}\big|_{\varepsilon^p = const} < 0 \text{ unloading}$$

3. ***Flow rule.*** An associated flow rule is employed and given as

$$\dot{\mathbf{E}}^p = \dot{\lambda}\frac{\partial \phi}{\partial \mathbf{T}} = \dot{\lambda}\frac{\mathbf{T}'}{\|\mathbf{T}'\|} = \dot{\lambda}\mathbf{n}'_T, \tag{19}$$

where $\dot{\lambda}$ and n'_T are the plastic multiplier and the flow surface normal or stress deviator direction, respectively.

4. ***Hardening rule.*** A linear isotropic hardening rule is considered, and the yield strength is stated as in [12]

$$\sigma_y(\varepsilon^p, \theta, f_j) = \sigma_{yo}(\theta, f_j) + H(\theta, f_j)\ \varepsilon^p \tag{20}$$

where σ_{yo} is the yield strength at the virgin state, and H is the plastic modulus. The hardening state variable is integrated from the plastic multiplier

$$\dot{\varepsilon}^p = \sqrt{\frac{2}{3}}\ \dot{\lambda} \tag{21}$$

5. ***Consistency condition.*** In general, the consistency condition $\dot{\phi}=0$ yields the value of the plastic multiplier $\dot{\lambda}$. The isothermal staggered algorithm [13] suggests that the temperature and phase fractions should be kept constant (i.e., $\dot{\theta}=0$ and $\dot{f}_j=0$), so that the plastic multiplier [11] becomes

$$\dot{\lambda}=\frac{2\mu\mathbf{n}_T'\cdot\left(\dot{\mathbf{E}}-\dot{\mathbf{E}}^{trip}\right)}{2\mu+\frac{2}{3}H} \tag{22}$$

3 Solution Methodology

The non-linear coupled simultaneous equations obtained through FEM are solved using the isothermal staggered algorithm [13]. Thermal, metallurgical and mechanical fields are sequentially solved in every time step in the following way: (a) the thermal field is solved at fixed configuration and phase fractions, (b) the metallurgical field is solved at fixed configuration and constant temperature, (c) the mechanical field is solved at constant temperature and phase fractions. In each time step, first the transient temperature field is solved iteratively, then the phase transitions are computed, and finally the displacement field is computed iteratively. The discrete form of all coupled equations are derived and discussed in detail in the following subsections.

3.1 Thermal Field Formulation

Using FEM, the final matrix form of thermal equilibrium is given as

$$\left\{\mathbf{K}^{\theta t+\Delta t}_{i}+\frac{1}{\Delta t}\mathbf{C}^{\theta t+\Delta t}_{i}\right\}(\Delta\boldsymbol{\Theta})=\mathbf{F}^{\theta t+\Delta t}-\mathbf{R}^{\theta t+\Delta t}_{i}, \tag{23}$$

where $\mathbf{K}^{\theta}$ is the global conductance matrix, $\mathbf{C}^{\theta}$ is the global capacitance matrix, $\mathbf{F}^{\theta}$ is the global thermal force vector, and $\mathbf{R}^{\theta}$ is the global residual thermal force vector. The elemental form of these matrices and vectors are given as in [14]

$$\begin{aligned}
\mathbf{K}^{\theta t+\Delta t}_{e\,i} &= \int_{\Omega}\left[\mathbf{H}^{\mathrm{T}}\,k_i^{t+\Delta t}\,\mathbf{H}\right]d\Omega+\int_{\Gamma_q}\left[\mathbf{N}^{\mathrm{S}}\alpha_i^{t+\Delta t}\left(\mathbf{N}^{\mathrm{S}}\right)^{T}\right]d\Gamma_q\\
\mathbf{C}^{\theta t+\Delta t}_{e\,i} &= \int_{\Omega}\left[\mathbf{N}\,\rho_i^{t+\Delta t}\,c_{pi}^{t+\Delta t}\,\mathbf{N}^{\mathrm{T}}\right]d\Omega\\
\mathbf{F}^{\theta t+\Delta t}_{e\,i} &= \int_{\Omega}\left[\mathbf{N}\,q_{vi}^{t+\Delta t}\right]d\Omega+\int_{\Gamma_q}\left[\mathbf{N}\,\alpha_i^{t+\Delta t}\,\theta_\infty\right]d\Gamma_q\\
\mathbf{K}^{\theta t+\Delta t}_{e\,i} &= \left\{\int_{\Omega}\left[\mathbf{H}^{\mathrm{T}}\,k_i^{t+\Delta t}\,\mathbf{H}\right]d\Omega\right\}\boldsymbol{\Theta}_{ei}^{t+\Delta t}+\mathbf{C}^{\theta t+\Delta t}_{e\,i}\left(\frac{\boldsymbol{\Theta}_{ei}^{t+\Delta t}-\boldsymbol{\Theta}_e^{t}}{\Delta t}\right)
\end{aligned} \tag{24}$$

where $\mathbf{N}$ is the element shape function and $\mathbf{H}$ is the element temperature-gradient interpolation operator.

3.2 Phase Field Formulation

At the end of thermal field computation, the current temperature $\Theta^{t+\Delta t}$ and the current temperature increment $\Delta\Theta = \Theta^{t+\Delta t} - \Theta^{t}$ are known at every integration point of the elements. The displacive and diffusive phase transitions are computed using these temperature details. Martensitic evolution can be directly determined from the Eq. (0.10). Pearlite is considered as the only product of diffusive transformation which is a reasonable simplification. Scheil's sum increment ΔS at the current time step can be computed using the IT diagram information [8]

$$\Delta S = \frac{\Delta t}{t_s^{\,t+0.5\Delta t}} \tag{25}$$

The current Scheil's sum can be updated to $S^{t+\Delta t} = S^{t} + \Delta S$. The general phase fraction evolved during the current time step can be given as

$$\Delta f = \frac{\zeta\ \Delta t}{t_e^{\,t+0.5\Delta t} - t_s^{\,t+0.5\Delta t}} \tag{26}$$

The following three possibilities arise in this calculation:

1. If $S^{t+\Delta t} < 1$, then $\Delta f = 0$.
2. If $S^{t+\Delta t} < 1$ and $S^{t+\Delta t} > 1$, the incubation time is reached during the current time step, and only a fraction ζ of Δt contributes to phase transition and $\zeta \approx \dfrac{s^{t+\Delta t} - 1}{\Delta s}$.
3. If $S^{t+\Delta t} > 1$ and also $S^{t} > 1$, the phase transition already started and ζ=1, since the full time step contributes to phase transition.

3.3 Displacement Field Formulation

The final global form of mechanical equilibrium equation becomes

$$\mathbf{K}_i^{ut+\Delta t}\ \Delta\hat{\mathbf{U}} = \mathbf{F}^{ut+\Delta t} - \mathbf{R}_i^{ut+\Delta t}, \tag{27}$$

where $\mathbf{K}^{u}$ is the global stiffness matrix, $\mathbf{F}^{u}$ is the global equivalent nodal load vector, $\mathbf{R}^{u}$ is the internal reaction vector taken from the previous iteration, and $\Delta\hat{\mathbf{U}}$ is the incremental global displacement vector. Elemental forms of matrices and vectors are given as in [14]

$$\begin{aligned}
\mathbf{K}_{ei}^{ut+\Delta t} &= \int_{\Omega}\left[\mathbf{B}^{\mathrm{T}}\ \mathbf{C}_i^{ept+\Delta t}\ \mathbf{B}\right] d\Omega \\
\mathbf{F}_{ei}^{ut+\Delta t} &= \int_{\Gamma_f}\left[\left(\mathbf{N}^{\mathrm{S}}\right)^{T}\ \mathbf{t}_e^{\,t+\Delta t}\right] d\Gamma_f + \int_{\Omega}\left[\mathbf{N}^{\mathrm{T}}\ \mathbf{b}_e^{\,t+\Delta t}\right] d\Omega \\
\mathbf{R}_{ei}^{ut+\Delta t} &= \int_{\Omega}\left[\mathbf{B}^{\mathrm{T}}\ \mathbf{T}_{ei}^{t+\Delta t}\right] d\Omega
\end{aligned} \tag{28}$$

where $\mathbf{t}_e$ is the boundary element traction vector, $\mathbf{b}_e$ is the element body force vector, $\mathbf{B}$ is the strain-displacement matrix which is unique for the particular structural problem which will be discussed in Section 3.4 $\mathbf{C}^{ep}$, is the elemental tangent elasto-plastic matrix [11].

3.4 Structural Application

Thermal, metallurgical and mechanical field computations which are discussed in Section 3.1, 3.2 and 3.3 are similar for any kind of a three dimensional metal quenching process except the calculation of strain-displacement matrix $\mathbf{B}$. The two-dimensional beam problem is considered in this section. Using an iso-parametric element formulation, the global co-ordinates and displacements are given in terms of local co-ordinates (ξ, η) by

$$\begin{aligned} x(\xi,\eta) &= \mathbf{N}^T\mathbf{X}, \quad y(\xi,\eta) = \mathbf{N}^T\mathbf{Y} \\ u(\xi,\eta) &= \mathbf{N}^T\mathbf{U}_e, \quad v(\xi,\eta) = \mathbf{N}^T\mathbf{V}_e \end{aligned} \tag{29}$$

The derivatives of the shape functions with respect to the global x and y co-ordinates are represented by the operator $\mathbf{H}$ of size 2x9 as

$$\mathbf{H} = \begin{bmatrix} \frac{\partial}{\partial x} \\ \frac{\partial}{\partial y} \end{bmatrix} \mathbf{N}^T = \mathbf{J}^{-T} \begin{bmatrix} \frac{\partial}{\partial \xi} \\ \frac{\partial}{\partial \eta} \end{bmatrix} \mathbf{N}^T . \tag{30}$$

The strain-displacement operator $\mathbf{B}$ (subscript 'e' is suppressed) for the plane stress case is the simplest one and it is refereed in this text as standard strain-displacement operator with size 3x18,

$$\mathbf{B} = \mathbf{B}_{std} = \left[\begin{array}{cc|c|cc} H_{x1} & 0 & \cdots & H_{x9} & 0 \\ 0 & H_{y1} & \cdots & 0 & H_{y9} \\ H_{y1} & H_{x1} & \cdots & H_{y9} & H_{x9} \end{array}\right], \tag{31}$$

where H_x and H_y are the elements of the first and second rows of derivative operator $\mathbf{H}$. Through the beam cross-sectional element the long profiles can be analyzed by introducing one extra global node with 3 degrees of freedom. The strain-displacement matrix for the beam case [8] has the size 4x21. There is one additional row and three additional columns. The introduced addition is named as $\mathbf{B}_{beam}$ and

$$\mathbf{B} = \begin{bmatrix} \mathbf{B}_{std} & \mathbf{0} \\ \mathbf{0} & \mathbf{B}_{beam} \end{bmatrix}, \quad \text{where } \mathbf{B}_{beam} = \tfrac{1}{l}[1 \quad y \quad -x] . \tag{32}$$

The additional operator $\mathbf{B}_{beam}$ is only for computing the strain in the axial direction, which is just related to axial elongation w and bending curvatures c_x and c_y. The standard elasto-plastic stress-strain operator is given as

$$\hat{\mathbf{C}}_{std}^{ep} = 3\kappa\hat{\mathbf{P}}_1 + 2\mu\hat{\mathbf{P}}_2 - \frac{2\mu}{1+\frac{H}{3\mu}}\hat{\mathbf{n}}'_T - \lambda\frac{4\mu^2}{\left\|(\mathbf{T}')^{trial}\right\|}\left(\hat{\mathbf{P}}_2 - \hat{\mathbf{n}}'_T\right), \tag{33}$$

where $\hat{\mathbf{P}}_1$ is the spherical projector, $\hat{\mathbf{P}}_2$ is the deviator projector, and $\hat{\mathbf{n}}_T'$ is the plastic flow direction projector. The projectors are of size 4x4

$$\hat{\mathbf{P}}_1 = \frac{1}{3}\begin{bmatrix} 1 & 1 & 0 & 1 \\ 1 & 1 & 0 & 1 \\ 0 & 0 & 0 & 0 \\ 1 & 1 & 0 & 1 \end{bmatrix}, \qquad \hat{\mathbf{P}}_2 = \frac{1}{2}\begin{bmatrix} \frac{4}{3} & \frac{-2}{3} & 0 & \frac{-2}{3} \\ \frac{-2}{3} & \frac{4}{3} & 0 & \frac{-2}{3} \\ 0 & 0 & 1 & 0 \\ \frac{-2}{3} & \frac{-2}{3} & 0 & \frac{4}{3} \end{bmatrix} \tag{34}$$

4 Results and Discussions

An **L**120x12 profile made up of 100Cr6 steel of unit length is modeled using the beam cross-sectional elements as discussed in the Section 3.4. The temperature-dependent material properties of the individual phases can be found in reference [6]. The distortion of the long profiles is represented by their curvature. The volume averages of the temperature and the effective stresses are considered for the comparison of different cooling strategies.

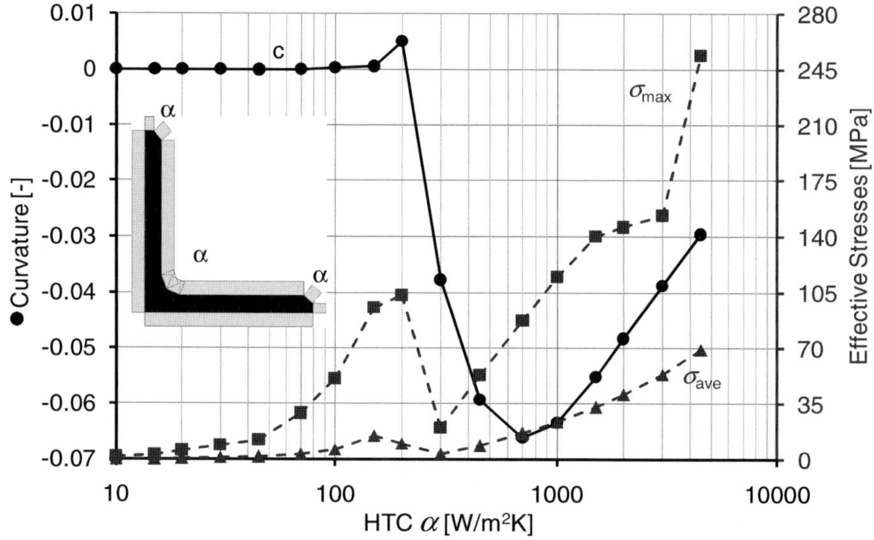

Fig. 2. HTC α vs. Curvature and Effective stresses in equal cooling of L-100Cr6

A series of simulations with equal HTC (α) ranging from 10-4500W/m^{2}K has been performed to find out the critical cooling regions. The computed final distortion, the average and the maximum effective stresses are plotted against α in Fig.2. In the low cooling range α < 200W/m^{2}K no distortion is observed, that it increases in the negative direction, reaches maximum at α = 700W/m^{2}K, and afterward changes its cooling range and they reaches a local maximum where the distortion of the profile is direction. The internal stresses gradually increase with the increasing α in the low first observed. Next, a local minimum of stress indicates that the different parts of the profile are plastified in the reverse direction which produces the distortion but at the same time relaxes the residual stress state.

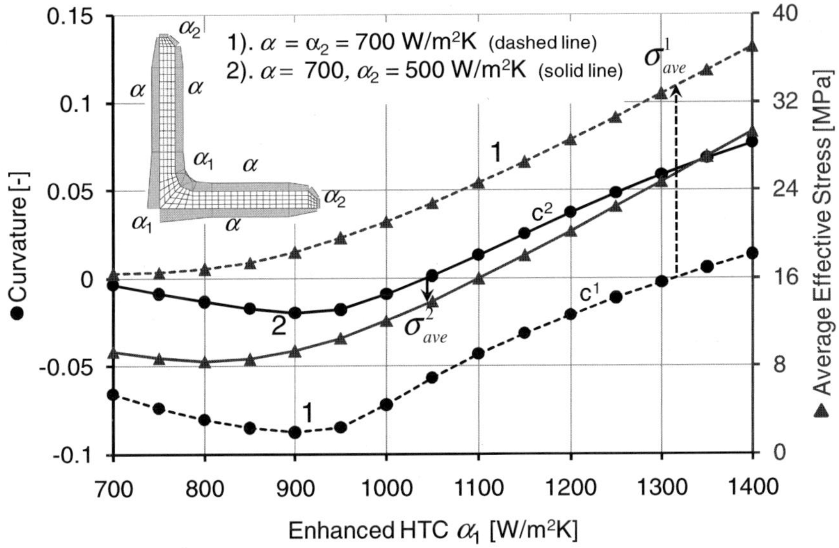

Fig. 3. Curvature and average effective stress as a function of enhanced HTC α_1 (L-100Cr6)

During an equal cooling, the temperature gradient is not uniform due to the mass distribution with respect to the locations of the boundaries. The distortion can be eliminated by increasing the local cooling at mass lumped regions [6]. This fact is verified in this section with the first cooling strategy as shown in Fig. 3. Firstly, increasing the HTC only at the mass lumped region moderately equalizes the temperature distribution. The HTC at the mass lumped region is designated by α. As the HTC α_1 is increased, the distortion gets reduced and totally eliminated at α_1 =1315W/m^{2}K. Further increase in the HTC α_1 produces a distortion in the reverse direction. However, this strategy increases the residual stress continuously as shown in Fig. 3. From this simulation result, one can come to the conclusion that increasing the HTC at the mass lumped region can only reduce the distortion but not the residual stresses.

Secondly, reducing the cooling at the edges and enhancing the cooling at the mass lumped region equalizes the temperature distribution inside the material to a greater extent [7]. To implement this, strategy 2 in which α_2 = 500W/m^{2}K is maintained at the edges along with uniform cooling α = 700W/m^{2}K is introduced. The curvature and the stresses are plotted for various values of enhanced cooling HTC α_1. The distortion is completely eliminated when α_1 reaches 1040W/m^{2}K. The residual stress at

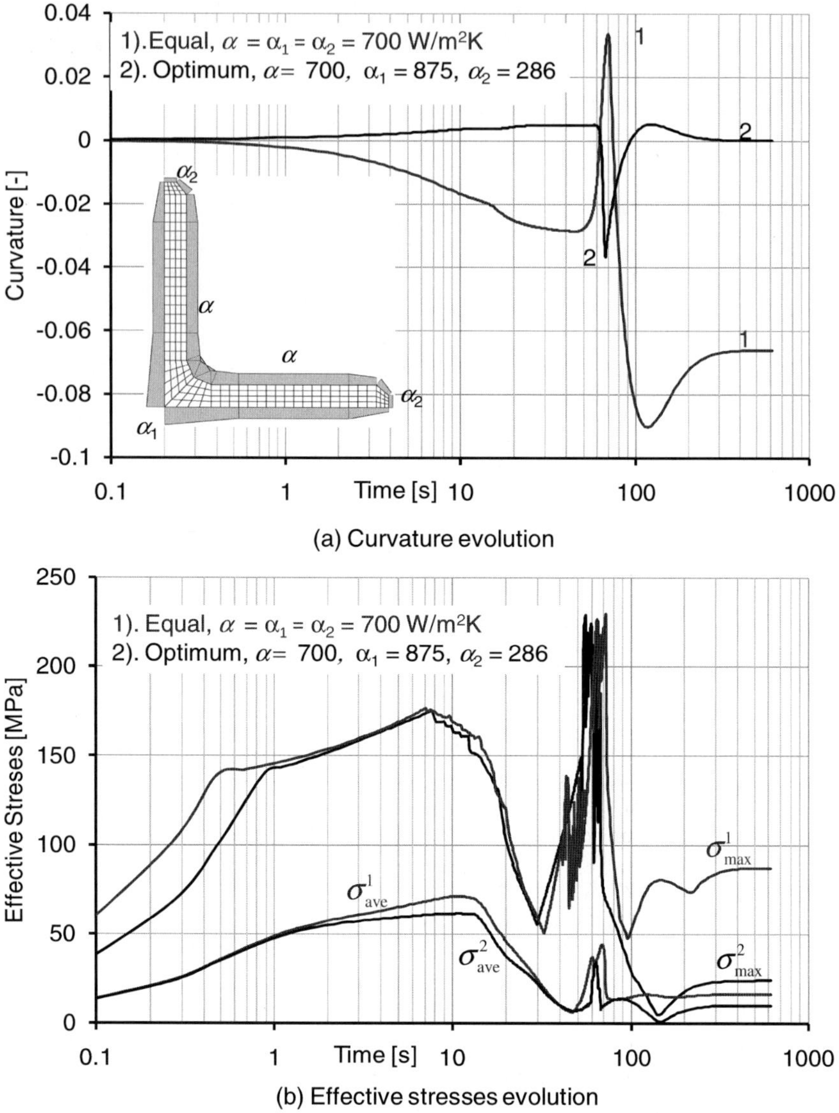

Fig. 4. Comparisons: equal and optimum cooling strategies - curvature and stresses evolution

zero curvature in strategy 2 is half the value of strategy 1. This fact indicates that with a continent combination of α_1 and α_2, it is a possible to reduce the residual stresses at a distortion free final state.

For $\alpha = 700\text{W/m}^2\text{K}$, the values of α_1 and α_2 are identified through a standard two parameter optimisation technique. In the case of the optimum cooling strategy, the distortion is much smaller during the cooling and it is finally eliminated as shown in Fig. 4a. The final maximum equivalent stress is reduced approximately from 87.2 to 24MPa. Similarly, the average effective stress is reduced from 16.4 to 10.1MPa as shown in Fig. 4b. During the phase transformation the maximum equivalent stress fluctuates a lot due to the transformation induced plasticity.

5 Concluding Remarks

The metal quenching process is analysed using a non-linear finite element technique which includes the coupling of the thermal, metallurgical and mechanical fields within the frame of the isothermal staggered approach. The distortion and residual stress evolution are calculated for long **L** profile made of steel. Along with an enhanced cooling at the mass lumped region, a reduced cooling at the edges and corners simultaneously reduces both the distortion and the residual stresses.

References

[1] Smoljan, B.: Numerical simulation of steel quenching. J. Mat. Engg. Perf. 11, 75–79 (2002)

[2] Lewis, R.W., Ravindran, K.: Finite element simulation of metal casting. Int. J. Num. Meth. Engg. 47, 29–59 (2000)

[3] Todinov, M.T.: Mechanism for formation of the residual stresses from quenching. Model. Sim. Mat. Sci. Engg. 6, 273–291 (1998)

[4] Pietzsch, R., Brzoza, M., Kaymak, Y., Specht, E., Bertram, A.: Minimizing the distortion of steel profiles by controlled cooling. Steel Res. Int. 76, 399–407 (2005)

[5] Brzoza, M., Specht, E., Ohland, J., Belkessam, O., Lübben, T., Fritsching, U.: Minimizing stress and distortion for shafts and discs by controlled quenching in a field of nozzels. Mat. Sci. Engg. Tech. 37, 97–102 (2006)

[6] Pietzsch, R., Brzoza, M., Kaymak, Y., Specht, E., Bertram, A.: Simulation of the distortion of long steel profiles during cooling. J. App. Mech. 74, 427–437 (2007)

[7] Kaymak, Y., Specht, E.: Strategies for controlled quenching to reduce stresses and distortion. Heat Proces. 5(3), 1–4 (2007)

[8] Kaymak, Y.: Modeling of metal quenching process and strategies to minimize distortion and stresses, PhD thesis, Otto-von-Guericke University, Magdeburg, Germany (2007), `http://diglib.uni-magdeburg.de/Dissertationen/2007/yalkaymak.pdf`

[9] Celentano, D., Orate, E., Oller, S.: A temperature-based formulation for finite element analysis of generalized phase-change problems. Int. J. Num. Meth. Engg. 37, 3441–3465 (1994)

[10] Kang, S.H., Im, Y.T.: Finite element investigation of multi-phase transformation within carburized carbon steel. J. Mat. Proce. Tech. 183, 241–248 (2007)

[11] Simo, J.C., Hughes, T.J.R.: Computational Inelasticity. Springer, New York (1997)

[12] Chen, W.-F.: Constitutive Equations for Engineering Materials. Plasticity and Modelling, vol. 2. Elsevier, Netherlands (1994)

[13] Armero, F., Simo, J.C.: A new unconditionally stable fractional step method for nonlinear coupled thermomechanical problems. Int. J. Num. Met. Engg. 35, 737–766 (1992)

[14] Bathe, K.J.: Finite Element Procedures. Prentice-Hall Inc., New Jersey (1996)

Micro Model for the Analysis of Spray Cooling Heat Transfer – Influence of Droplet Parameters

M. Nacheva and J. Schmidt

Institut für Strömungstechnik und Thermodynamik,
Otto-von-Guericke-Universität Magdeburg

Abstract. The quenching process of hot metal surfaces is studied on the micro level based on the behaviour of impinging single droplets. As an application, water cooling of hot steel sheet at initial surface temperatures above the Leidenfrost point is regarded. The heat transfer is investigated for a Representative Volume Element. It is assumed that the direct liquid-wall contact and the kinetics of the droplet spreading dominate the heat transfer. After a numerical solution of the energy balance, the heat transferred from the wall to the droplet is calculated as a function of spray parameters. On the macro scale, the heat transfer coefficient of the spray can be determined by homogenisation using the results of the micro model. Simulations are carried out under different assumptions for the droplet spreading and the model parameters for a wide range of droplet diameters, velocities and mass fluxes. The results are presented and discussed in dependence on the varied parameters.

1 Introduction

Spray cooling is used in many industrial applications, for example in the process engineering, metallurgy, and certain finishing processes. One of the important advantages is the good adjustment of the local cooling conditions on the cooling demands. Often, defined cooling conditions are required to guarantee specific pro-cess conditions or specific product properties. In heat treatment processes, the cooling rate influences the micro structure of the materials, and certain properties of the material or finished product can be obtained such as strength, hardness or machinability. In other applications, the aim is to avoid distortion, and to minimise the stresses induced during the quenching process [1]. The necessary quality can only be achieved in all processes by certain cooling strategies with defined cooling conditions.

The spray cooling is investigated in a large number of scientific works, and many researchers give empirical, semi-empirical or numerical models to describe the process and to calculate the corresponding heat transfer coefficient (HTC). A comparison of the HTC calculated by some empirical models is presented in Fig. 1. The HTC depends on the mass flux of the cooling fluid, which is the most important spray parameter of the cooling. Only few of the studies correlate the HTC with the droplet velocity and rarely with the droplet diameter. The great discrepancies in the different experimental values of the HTC can be seen clearly in Fig. 1. Furthermore, different trends of the dependence of the HTC on the droplet parameter, diameter and velocity are described.

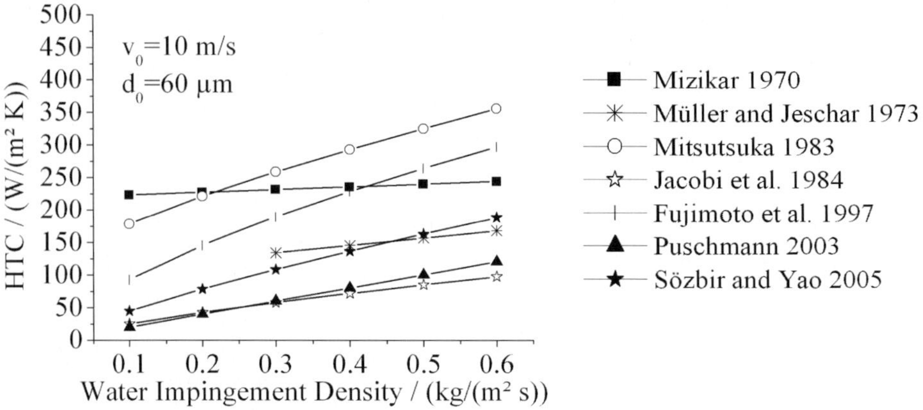

Fig. 1. Comparison of HTC calculated with different correlations [2-8]

The design of the cooling process requires the understanding of the heat transfer mechanism and the determination of the HTC. In the case of spray cooling at high surface temperatures, the Leidenfrost phenomenon arises. In dilute sprays, the heat transfer can be dominated by the direct liquid-solid contact of the droplets.

The influence of the droplet parameters on the cooling process can be explained based on the behaviour of single droplets (impacting regimes, spreading, etc.). The impacting regimes and the deformation of the droplet at different Weber numbers are investigated and described by many authors [9-11]. The impact of water droplets on aluminium surfaces heated up to different temperatures is investigated with a photographic technique by Bernardin et al. [12]. The cooling process above the Leidenfrost point is also considered by Suzuki and Mitachi [13]. They consider the impact of single water drops on a brass surface heated from 200 to 450°C. Contrary to [12] and [13], Šikalo [14] and Roisman et al. [15] describe the droplet spreading on cold solid surfaces with different surface properties. The experimental results for the droplet spreading are used in the presented model for the description of the droplet kinetics.

2 Modelling of Heat Transfer

2.1 Micro-macro Interactions

The heat transfer and the cooling of the work piece can be modelled on the micro and the macro level [16]. On the macro scale, the spray and the total hot surface are considered as shown in Fig. 2. The cooling process is described by the HTC α_{sp} which is locally distributed and time averaged. The HTC is measured and correlated to the local liquid mass flux of the spray, as mean parameter, and with the droplet parameters, diameter and velocity. In real sprays, the droplet diameter and velocity are distributed, but in the correlations, mean values, such as the Sauter diameter d_{32} and mean velocity $\bar{v}$ are used.

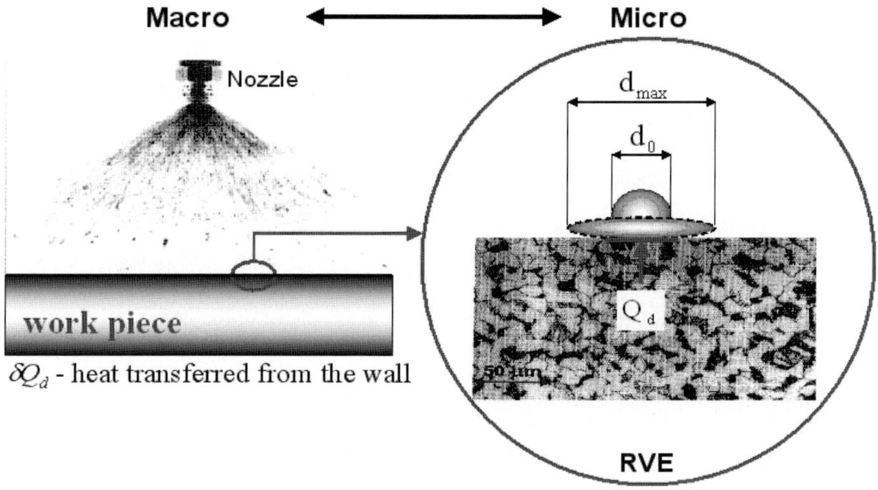

Fig. 2. Micro-macro interactions in spray cooling

In particular, concerning the influence of the diameter and the velocity on the HTC, no clear dependence is described in the literature. For dilute sprays, the effect of the velocity and the diameter is related to the droplet-wall interactions, the contact and the spreading of the droplets on the wall. The interaction kinetics also influences the heat transfer phenomena, such as direct liquid-wall contact, superheating of the liquid, and vapour film formation. In this way, the investigation of single droplets on the micro scale, Fig. 2, can disclose important information concerning the influence of the spray parameters on the heat transfer. Simultaneously, the modelling of the heat transfer on the micro level is a supposition for the investigation of thermally caused changes in the microstructure and possible micro scale damages. On the micro scale, the expected temperature gradients are much higher as on the macro level, and they could cause different effects in comparison to the calculated temperature field using the homogenised HTC.

The modelling on the micro scale uses a Representative Volume Element (RVE) formed by a single droplet and a characteristic part of the work piece as shown in Fig. 2. In the RVE, the droplet spreads out to the maximum diameter and causes a local cooling of the work piece. The heat Q_d transferred from the work piece to the droplet depends on the complex and coupled transport processes in the droplet and the work piece and is determined by the kinetics of the droplet deformation. In a first approximation, the heat transfer in the droplet and in the work piece is considered to be decoupled. During a parallel work, the droplet spreading and deformation are considered using CFD-methods, Fig. 3 [17].

This paper is concentrated on the solid side transport processes. Droplet spreading and heat transfer are described by suitable boundary conditions. The experimental results of many authors indicate that the amount of heat transferred from the wall to the droplet after the impact is determined by the liquid-solid contact in the first phase and

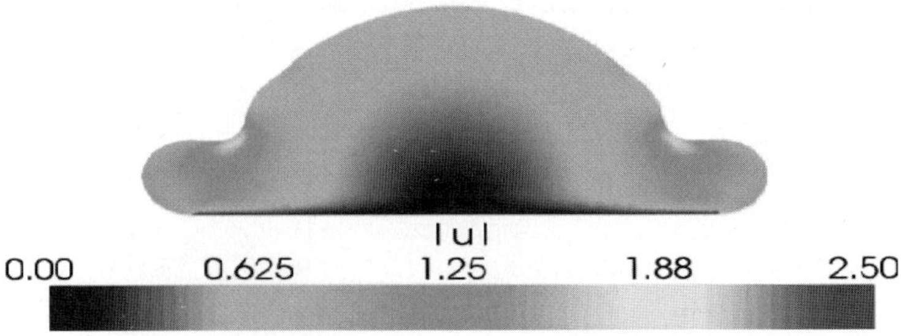

Fig. 3. Droplet deformation and velocity distribution [17]

it is an order of magnitude higher than the amount of heat transferred through the vapour layer, which forms later on the surface. Because of this, the model characterises the heat transfer by direct contact in relation to the time-dependent droplet spreading which determines the area of the heat transfer.

For the modelling of RVE, the following assumptions are made.

- The considered spray is dilute and monodisperse concerning the homogenisation of the single droplet results.
- The droplets in the spray are homogeneously distributed. In this case, the distance between two droplets can nearly be described in all directions by the same characteristic length δ_d and the time between two droplets impact in the RVE is

$$t_\delta = \delta_d / v_0 . \tag{1}$$

- The droplet impact takes place at Weber numbers $We > 30$.
- The initial temperature distribution in the wall at the droplet impact is uniform.

2.2 Droplet Spreading with Time

At first, the time-dependent droplet deformation is analysed and applied later in the boundary conditions for the heat transfer. After impact, the droplet spreads out on the surface to its maximum diameter d_{max} in a couple of microseconds. The corresponding time, in which the cooling takes place, is represented by the contact time $t_{cont} = t(d_{max})$. The modelling of the spreading function is already described in [18] and is presented here briefly.

The maximum spreading diameter d_{max} for droplet impingement on super-heated walls is described in dependence on the droplet Weber number using the empirical correlation of Suzuki and Mitachi [13]

$$\frac{d_{max}}{d_0} = \mathrm{a}\, We^{\mathrm{n}}. \tag{2}$$

In comparison to further experiments, i.a. of Sikalo [14], the coefficients a and n are adapted:

$$\mathrm{a} = 1{,}18, \quad \mathrm{n} = 0{,}24 \quad 30 \le We < 1000 . \tag{3}$$

For the contact time, different formulations are given in the literature [11, 13]. For lower *We*-numbers, the Rayleigh time is often used

$$t_{\mathrm{Ray}} = \frac{\pi}{4}\sqrt{\frac{\rho_{\mathrm{fl}} d_0^3}{\sigma_{\mathrm{fl}}}}\,. \tag{4}$$

In the experiments of Inada [19], a vapour film could only be measured in the second half of the interaction time, when the droplet recoils. Therefore, the contact time $t_{\mathrm{cont}} = t(d_{\mathrm{max}})$ is only related to the spreading phase. In the model, a correlation proposed by Suzuki and Mitachi [13], is used. This correlation to the dimensionless contact time t^*_{cont}

$$t^*_{\mathrm{cont}} = \frac{t_{\mathrm{cont}}}{d_0 / v_0} = 1.85\, We^{0.16} \qquad 10 \le We \le 1000 \tag{5}$$

is valid in a large range of the *We*-number. The spreading diameter $d(t^* = t\, v_0 / d_0)$ can be calculated from the maximum diameter d_{max} and the dimensionless contact time t^*_{cont} using as a first approximation a linear function with constant spreading velocity v_{spread}

$$\frac{d(t^*)}{d_0} = 2\frac{v_{\mathrm{spread}}}{v_0} t^*\,, \tag{6}$$

$$\frac{v_{\mathrm{spread}}}{v_0} = \frac{1}{2}\frac{d_{\mathrm{max}}}{d_0}\frac{1}{t^*_{\mathrm{cont}}} = \mathrm{c}\, We^{\mathrm{m}} = const\,. \tag{7}$$

The constants c and m in Eq. (7) are determined by Eqs. (2), (3) and (5). The used approach describing the maximum diameter differs to previous publications [20]. With the new description, a smooth function of the spreading velocity v_{spread} (We) and a better approximation are obtained.

2.3 Temperature Field

Fourier's differential equation as energy balance is used to determine the temperature distribution $\vartheta(r,z,t)$ and the heat transferred to the droplet Q_{d}. For the considered case, a 2 D-formulation of the heat conduction equation

$$\rho c_p \frac{\partial \vartheta}{\partial t} = \frac{1}{r}\frac{\partial}{\partial r}\left(r\lambda \frac{\partial \vartheta}{\partial r}\right) + \frac{\partial}{\partial z}\lambda \frac{\partial \vartheta}{\partial z} \tag{8}$$

is applicable, which is solved numerically by means of the explicit Finite Diffe-rence Method. Since boundary conditions are considered as time- and space-dependent, a numerical solution is necessary. In the case of changes in the microstructure, the temperature-dependence of the thermophysical material properties and the phase change enthalpy have to be taken into account. In this case, the use of the energy balance in the enthalpy formulation

$$\frac{\partial H_v}{\partial t} = \frac{1}{r}\frac{\partial}{\partial r}\left(r \frac{\partial \Lambda}{\partial r} \right) + \frac{\partial^2 \Lambda}{\partial z^2} \tag{9}$$

is appropriated. H_v and Λ present the volumetric enthalpy

$$H_v(\vartheta) - H_{v,0}(\vartheta_0) = \int_{\vartheta_0}^{\vartheta} \rho_{\mathrm{w}}\, c_{p,\mathrm{w}}(\vartheta)\, d\vartheta \tag{10}$$

and the integral thermal conductivity

$$\Lambda(\vartheta) - \Lambda_0(\vartheta_0) = \int_{\vartheta_0}^{\vartheta} \lambda_{\mathrm{w}}(\vartheta)\, d\vartheta\,, \tag{11}$$

respectively. The advantages of the Enthalpy Method can especially be used in combination with the explicit Finite Difference Method.

For the solution of the energy balance, initial and boundary conditions are necessary. According to the model assumptions, the wall temperature at time $t = 0$ lies above the Leidenfrost temperature $\vartheta_{\mathrm{Leid}}$ and is uniform

$$t = 0: \qquad \vartheta(r, z, t = 0) = \vartheta_{\mathrm{w},0} > \vartheta_{\mathrm{Leid}}\,. \tag{12}$$

A boundary condition of first kind is chosen at the direct contact area $r \leq r_{\mathrm{spread}}(t)$ and in a first approximation, adiabatic conditions are supposed for the other boundaries of the RVE.

$$z = 0,\ 0 \leq r \leq r_{\mathrm{spread}}(t): \quad \vartheta_{\mathrm{w}} = \vartheta^*_{\mathrm{cont}}(t) \tag{13}$$

$$z = 0,\ r_{\mathrm{spread}} < r \leq r_{\mathrm{RVE}}: \quad \frac{\partial \vartheta}{\partial z} = 0 \tag{14}$$

In difference to a commonly used constant wall temperature, for example by Bolle, Moureau [10] or Wruck [11], the wall temperature ϑ_{w} in Eq. (13) is considered as time-dependent. The temperature $\vartheta^*_{\mathrm{cont}}$ is comparable to the contact temperature $\vartheta_{\mathrm{cont}}$ but it changes with time.

According to the definition, the contact temperature is a function of wall and liquid temperatures and the thermal effusivities $b_{\mathrm{w},0}$ and b_{fl}

$$\vartheta_{\mathrm{cont}} = \vartheta_{\mathrm{w},0} - \frac{\vartheta_{\mathrm{w},0} - \vartheta_{\mathrm{fl},0}}{1 + \dfrac{b_{\mathrm{w}}}{b_{\mathrm{fl}}}}\,. \tag{15}$$

This temperature corresponds to the surface temperature when two semi-infinite bodies with different initial temperatures come into contact with each other. Due to the small droplet sizes, the heating of the liquid film on the wall occurs very fast and demands the consideration of the liquid temperature as time-dependent. Using this temperature $\vartheta_{\mathrm{fl}}(t)$ and an effective thermal effusivity of the fluid, $b_{\mathrm{fl,eff}} = \varphi_{\mathrm{M}}\, b_{\mathrm{fl}}$, with a model parameter φ_{M}, the temperature $\vartheta^*_{\mathrm{cont}}$ is calculated analogously to the definition of contact temperature [18]

$$\vartheta^*_{\mathrm{cont}}(t) = \vartheta_{\mathrm{w},0} - \frac{\vartheta_{\mathrm{w},0} - \vartheta_{\mathrm{fl}}(t)}{1 + \dfrac{b_{\mathrm{w}}}{\varphi_{\mathrm{M}} b_{\mathrm{fl}}}} . \tag{16}$$

The temperature $\vartheta^*_{\mathrm{cont}}$ is calculated in dependence on the liquid temperature $\vartheta_{\mathrm{fl}}(t)$ for each time step and used in the boundary condition, Eq. (13). In the energy balance, from which $\vartheta_{\mathrm{fl}}(t)$ is determined, only liquid phase heat transfer is considered. Therefore, the liquid temperature is not limited by the boiling temperature, and a super heating could arise. The maximum superheating temperature $\vartheta_{\mathrm{sh,max}}$ depends on different influence parameters [18] and is considered as model parameter. If the liquid temperature reaches $\vartheta_{\mathrm{sh,max}}$, a change in the boundary conditions is assumed, and Eq. (13) is replaced by a boundary condition of second kind. The corresponding heat flux depends on the wall temperature and is determined according to the Nukiyama boiling curve [18].

The heat Q_{d}, transferred from the hot surface to the droplet, can be calculated from the temperature field $\vartheta(r,z,t)$

$$Q_{\mathrm{d}} = \int_{t=0}^{t_\delta} \int_{r=0}^{r_{RVE}} 2\pi r \lambda_{\mathrm{w}} \left(\frac{\partial \vartheta}{\partial z} \right)_{z=0} dr\, dt\ . \tag{17}$$

This heat is typical for a certain pair of droplet diameter and velocity and permits the estimation of the evaporation efficiency and the HTC for dilute sprays.

2.4 Evaporation Efficiency and HTC

The calculated heat Q_{d} depends strongly on the droplet diameter. Therefore, the evaporation efficiency ε_{d} of a single droplet is better suited for the comparison of heat transfer of different droplet sizes. The evaporation efficiency

$$\varepsilon_{\mathrm{d}} = \frac{Q_{\mathrm{d}}}{\rho_{\mathrm{fl}} \dfrac{\pi}{6} d_0^3 \Delta h_{\mathrm{v}}} \tag{18}$$

presents the ratio of the absorbed heat Q_{d} to the heat required for the complete evaporation of the droplet.

The transferred heat Q_{d} can also be used for the determination of the HTC of dilute sprays. If the spray is homogeneous and monodisperse, a simple possibility for the homogenisation is given by the characteristic length δ_{d}. The parameter δ_{d} describes the distance between each pair of neighbour drops

$$\delta_{\mathrm{d}} = \sqrt[3]{\frac{M_d v_0}{\dot{m}_{\mathrm{sp}}}} = \sqrt[3]{\frac{\pi}{6} \frac{d_0^3 \rho_{\mathrm{fl}} v_0}{\dot{m}_{\mathrm{sp}}}}, \tag{19}$$

which depends on the droplet diameter d_0 and velocity v_0, the droplet mass M_{d} and the liquid mass flux $\dot{m}_{\mathrm{sp}}$. The HTC can be obtained by coupling Eq. (19) with the heat flux definition

$$\dot{q}_{\mathrm{sp}} = \alpha(\vartheta_{\mathrm{w}} - \vartheta_{\mathrm{fl}}) = \frac{Q_{\mathrm{d}}}{\delta_{\mathrm{d}}^2} \frac{v_0}{\delta_{\mathrm{d}}}. \tag{20}$$

The HTC can be written as

$$\alpha_{\mathrm{sp}} = \frac{6\dot{m}_{\mathrm{sp}}}{\pi d_0^3 \rho_{fl}} \frac{Q_{\mathrm{d}}(v_0, d_0, \vartheta_{\mathrm{w}} - \vartheta_{\mathrm{fl}})}{\vartheta_{\mathrm{w}} - \vartheta_{\mathrm{fl}}}. \tag{21}$$

Eq. (21) clarifies that the HTC is a linear function of the mass flux $\dot{m}_{\mathrm{sp}}$ and also depends on the droplet parameter d_0 and v_0.

3 Results and Discussion

The objective of the current investigation is the theoretical analysis of the heat transfer resulting during the spray cooling of the work piece at different spray impingement densities, droplet diameters and droplet velocities. Furthermore, the influence of the thermal boundary conditions (time-dependent temperature at the liquid-solid contact area, maximum limit for the superheating of the liquid, effective thermal effusivity of the liquid) on the heat transfer is considered. The presented results are obtained with constant thermophysical properties of the metal and the liquid under the following conditions:

- Initial surface temperature: $\vartheta_{\mathrm{w},0} = 500°C$
- Initial liquid temperature: $\vartheta_{\mathrm{fl},0} = 20°C$
- Water impingement density: $0.1 \le \dot{m}_{\mathrm{sp}} \le 0.6\ kg/(m^2\ s)$
- Droplet diameter: $40 \le d_0 \le 300\ \mu m$
- Impact droplet velocity: $7 \le v_0 \le 15\ m/s$
- Droplet Weber number: $27 \le We \le 927$
- Maximum superheating of the liquid: $\vartheta_{\mathrm{sh,max}} = 100,\ 150\ °\mathrm{C}$
- Parameter for the effective thermal effusivity of the liquid: $\varphi_{\mathrm{M}} = 1, 2, 4, 6$

The boundary temperature at the liquid-solid contact area strongly influences the simulation results. Some researchers [10, 11] define it as constant in time, which offers the advantage, that a similarity exists in relation to the dimensionless temperature $\theta = (\vartheta - \vartheta_{\mathrm{fl}})/(\vartheta_{\mathrm{w},0} - \vartheta_{\mathrm{fl}})$.

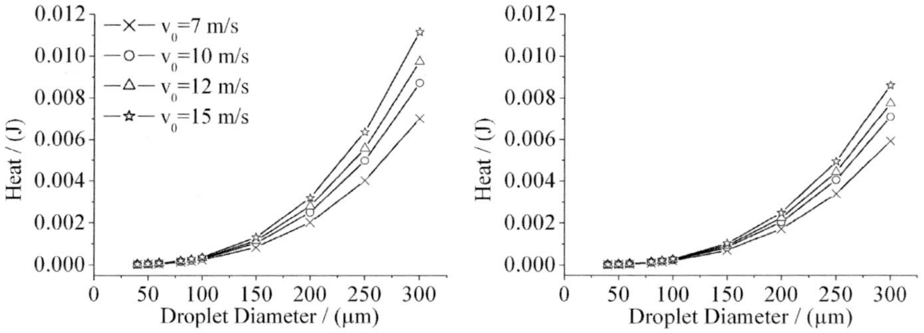

Fig. 4. Comparison of the transferred heat Q_d for $\vartheta_{cont} = const$ (left) and $\vartheta_{cont} = (t)$ (right)

In the present study, a time-dependent contact temperature is assumed. Its effect on the heat transferred from the wall to the droplet is compared in Fig. 4 to the results with a constant contact temperature. Concerning the liquid superheating, no limit is considered.

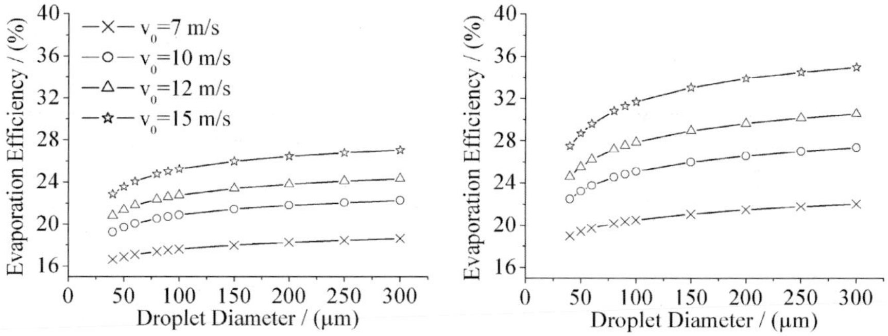

Fig. 5. Comparison of the evaporation efficiency for $\vartheta_{cont} = const$ (left) and $\vartheta_{cont} = (t)$ (right)

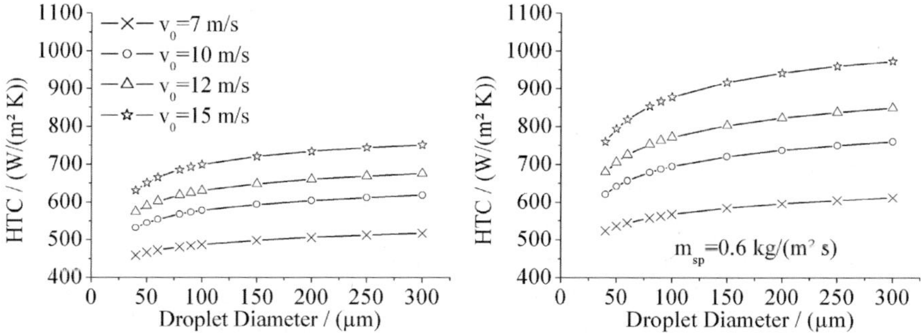

Fig. 6. Comparison of the HTC $\dot{m}_{sp} = 0.6\,kg/(m^2 s)$ for $\vartheta_{cont} = const$ (left) and $\vartheta_{cont} = (t)$ (right)

Both cases (time-independent and time-dependent ϑ_{cont}) show the same trend: the heat transferred from the wall to the droplet increases with increase of the droplet diameter at constant velocity. A similar trend can be seen for the estimated evaporation efficiency of the droplet, Fig. 5, and for the HTC, Fig 6.

In comparison to the values with $\vartheta_{\text{cont}} = const$, the amount of heat transferred at a time-dependent contact temperature is clearly smaller than those at a constant contact temperature. Furthermore, the smaller amount of transferred heat leads to smaller HTC, Fig 6, and to a smaller evaporation efficiency of the droplet, Fig 5.

The liquid temperature increases by the transferred heat during the droplet deformation up to a maximum value. Figure 7 shows this maximum increase of li-quid temperature for certain combinations of droplet parameters.

In Figs. 8 - 10, the influence of the assumed liquid superheating limit $\vartheta_{\text{sh,max}}$ on the transferred heat, the evaporation efficiency and the HTC is shown in depen-dence on both droplet size and velocity.

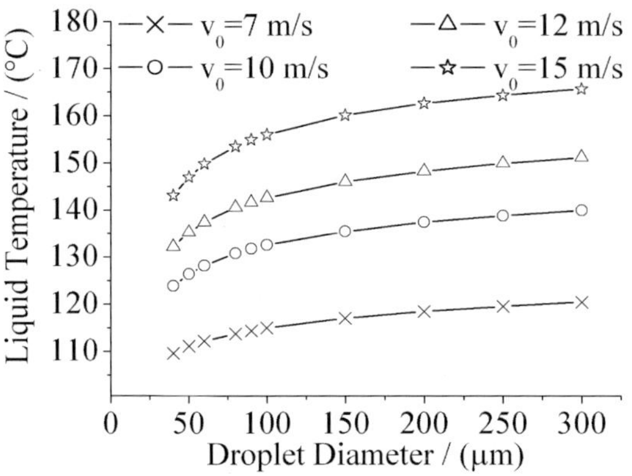

Fig. 7. Maximum liquid temperatures of droplets

After reaching the limit for the superheating, the liquid temperature remains constant and the heat flux is determined in accordance with the Nukiyama boiling curve. As shown in Fig. 7, the maximum liquid superheating limit depends on the droplet parameter. The limit of 100°C is reached for all tested droplet velocities and diameters.

HTC as well as evaporation efficiency depend mainly on the droplet velocity, Fig. 9 and 10. This prediction conforms to experimental results of Jacobi et al. [4], Puschmann [3], Schmidt and Boye [21]. The second case of superheating limit is set to 150°C. As shown in Fig. 7, this limit is only attained for higher droplet velocities and the corresponding results are indicated by filled markers in Fig. 9 and 10. The HTC and the evaporation efficiency depend on both droplet diameter and velocity below the superheating limits. This validates the experimental conclusion of Choi and Yao [22], Deb and Yao [23] that the HTC of dilute and monodisperse sprays increases with increasing droplet parameters. In summary, the influence of different parameters

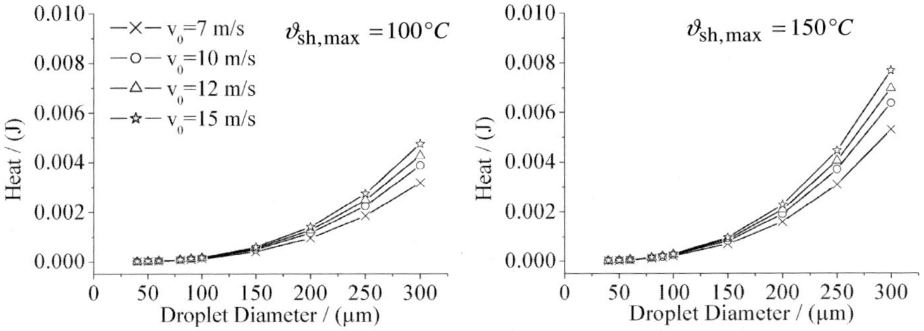

Fig. 8. Transferred heat Q_d for different limits of superheating $\vartheta_{sh,max}$

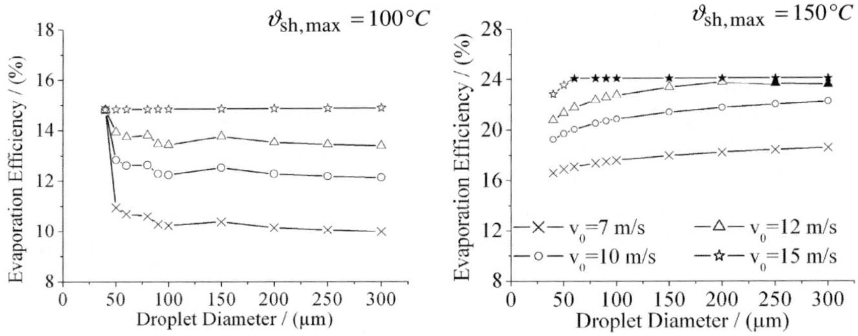

Fig. 9. Evaporation efficiency ε_v for different limits of superheating $\vartheta_{sh,max}$

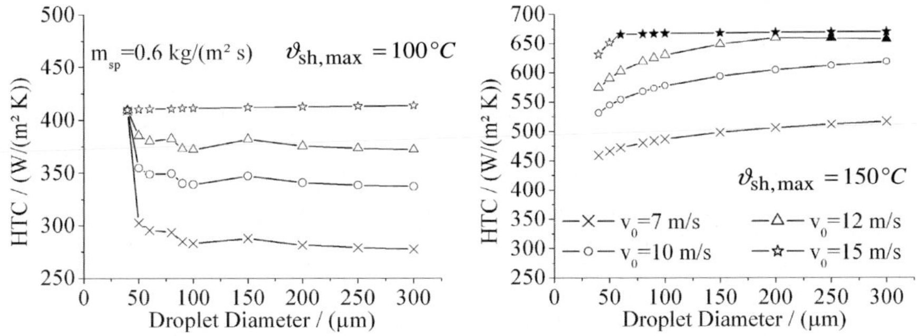

Fig. 10. HTC for different limits of superheating $\vartheta_{sh,max}$

on heat transfer depends on the liquid superheating limit. If the limit is attained, only droplet velocities influence the heat transfer. Otherwise, the heat transfer is affected by droplet velocity and diameter.

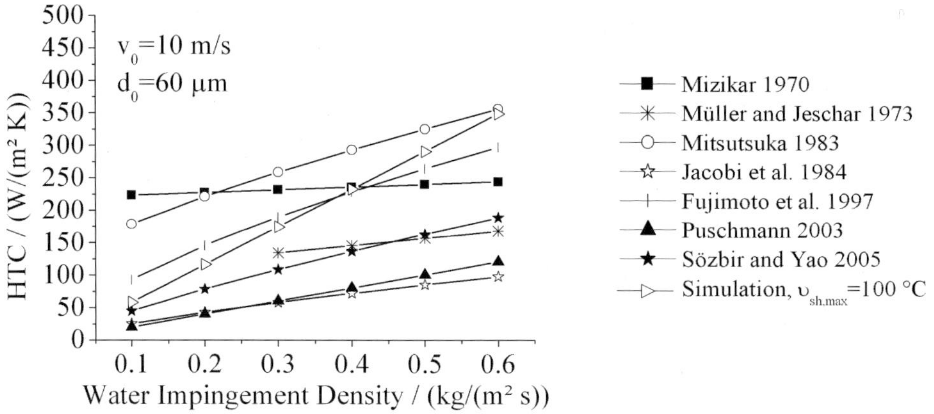

Fig. 11. Comparison of simulation and experimental results

The heat penetration into the droplet is affected by heat conductivity as well as by convection. The intensification due to the convective heat transport is considered by the effective thermal effusivity of the liquid. Generally, the consideration of effective thermal effusivity of liquid in the present modelling improves the spray cooling heat transfer and accelerates the reaching of the superheating limit. Below the superheating limit, increasing droplet parameters and increasing effective thermal effusivity yield higher HTC.

The calculated HTC according to the presented model are compared with experiments taken from the literature at $d_0 = 60\mu m, v_0 = 10 m/s$ and superheating limit of 100 °C, Fig. 11. The simulation results lie in the middle range of the HTC and suggest the values of Fujimoto et al. [6].

4 Conclusions

A new model for spray cooling heat transfer based on the droplet-wall interactions is presented. In comparison to other models in the literature, a time- and space-dependent wall temperature in the boundary condition and a numerical solution, which is also applicable in the case of changes in the microstructure, are used. The spreading is described in dependence on the *We*-number and direct liquid-wall contact is assumed as dominating heat transfer mechanism in the spreading phase. The maximum superheating temperature and an effective thermal effusivity of the liquid are used as model parameters.

A numerical study is performed to investigate the influence of different spray and model parameters. The results of the simulations show the dependence of the evaporation efficiency and the HTC on the mass flux, the droplet diameter and velocity. Especially, the maximum superheating temperature influences strongly the results and leads to a change in the dependence of the HTC on the droplet parameters.

For dilute sprays, the evaporation efficiency and the HTC show the same qualitative dependence on the droplet diameter d_0 and velocity v_0. Both increase with increase of d_0 and v_0. If the droplets reach the super heating limit, the evaporation efficiency and the HTC are nearly independent of the droplet diameter. The HTC is directly proportional to the mass flux $\dot{m}$, Eq. (21), and $\dot{m}$ does not change the influence of v_0 and d_0, which is given by $Q_d(v_0, d_0)$.

References

[1] Pietzsch, R., Brzoza, M., Kaymak, Y., Specht, E., Bertram, A.: Minimizing of distortion of steel profiles by controlled cooling. Steel Research Int. 76, 399–407 (2005)

[2] Müller, H., Jeschar, R.: Untersuchung des Wärmeübergangs an einer simulierten Sekundärkühlzone beim Stranggießverfahren. Eisenhüttenwesen 44, 589–594 (1973)

[3] Puschmann, F.: Experimentelle Untersuchung der Sprühkühlung zur Qualitätsverbesserung durch definierte Einstellung des Wärmeübergangs, PhD Thesis, University of Magdeburg, Magdeburg (2003)

[4] Jacobi, H., Keastle, G., Wünnenburg, K.: Heat transfer in cyclic secondary cooling during solidification of steel. Ironmaking and Steelmaking 11, 132–145 (1984)

[5] Mizikar, E.A.: Spray cooling investigation for continous casting of billets and blooms. Iron and Steel Engineer, 53–60 (1970)

[6] Fujimoto, H., Hatta, N., Asakawa, H., Hashimoto, T.: Predictable modeling of heat transfer coefficient between spraying water and a hot surface above the Leidenfrost temperature. ISIJ Int. 37(5), 492–497 (1997)

[7] Mitsutsuka, M.: Heat transfer coefficients in the surface temperature range of 400 °C to 800 °C during water-spray cooling of Hot Steel Product. Tetsu-to-Hagane 69, 268–274 (1983)

[8] Sözbir, N., Yao, S.C.: Experimental investigation of spray cooling on high temperature metal surfaces. In: Proc. Int. Symp. Heat Mass Transfer in Spray Systems, Antalya, Turkey, June 5-10 (2005)

[9] Wachters, L.H., Westerling, N.A.: The heat transfer from a hot wall to impinging water drops in the spheroidal state. Chem. Eng. Sc. 21, 1047–1056 (1966)

[10] Bolle, L., Moureau, J.C.: Spray cooling of surfaces. In: Multiphase Science and Technology, pp. 1–97. McCraw-Hill, New York (1982)

[11] Wruck, N.M.: Transientes Sieden von Tropfen beim Aufprall, PhD Thesis, Shaker Verlag, Aachen (1999)

[12] Bernardin, J.D., Stebbins, C.J., Mudawar, I.: Mapping of impact and heat regimes of water drops impinging on a polished surface. Int. J. Heat Mass Transfer 40(2), 247–267 (1997)

[13] Suzuki, T., Mitachi, K.: Experimental observation of liquid droplet impingement onto super-heated wall with high Weber numbers. In: Proc. Int. Symp. Heat and Mass Transfer in Spray Systems, Antalya, Turkey, June 5-10 (2005)

[14] Šikalo, Š.: Analysis of droplet impact onto horizontal and inclined surfaces, PhD Thesis. Shaker Verlag, Aachen (2003)

[15] Roisman, I.V., Rioboo, R., Tropea, C.: Normal impact of a liquid drop on a dry surface: model for spreading and receding. Proc. R. Soc. Lond. A 458, 1411–1430 (2002)

[16] Nacheva, M., Dontchev, D., Todorov, T., Schmidt, J.: Experimental and theoretical investigation of spray cooling heat transfer on macro and micro level. In: Proc. Int. Symp. Heat and Mass Transfer in Spray Systems, Antalya, Turkey, June 5-10 (2005)

[17] Sashikumaar, G.: Finite Element Methods on Moving Meshes for Free Surface and Interface Flows, PhD Thesis, University of Magdeburg, Magdeburg (2006)
[18] Nacheva, M., Schmidt, J.: Theoretical investigation of spray cooling heat transfer using a micro model for single droplets. In: Proc. 6th Int. Boiling Heat Transfer Conf., Spoleto, Italy, May 6-11 (2006)
[19] Inada, S., Miyasaka, Y., Sakamoto, K., Hojo, K.: Liquid-solid contact state and fluctuation of the vapor film thickness of a drop impinging on a heated surface. J. Chem. Eng. Japan 21(5), 463–468 (1988)
[20] Nacheva, M., Schmidt, J.: Micro modelling of the spray cooling of hot metal surfaces above the Leidenfrost Temperature. In: Proc. 10th Int. Congress on Liquid Atomization and Spray Systems ICLASS 2006, Kyoto, Japan, August 27 – September 1 (2006)
[21] Schmidt, J., Boye, H.: Influence of velocity and size of the droplets on the heat transfer in spray cooling. Chem. Eng. Technol. 24(3), 255–260 (2001)
[22] Choi, K.J., Yao, S.C.: Mechanisms of film boiling heat transfer of normally impacting spray. Int. J. Heat Mass Transfer 30, 311–318 (1987)
[23] Deb, S., Yao, S.C.: Analysis on film boiling heat transfer of impacting sprays. Int. J. Heat Mass Transfer 32, 2099–2112 (1989)

Finite Element Simulation of an Impinging Liquid Droplet

S. Ganesan and L. Tobiska

Institut für Analysis und Numerische Mathematik,
Otto-von-Guericke-Universität Magdeburg

Abstract. Numerical investigations of the deformation of a single droplet are useful for better understanding of the macro behaviour of spray cooling. In this chapter we present a finite element scheme, which is developed for computations of a single liquid droplet deformation on a horizontal surface. The numerical scheme is based on an Arbitrary Lagrangian Eulerian approach in a moving mesh. A brief overview of a few interface capturing and tracking techniques is also presented.

1 Introduction

Fluid flows with free surfaces and interfaces are encountered in many industrial and scientific applications such as spray cooling, film boiling, coating a solid substrate with liquids and inkjet printing. In many fluid flows, it is essential to study the micro properties of the flow to understand their macro behaviour. For instance to study the spray cooling effect in a macro scale, we must know the wetting behaviour of a single droplet in the spray, the influence of the droplet size and the droplet impinging velocity on the wetting rate, and the maximal wetting diameter of the droplet. It is the purpose of this work to develop an accurate and robust finite element scheme for computing the deformation of a single liquid droplet on a horizontal solid surface.

Apart from other difficulties associated with the computation of free surface and interface flows, computations of droplet deformations exhibit two main challenges: (i) the inclusion of the dynamic contact angle, (ii) the handling the moving contact line. These two problems are handled in our numerical scheme by using the Navier-slip boundary condition on the solid surface, and the Laplace Beltrami operator technique. In this article, we briefly recall our finite element scheme, for more details we refer to [7]. This article is organized as follows. In Section 2 we present a mathematical model for a single droplet deformation and their variational form. In Section 3 we present a brief overview of the commonly used interface capturing and tracking methods. A few numerical results of an impinging liquid droplet are provided in Section 4.

2 Model for an Impinging Droplet

In a given time interval $(0, T_g]$, the deformation of an incompressible impinging liquid droplet on a horizontal solid surface is described by the time-dependent incompressible Navier Stokes equations

$$\nabla \bullet u = 0; \qquad \rho\left(\frac{\partial u}{\partial t} + u \bullet \nabla u\right) - \nabla \bullet \mathrm{S}(u,p) = \rho g e \tag{1}$$

in a time dependent domain $\Omega(t) \subset \Re^3, t \in [0, \mathrm{T_g}]$. The model equations are completed by the initial condition, $u(0)=u_0$, the kinematic and force balance conditions

$$u \bullet n = w \bullet n, \qquad n \bullet \mathrm{S}(u,p) = \sigma \kappa n$$

on the free surface $\Gamma_F(t)$, and by the Navier slip boundary condition

$$u \bullet n = 0, \qquad u \bullet t + \varepsilon(n \bullet \mathrm{S}(u,p) \bullet t) = 0$$

on the solid surface $\Gamma_S(t) := \partial\Omega \backslash \Gamma_F(t)$. Here, u is the fluid velocity, p the pressure, w the free surface velocity, $\mathrm{T_g}$ the given time, ρ is the density of the fluid, κ the curvature of the free surface, σ the coefficient of the surface tension, ε the slip coefficient, g is the gravitational constant, and n, t are the normal and tangential vectors on their corresponding boundary, respectively. The stress tensor $\mathrm{S}(u, p)$ for the Newtonian fluid is given by

$$\mathrm{S}(u,p) = 2\mu\big(D(u) - \mathrm{I}p\big), \qquad D(u) = \frac{1}{2}\big(\nabla u + (\nabla u)^{\mathrm{T}}\big)$$

where μ is the dynamic viscosity, and $D(u)$ is the velocity deformation tensor. Further, the superscript T denotes the transpose, and I denotes the identity tensor.

2.1 Dynamic Contact Angle

An important property in a liquid droplet deformation on a solid surface is the dynamic contact angle θ_d, which is measured between the free surface and the solid surface. The contact angle plays a significant role in the flow dynamics. The droplet deformations with the same liquid but with different dynamic contact angles induce different flow dynamics. Thus to obtain a physically acceptable solution, the contact angle must be included into the model. The detailed description of the contact angle inclusion using the Laplace Beltrami operator technique is given in [11]. Here, we briefly recall it for completeness.

Let $\mathrm{V} := (\mathrm{H}^1(\Omega(t)))^3$ and $\mathrm{Q} := \mathrm{L}^2(\Omega(t))$ be the usual Sobolev spaces. We multiply the Navier Stokes equations (2) (after making it into a dimensionless form) with test functions $\mathrm{v} \in \mathrm{V}$ and $\mathrm{q} \in \mathrm{Q}$, and integrate over $\Omega(t)$. After applying integration by parts to the integral, which contains the stress tensor, and incorporating the boundary conditions, the variational form the Navier Stokes equation (1) reads: For a given $\Omega(0)$ and u_0, find $(u(t), p(t)) \in \mathrm{V} \times \mathrm{Q}$ such that

$$\left(\frac{\partial u}{\partial t}, v\right) + a(u;u,v) - b(p,v) + b(p,u) = f(\kappa, v) \tag{2}$$

for all $v \in V$ and $q \in Q$. Here,

$$a(\bar{u};u,v) = \frac{2}{\text{Re}} \int_{\Omega(t)} D(u) : D(v) + \int_{\Omega(t)} (\bar{u} \bullet \nabla) u \, v + \frac{1}{\beta} \int_{\Gamma_S(t)} (u \bullet t)(v \bullet t) \, ;$$

$$b(q,v) = \int_{\Omega(t)} q \nabla \bullet v \, ; \qquad f(\kappa,v) = \frac{1}{\text{Fr}} \int_{\Omega(t)} e \bullet v + \frac{1}{\text{We}} \int_{\Gamma_F(t)} (v \bullet n) \kappa.$$

After replacing the curvature in $f(\kappa,v)$ by the Laplace Beltrami operator of the identity, and integration by parts, the source term with the contact angle θ becomes:

$$f(\kappa,v) = \frac{1}{\text{Fr}} \int_{\Omega(t)} e \bullet v - \frac{1}{\text{We}} \int_{\Gamma_F(t)} \underline{\nabla} id_{\Gamma_F} : \underline{\nabla} v + \frac{1}{\text{We}} \int_{\Phi(t)} \cos(\theta) \, v \bullet t \qquad (3)$$

where $\Phi(t)$ is the moving contact line, for more details of this derivation see [11]. Here, $\underline{\nabla}$ denotes the tangential gradient, id_{Γ_F} is the identity mapping, and the dimensionless numbers (Reynolds, Weber and Froude) are defined as

$$\text{Re} = \frac{\rho \text{UL}}{\mu}, \qquad \text{We} = \frac{\rho \text{U}^2 \text{L}}{\sigma}, \qquad Fr = \frac{\text{U}^2}{\text{L}g}.$$

Here, U is the characteristic velocity and L is the characteristic length. To discretise and solve the variational Eq. (2), the free surface has to be captured or tracked over the time. In the next Section, we discuss a few commonly used interface capturing and tracking methods.

3 Interface Capturing and Tracking Methods

Numerical schemes for computing the dynamics of incompressible fluids in fixed domains are well established. However, the situation changes when the entire or a part of the domain is *a priori* unknown. This is the case in computations of fluid flows with free boundaries and interfaces. Three main challenges in a numerical scheme for handling these moving boundaries are their discrete representation, the inclusion of the boundary/interfacial condition on them, and their evolution in time. Based on how a numerical scheme handles these three challenges, it can be classified as an Eulerian or a Lagrangian scheme. In an Eulerian scheme the grid remains fixed throughout the computation, and the fluid flows through the grid. Thus, a cell in the fixed grid may contain two fluids, and therefore discontinuities in the fluid properties between these fluids occur within a cell. Further, an additional numerical scheme must be provided to impose surface or interface forces into the Eulerian model. The Lagrangian representation of fluid flows is conceptually simple, since the interface between two fluids is resolved by edges/faces of moving cells, and a single cell contains only one fluid phase for all time. Thus, the inclusion of surface/interfacial forces is easy in Lagrangian schemes, and these schemes offer potentially the highest accuracy.

3.1 Volume of Fluid (VOF) Method

The Volume of fluid method is based on the Eulerian representation, and is a popular method for capturing the interface in computations of interface flows [14, 25, 26, 27]. In the volume-of-fluid method (VOF), a *volume-of-fluid function* C (also called *marker function*) is used to identify each fluid phase in a cell. For a two-phase flow containing fluid-A and fluid-B, the value of the function C would be unity at any point which is occupied by fluid-A, and zero otherwise. The average value of C in a cell would then represent the fractional volume of the fluid-A in the cell. A cell with the average value of unity would corresponds to the fluid-A, while a zero average C value of a cell corresponds to fluid-B. In particular, cells with average C values between zero and one must then contain an interface. The time-dependency of the function C is governed by the transport equation

$$\frac{\partial C}{\partial t} + u \bullet \nabla C = 0.$$

Here, u is the fluid velocity. Even though the volume fraction is unique in each cell, the representation of the interface is not unique. Several interface reconstruction algorithms such as SLIC, piecewise linear approximations are proposed in the literature to represent the interface, (see [23, 24] for more details of these algorithms).

3.2 Level Set Method

The level set method is another popular and flexible method for computations of interface flows in the Eulerian description. This method has been introduced by Osher and Sethian [22]. A continuous zero level set function is used to represent the interface in this method. The level set function $\phi(x,t)$ is constructed as a signed distance function in such a way that $\phi(x,t) > 0$ holds on one side of the interface, and $\phi(x,t) < 0$ on the other. Similarly, several interfaces can be represented by the same level set function.

Let us consider an interface of a unit circle centred at the origin in a [-2, 2] square domain at time t=0. A signed distance level set function, which has positive values inside the circle and negative values outside the circle, is defined by

$$\phi(x,0) = 1 - \sqrt{x^2 + y^2}.$$

Here, the zero level set function $\phi(x,t) = 0$ represents the interface, see Fig. 1. Given the fluid velocity u, the new position of the interface is captured by solving the advection equation

$$\frac{\partial \phi}{\partial t} + u \bullet \nabla \phi = 0$$

and setting $\phi(x,t) = 0$.

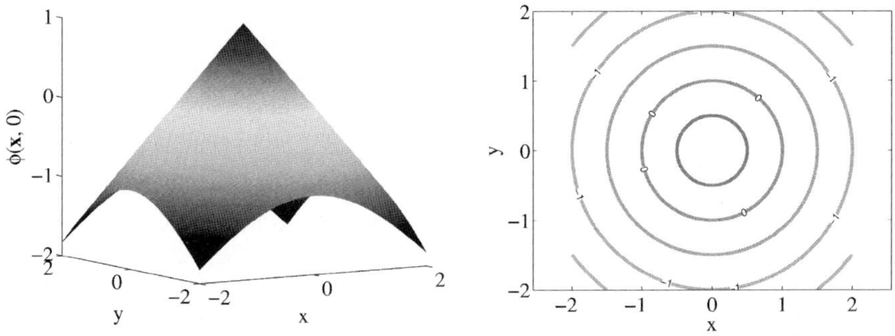

Fig. 1. The level set function (left) and their contours (right)

3.3 Front Tracking Method

The basic idea in the front tracking method is to describe the interface using points connected by line segments and triangles in two- and three dimensions, respectively, see for example [31]. These points are called front points, and for accuracy it is best to limit the distance between the neighbouring front points to less than the cell size. The evolution of the interface is accomplished by simply moving these front points with the fluid velocity determined by an interpolation in the surrounding fixed grid. Since the front points are moved in a Lagrangian manner, these points may accumulate or the resolution becomes inadequate at some parts of the interface. These difficulties can be handled by adding and deleting or redistributing the front points. In the front tracking method there is no need to solve any additional equation to advect the interface. However, one has to solve some algebraic equations to construct a parameterised curve to calculate the curvature and redistribute the front points.

3.4 Arbitrary Lagrangian Eulerian (ALE) Approach

The ALE approach is similar to the Lagrangian approach, but the only difference lies in the mesh movement. In the ALE approach the mesh movement is done in two steps. First, the interface and the boundary points are moved with their corresponding fluid velocity in a Lagrangian manner. To move these points, we can use either the full velocity of the fluid or the fluid velocity only in the normal direction $(u \bullet n)n$ to avoid accumulation of points at one position of the interface [1]. Apart from these, we can also move the boundary points in a prescribed direction by solving an additional equation to satisfy kinematic condition [2]. In the second step, the inner mesh points are displaced according to the displacement of the interface and the boundary points by solving a linear elasticity equation. This technique is called *elastic solid update*.

To rewrite the variational form of the Navier-Stokes equation (2), which is defined in a time-dependant domain, the term $a(\overline{u-w};u,v)$ has to be added instead of the term $a(\overline{u};u,v)$ in the equation (2), see [18]. Here, w is the mesh velocity. The solution can be approximated more accurately with the ALE approach in comparison to all other interface tracking and capturing methods for deforming domains, which do not require re-meshing very often. However, if the re-meshing of the domain is needed very often or any topological change in the domain occurs, the accuracy of the approximated solution might be reduced due to the re-meshing and interpolation.

4 Eulerian vs. Lagrangian

In the previous Section a collection of interface capturing (VOF and Level Set) and tracking methods (Front Tracking and ALE approach) have been discussed. There is no general method, which can be used for all types of free surface and multiphase flows. Each method has its own advantages and disadvantages. Therefore, the choice of an interface capturing or a tracking method depends on the type of flow is considered and the type of information we are interested in to obtain from the solution? Thus, instead of recommending a particular method, we discuss what has to be considered before choosing an interface capturing or a tracking method for a particular free surface or interface flow.

4.1 Topological Issues

Topological changes such as merging and splitting of one fluid phase during the computations is handled automatically by the Volume-of-Fluid and the Level Set method. Sometimes this is considered as an essential advantage of these methods. However, merging or splitting depends on the physics of the fluids and is not only a geometric property. In the Front Tracking and the Arbitrary Lagrangian Eulerian approaches merging and splitting will never happen unless we implement a special numerical procedure on the basis of a certain *a priori* criteria. These criteria can be driven by the spatial location of both phases to each other but also by other means.

4.2 Spurious Velocities

One of the undesirable effects in computations of interface flows is the occurrence of spurious velocities [8, 12]. These non-physical velocities (also called *parasitic currents*) occur in computations due to an improper handling of the discontinuity across the interface in the pressure and in the flow properties, and due to numerical errors such as the approximation errors in the curvature and in the interface. In particular, in Eulerian methods spurious velocities appear. The main reason is that the interface is not resolved in Eulerian methods, *i.e.*, two phases of fluid may be contained in a single cell. In general, the pressure across the interface has jumps, and if we approximate the discontinuous pressure in a cell with a continuous function it will induce some spurious velocities, see Figure 2.

To overcome this difficulty, a discontinuity is induced in the finite element pressure space for cells which contain the interface. However, it is not clear whether the finite element spaces satisfy the "inf-sup" stability condition, and how to use a quadrature rule for a cell with discontinuous functions. It has been shown in [8] that the spurious velocities can be suppressed in computations by using an interface resolved mesh with a discontinuous pressure approximation and a proper approximation of the surface force.

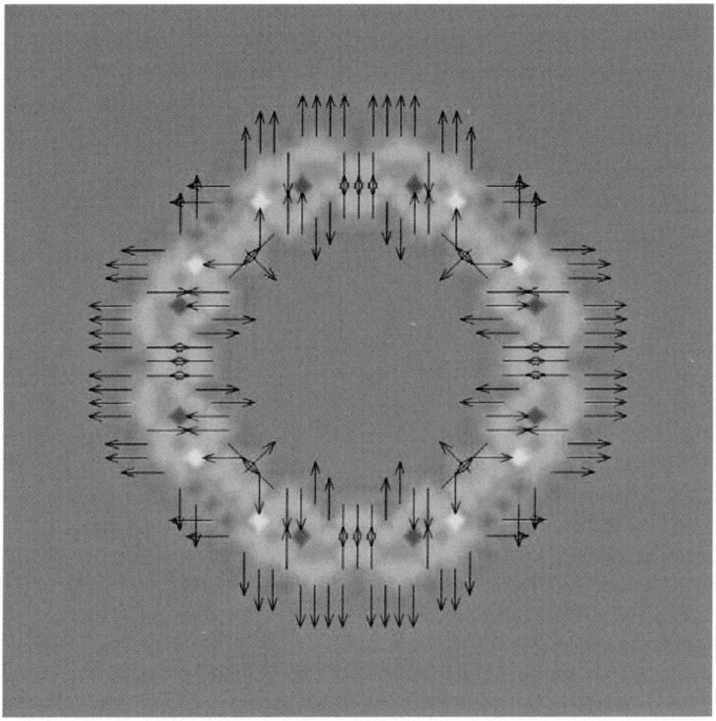

Fig. 2. Spurious velocities generated in the computation of the static bubble problem with the continuous pressure approximation in the interface unresolved mesh

4.3 Algorithmic and Computational Issues

The main challenging part in a free surface and interface flow programming is the inclusion of the surface force and the fluid properties such as the density and the viscosity which contain jumps across the interface. This challenge arises only in fixed grid methods, since the interface is not resolved by the mesh. The Continuum Surface Force (CSF) technique is often used to include the surface. The CSF technique was first introduced in [3]. The basic idea behind this technique is to include the surface force over a small region near the interface instead of only over the interface. To include the fluid properties such as the density and the viscosity, several techniques such as defining the material property as a function of shortest distance from the interface, using a steep gradient to translate the jumps and defining the material parameters by a smoothed Heaviside function, have been proposed in the literature, see for instance [31]. However, these smoothing techniques often induce additional numerical errors in the solution.

In the Volume-of-Fluid method, the main challenging part lies in the calculation of the local curvature of the interface from the volume fraction. A number of recent developments, including a technique to improve the resolution of the interface by high order schemes [13, 16, 17, 23, 29], and local mass preserving schemes [25], have increased the applicability of this method. In the level set method, it is essential that the level set function remains a signed distance function. However, this property is not preserved during the advection in general. Thus, a re-initialisation of the level set function has to be done regularly to guarantee the mass conservation (for recent improvement in level set method see for example [19, 20, 21, 28, 32]). A coupled level set and Volume-of-Fluid method has also been proposed to overcome these difficulties in both of these methods [30]. Both in the Volume-of-Fluid and in the level set methods, solving the advection equation, especially in the finite element context is not a trivial task. Additionally, a stabilization technique such as the streamline diffusion (SDFEM) or the Local Projection Stabilization (LPS) has to be used for the advection equation [9]. In the front tracking method, since the interface is tracked explicitly, topological changes such as breaking and merging of interfaces have to be done manually. If such intersections are anticipated, the reordering of these front points may not be difficult. However, it is difficult to formulate a criterion for breaking and merging of interfaces since the physical behaviour of the fluid has to be taken into account. Furthermore, the interface is advected by an interpolated velocity, and, in general, this interpolated velocity is not divergence free at the front points. Therefore, it is very difficult to guaranty the mass conservation in each phase is very difficult.

These difficulties in the fixed grid methods do not exist in moving grid methods. However, the main challenge in moving grid methods is the handling of the mesh movement. If the distortion of the mesh is large, then a re-meshing and an interpolation of the solution have to be done. Nevertheless, the mesh distortion and the re-meshing can be avoided almost completely by using the Arbitrary Lagrangian Eulerian method. However, if the topology of the domain changes, it is not an easy task to handle it by a moving grid method.

4.4 Verification and Validation

One of the important steps in computations of free surface and interface flows, irrespective of the choice of an interface capturing or a tracking method, is to verify the computed solution. One possibility is to compare the computed solution with the experimentally obtained solution with the same configuration. However, the accuracy of the experimental results depends on the experimental setup, the experimental procedure, and the instruments used in the experiments. Therefore, experimental comparisons are never reliable, and thus it is essential to verify the computed solution with the analytical solution or with a reference value.

For fluid flows in a fixed domain, there are several flow problems in 2D and 3D with analytical solutions to verify the computed solution. However, this is not the case for flows with free surfaces and interfaces. A static bubble example [8] can be used to verify the implementation of the surface force and the material properties of the fluid. However, there is no example in interface flows with the analytical solution to verify the dynamics of the interface. Nevertheless, benchmark reference values for a 2D rising bubble are available in [15] for comparisons. The reference values obtained from three different research groups with the Level Set and the Arbitrary Lagrangian Eulerian methods are provided in [15].

5 ALE Approach for Computations of an Impinging Droplet

We prefer the ALE approach for computation of the considered droplet deformation problem. In this example, we study the flow parameters such as wetting rate, the maximum wetting diameter, and the dynamic contact angle. Therefore, we need a very accurate and spurious-free solution to study these micro properties. Further, topological changes are not considered in this study.

In our finite element scheme, we derive the 3D-axisymmetric form [10] of the variation form of the model equation. Then, we discretise the 3D-axisymmetric variational form in time by the second order strongly A-stable *fractional-step-*ϑ scheme and in space with second order "*inf-sup*" stable finite element pair. The used mixed finite element pair is $P_2^{bubble} / P_1^{disc}$, *i.e.*, a continuous piecewise quadratic enriched with

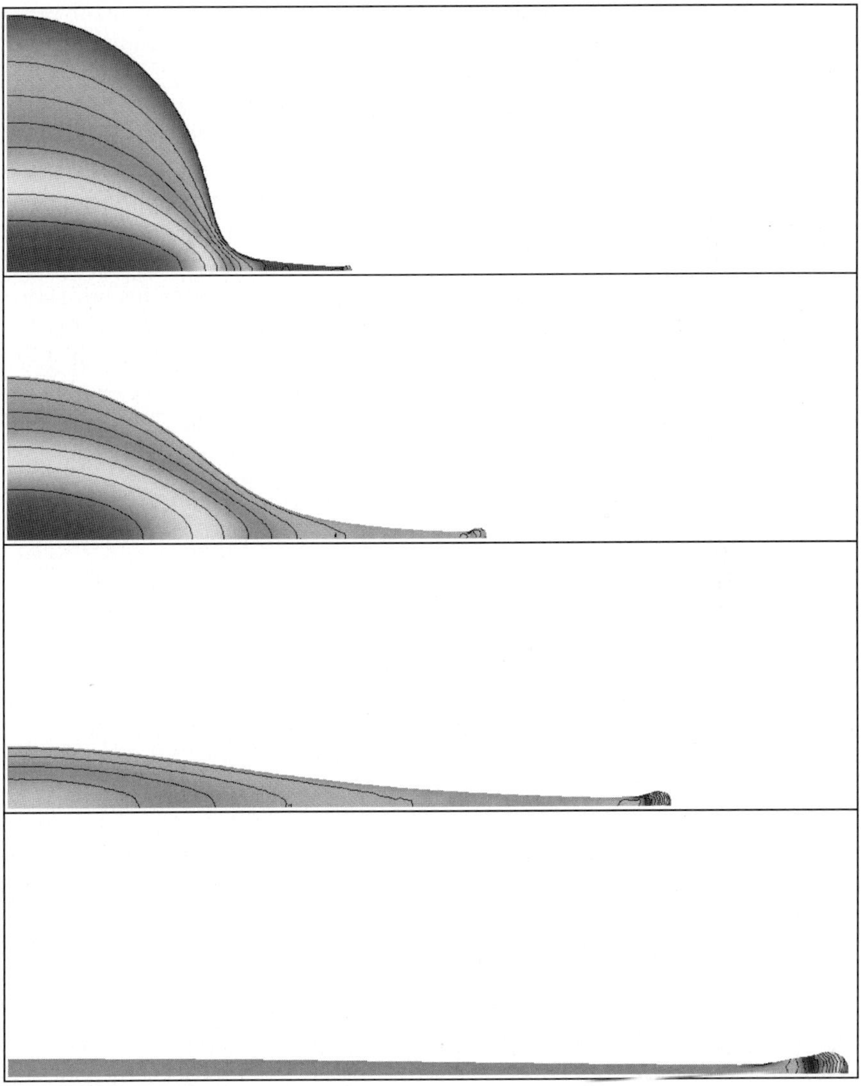

Fig. 3. The shape of a droplet at t = 0.25, 0.5, 1, 2 ms. Re=8761, We=391, Fr=401, θ_e=100°

a cubic bubble function for the velocity, and a discontinuous, piecewise linear function for the pressure. This finite element pair guaranties the mass conservation cellwise and provides a stable solution [4]. We solve the non-linear discrete system by an iteration of fixed point type, and in each iteration we solve the linear algebraic system by the direct solver UMFPACK [5, 6]. At each time step, the domain is moved using the elastic mesh update.

Impinging water droplet

Here, we provide a few numerical results for a water droplet impinging on a horizontal solid surface. For a detailed numerical study we refer to [7]. Here, in all our computations we use the equilibrium contact angle as the input parameter in the equation (3).

In the first example, we consider a water droplet of diameter $d_0=2.7\times10^{-3}$m impinging on a wax surface with the impact velocity $u=(u_r, u_z)=(0, 3.26)$m/s. The following parameters are used for the water: the density $\rho=996$kg/m^3, the dynamic viscosity $\mu=1\times10^{-3}$Ns/m^2, the surface tension $\sigma=7.3\times10^{-2}$N/m. Further, the equilibrium contact angle $\theta_e=100°$ and the slip number $\beta=2$ are used. For these data, we get Re=8761, We=391, Fr=401 with $U=u_z$ and $L=d_0$.

A sequence of images obtained at times t = 0.25, 0.5, 1 and 2 ms are shown in Figure 3. Further the scaled wetting diameter (d/d_0) and the dynamic contact angle obtained by the computations are presented in Figure 4. For the same set of data, the wetting rate, the wetting behaviour and the maximum wetting diameter obtained in experiments [27] are in good agreement with our numerical results.

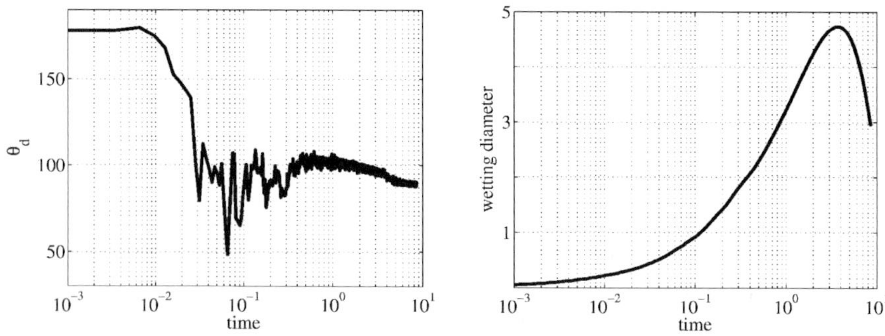

Fig. 4. The scaled wetting diameter (left) and the dynamic contact angle (right) over the dimensionless time for a droplet with Re=8761, We=391, Fr=401, $\theta_e=100°$

Table 1. Values of the dimensionless numbers for the water droplet of diameter $d_0=2.5\times10^{-3}$m

Re	3014	4264	5222	6030	7384	8526
We	50	100	150	200	300	400
Fr	60	120	180	239	359	479

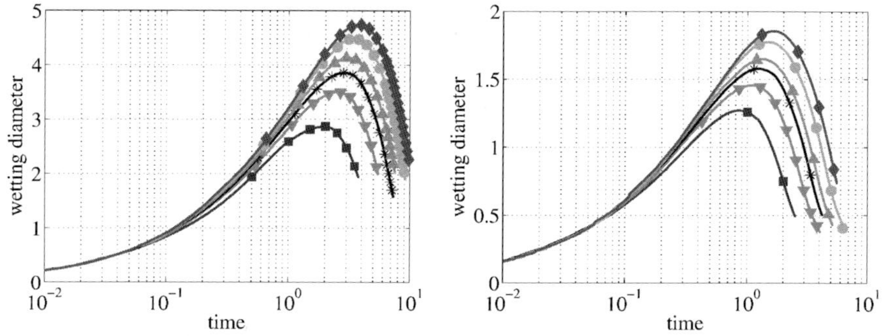

Fig. 5. The scaled wetting diameter over the dimensionless time for a droplet with θ_e=100°

Next, we consider an impinging water droplet of diameter d_0=2.5×10^{-3}m, and perform an array of computations for the dimensionless numbers given in Table 1. Further, for the same array of We and Fr, but with one fifth reduced Re, *i.e.*, Re*=0.2Re, we perform another set of computations. The scaled wetting diameter (d/d_0) obtained in computations over the dimensionless time is shown in Figure 5. In both cases the wetting behaviour is similar, but the scaled maximum wetting diameter is relatively less in the reduce Reynolds number case.

Acknowledgments. This research work has been supported by the Deutsche Forschungsgemeinschaft (DFG) within the graduate program Micro-Macro-Interactions in Structured Media and Particle Systems (GK 828) and partially through the research grant TO143/9. Further, the authors would like to thank Prof. Jürgen Schmidt for his suggestions.

References

[1] Bänsch, E.: Finite element discretization of the Navier-Stokes equations with a free capillary surface. Numer. Math. 88, 203–235 (2001)

[2] Behr, M., Abraham, F.: Free-surface flow simulations in the presence of inclined walls. Comp. Meth. App. Mech. Engg. 191, 5467–5483 (2002)

[3] Brackbill, J.U., Kothe, D.B., Zemach, C.: A Continuum method for modeling surface tension. J. Comp. Phys. 100, 335–354 (1992)

[4] Crouzeix, M., Raviart, P.-A.: Conforming and nonconforming finite element methods for solving the stationary Stokes equations. I. R.A.I.R.O. Anal. Numér 7, 33–76 (1973)

[5] Davis, T.A.: Algorithm 832: UMFPACK, an unsymmetric-pattern multifrontal method. ACM Trans. Math. Soft. 30, 196–199 (2004)

[6] Davis, T.A., Duff, I.S.: An unsymmetric-pattern multifrontal method for sparse LU factorization. SIAM J. Mat. Analy. App. 18, 140–158 (1997)

[7] Ganesan, S.: Finite element methods on moving meshes for free surface and interface flows. PhD Thesis, Faculty of Mathematics, Otto-von-Guericke-University. Docupoint Verlag, Magdeburg (2006)

[8] Ganesan, S., Matthies, G., Tobiska, L.: On spurious velocities in incompressible flow problems with interfaces. Comp. Meth. App. Mech. Engg. 196, 1193–1202 (2007)

[9] Ganesan, S., Matthies, G., Tobiska, L.: Local projection stabilization of equal order interpolation applied to the Stokes problem. Math. Comp (article in press, 2008)

[10] Ganesan, S., Tobiska, L.: An accurate finite element scheme with moving meshes for computing 3D-axisymmetric interface flows. Int. J. Num. Meth. Fluids 57, 119–138 (2008)

[11] Ganesan, S., Tobiska, L.: Modelling and simulation of moving contact line problems with wetting effects. Comput. Visual. Sci. (2008) doi: 10.1007/s00791-008-0111-3

[12] Gerbeau, J.-F., le Bris, C., Bercovier, M.: Spurious velocities in the steady flow of an incompressible fluid subjected to external forces. Int. J. Num. Meth. Fluids. 25, 679–695 (1997)

[13] Hernández, J., López, J., Gómez, P., Zanzi, C., Faura, F.: A new volume of fluid method in three dimensions - Part I: Multidimensional advection method with face-matched flux polyhedra. Int. J. Num. Meth. Fluids (2008) doi: 10.1002/fld.1776

[14] Hirt, C.W., Nichols, B.D.: Volume of fluid (VOF) method for the dynamics of free boundaries. J. Comp. Phys. 39, 201–225 (1981)

[15] Hysing, S., Turek, S., Kuzmin, D., Parolini, N., Burman, E., Ganesan, S., Tobiska, L.: Proposal for quantitative benchmark computations of bubble dynamics. Faculty of Mathematics, University of Dortmund, Preprint Number 351 (2007)

[16] López, J., Hernández, J., Gómez, P., Faura, F.: A volume of fluid method based on multidimensional advection and spline interface reconstruction. J. Comp. Phys. 195, 718–742 (2004)

[17] López, J., Zanzi, C., Gómez, P., Faura, F., Hernández, J.: A new volume of fluid method in three dimensions - Part II: Piecewise-planar interface reconstruction with cubic-Bézier fit. Int. J. Num. Meth. Fluids (2008) doi: 10.1002/fld.1775

[18] Nobile, F.: Numerical approximation of fluid-structure interaction problems with application to haemodynamics, PhD Thesis, Department of Mathematics, École Polytechnique Fédérale de Lausanne (2001)

[19] Olsson, E., Kreiss, G.: A conservative level set method for two phase flow. J. Comp. Phys. 210, 225–246 (2005)

[20] Olsson, E., Kreiss, G., Zahedi, S.: A conservative level set method for two phase flow II. J. Comp. Phys. 225, 785–807 (2007)

[21] Osher, S., Fedkiw, R.P.: Level Set Methods: An overview and Some Recent Results. J. Comp. Phys. 169, 463–502 (2001)

[22] Osher, S., Sethian, J.A.: Fronts propagating with curvature dependent speed: Algorithms based on Hamilton-Jacobi formulations. J. Comp. Phys. 79, 12–49 (1988)

[23] Pilliod Jr., J.E., Puckett, E.G.: Second-order accurate volume-of-fluid algorithms for tracking material interfaces. J. Comp. Phys. 199, 465–502 (2004)

[24] Renardy, Y., Renardy, M.: PROST: A Parabolic Reconstruction of Surface Tension for the Volume-of-Fluid Method. J. Comp. Phys. 183, 400–421 (2002)

[25] Renardy, M., Renardy, Y., Li, J.: Numerical simulation of moving contact line problems using a volume-of-fluid method. J. Comp. Phys. 171, 243–263 (2001)

[26] Rider, W.J., Kothe, D.B.: Reconstructing Volume Tracking. J. Comp. Phys. 141, 112–152 (1998)

[27] Šikalo, Š.: Analysis of droplet impact onto horizontal and inclined surfaces, PhD Thesis, Technische Universität, Darmstadt (2003)

[28] Spelt, P.D.M.: A level-set approach for simulations of flows with multiple moving contact lines with hysteresis. J. Comp. Phys. 207, 389–404 (2005)

[29] Sussman, M.: A second order coupled level set and volume-of-fluid method for computing growth and collapse of vapor bubbles. J. Comp. Phys. 187, 110–136 (2003)

[30] Sussman, M., Smith, K.M., Hussaini, M.Y., Ohta, M., Zhi-Wei, R.: A sharp interface method for incompressible two-phase flows. J. Comp. Phys. 221, 469–505 (2007)
[31] Tryggvason, G., Bunner, B., Esmaeeli, A., Juric, D., Al-Rawahi, N., Tauber, W., Han, J., Nas, S., Jan, Y.-J.: A Front-Tracking Method for the Computations of Multiphase Flow. J. Comp. Phys. 169, 708–759 (2001)
[32] van der Pijl, S.P., Segal, A., Vuik, C., Wesseling, P.: A mass-conserving Level-Set method for modeling of multi-pahse flows. Int. J. Num. Meth. Fluids 47, 339–361 (2005)

Pore-Scale Modelling of Transport Phenomena in Drying

T. Metzger[1], T.H. Vu[2], A. Irawan[3], V.K. Surasani[1], and E. Tsotsas[1]

[1] Institut für Verfahrenstechnik, Otto-von-Guericke-Universität Magdeburg
[2] Dept. Machinery & Equipment of Chemical Engineering, Hanoi University of Technology
[3] Dept. of Chemical Engineering, Sultan Ageng Tirtayasa University, Banten

Abstract. The work in this chapter is concerned with drying of capillary porous media and investigates how material properties characterizing the pore scale influence macroscopic process behaviour. A first approach to the problem takes a bundle-of-capillaries representation of pore space to parameterise a traditional continuous model of drying. In a second approach, the porous structure is represented by a network of pores, and transport is described by discrete rules at the pore level. By applying these two methods, the influence of pore structure, namely pore volume distribution and spatial correlations of pore size, is studied. Additionally, the role of individual transport phenomena, namely liquid viscosity and heat transfer, for drying behaviour is investigated.

1 Introduction

Traditionally, drying is described by continuous models treating the partially saturated porous medium as homogeneous. In such a description, state variables like liquid saturation are spatially averaged quantities that are changing smoothly with time and position, and transport is driven by gradients in these quantities and controlled by effective parameters. The most advanced literature model for heat and mass transfer during drying has been proposed by Perré (1999). This model is not phenomenological but has been derived by the use of volume averaging, and all its parameters have a physical meaning. The major advantage of continuous models is that they may be solved by efficient numerical techniques at a scale that is large as compared to pore size (Nasrallah and Perré 1988; Turner and Perré 1996). At the same time, this may be a fundamental problem if the micro and the macro scale cannot be separated, as in the case of percolation phenomena where the filling state of single pores decides over the transport behaviour of the whole porous medium. Another difficulty of continuous models is to determine, for a given porous structure, the effective parameters for transport (vapour diffusivity, absolute permeability, relative liquid and gas permeability and thermal conductivity) and for equilibrium (capillary pressure and sorption isotherm). This has to be done by experiments which are time-consuming since the parameters are functions of liquid saturation.

In the present work, two routes have been taken to link pore-level information and phenomena to macroscopic drying behaviour. The first approach stays with the classical continuous model but obtains effective parameters theoretically by representing the pore structure by a bundle of capillaries (Vu 2006a/b). An advantage is that, for this simplified geometry, both the liquid and the gas are continuous. But the only pore scale information that can be studied concerning its role for drying behaviour is the pore size distribution. Other structural properties characterizing a real three-dimensional pore space, such as pore coordination number, cannot be investigated.

The second route takes up a recent discrete modelling approach, in which the porous medium is represented by a network of pores and all relevant phenomena are modelled directly at the pore scale. This ensures that no effect is lost or hidden during up-scaling to a macroscopic model. Several research groups have developed pore network models – mainly two-dimensional and isothermal ones – addressing different aspects of drying: effective parameters for continuous models (Nowicki 1992), liquid phase patterns and drying curves with and without gravity, also by experiments (Prat 1993; Laurindo and Prat 1996, 1998), three-dimensional simulations (Le Bray and Prat 1999; Yiotis et al. 2006), viscous effects in the gas phase (Yiotis et al. 2001), liquid films (Yiotis et al. 2004; Prat 2007), and influence of temperature gradients (Huinink et al. 2002; Plourde and Prat 2003). It may be stated that literature work has focused on the role of individual transport phenomena – without having completed this investigation – and that the ability of pore networks to study the influence of pore structure has not yet been exploited. Only very recent work addresses the influence of pore shape (Segura 2005; Prat 2007). Therefore, this investigation has started exploring the role of pore structure by comparing drying behaviour of pore networks with mono- and bimodal pore size distributions and different spatial correlations of pore sizes. Additionally, we have addressed the effect of liquid viscosity on phase distributions and drying rates. And finally, a non-isothermal drying model has been developed by incorporating the free evolution of temperature. These two latter efforts have to be understood as crucial steps towards a pore network model that accounts for all effects that are incorporated in the continuous model. When this model is complete, it can be used to assess the continuous model, i.e. to define its range of validity, and to parameterise it for a given pore structure.

2 Account of Pore Size Distribution in Continuous Modelling

In this section, the continuous drying model is solved for a bundle of capillaries. Such pore geometry is strictly one-dimensional: the capillaries are set perpendicular to the product surface and connected without resistance. This ensures that the liquid phase is continuous – an important assumption of the drying model. Furthermore, at any position, liquid is contained in the smallest pores due to capillary pumping; this implies that effective parameters are unique functions of saturation. Figure 1 illustrates the distribution of liquid in a partially saturated bundle of capillaries during drying. This approach allows investigating the influence of pore volume distribution, represented here as the density function for liquid saturation of free water S_{fw}, but no other structural properties. (A discrete model for a bundle-of-capillaries geometry has been proposed previously, Metzger and Tsotsas 2005).

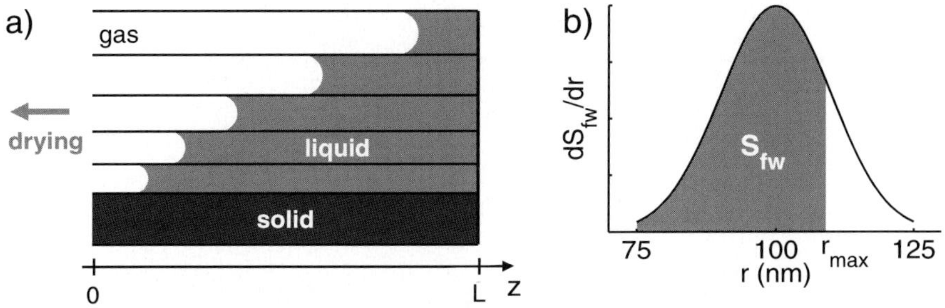

Fig. 1. Representation of pore space by a bundle of capillaries with distributed radii: during drying, liquid saturation may be non-uniform in space coordinate z (a); but at any position, liquid is contained in the smallest pores up to a maximum radius r_{max} which depends on pore volume distribution and local saturation of free water S_{fw} (b)

The general equations of the continuous drying model are given in Table 1. The evolution of liquid saturation S, temperature T and gas pressure P_g during drying is determined by equations (1-3) for conservation of water, air and enthalpy. Heat and mass transfer are by vapour diffusion, liquid or gas convection with velocity v_i as computed from the generalized Darcy law (4), and, additionally, heat conduction. The respective transport parameters are effective diffusivity D_{eff}, absolute permeability K, relative permeability k_i (taking values between 0 and 1) and effective thermal conductivity λ_{eff}.

Liquid and (partial) gas pressures P_i are obtained from capillary pressure P_c (5) and ideal gas laws (6-7), respectively. Additionally, P_v is related to saturation S by sorption equilibrium. Major variables are the densities ρ_i of solid (s), liquid (w), vapour (v), air (a) and gas (g), the mass fractions y_i (8) and the specific enthalpies h_i (9), with specific heat capacities c_{pi} and T_{ref} = 0°C. Further, ψ is porosity and $\tilde{R}$ ideal gas constant; $\tilde{M}_i$ are molar masses of gas components or mixture.

To solve this set of equations, three boundary conditions for the porous body are required. At the open surface ($z = 0$), vapour and heat flux are prescribed for convective drying by boundary layer theory (with Stefan correction) in Eqs. (10-11) with transfer coefficients for mass β and heat α, evaporation enthalpy Δh_v and vapour pressure $P_{v,\infty}$ and temperature T_∞ of bulk drying air, respectively; additionally, gas pressure is imposed as $P_g = P_\infty$. The other side ($z = L$) is impervious to heat and mass transfer and can represent the centre of a symmetrically dried material. Finally, initial values of saturation S_0 and product temperature T_0 must be given, while $P_{g,0} = P_\infty$.

This general model must now be parameterised with product-specific properties, either describing transport or equilibrium. These parameters depend on the structural and surface properties of the material and are functions of saturation. For the bundle of capillaries, all required parameters are given in Table 2. Due to the parallel arrangement, the diffusivity (12) and the conductivity (13) only depend on poroity ψ and saturation S whereas pore volume distribution dS_{fw}/dr plays a role in the permeability (14-16) and the capillary pressure (17). For these, only free water is accounted for and the relationship between free water saturation S_{fw} and the maximum radius

Table 1. Continuous drying model (Perré 1999)

$$\frac{\partial}{\partial t}\left(S\psi\rho_w+(1-S)\psi\rho_v\right)=\nabla\cdot\left(\rho_g D_{eff}\nabla y_v\right)-\nabla\cdot\left(\rho_w v_w+\rho_v v_g\right) \quad (1)$$

$$\frac{\partial}{\partial t}\left((1-S)\psi\rho_a\right)=\nabla\cdot\left(\rho_g D_{eff}\nabla y_a\right)-\nabla\cdot\left(\rho_a v_g\right) \quad (2)$$

$$\begin{aligned}\frac{\partial}{\partial t}&\left(S\psi\rho_w h_w+(1-S)\psi(\rho_v h_v+\rho_a h_a)+(1-\psi)\rho_s h_s-(1-S)\psi P_g\right)\\ &=\nabla\cdot\left(\rho_g D_{eff}\left(h_v\nabla y_v+h_a\nabla y_a\right)\right)\\ &\quad-\nabla\cdot\left(\rho_w h_w v_w+(\rho_v h_v+\rho_a h_a)v_g\right)+\nabla\cdot\left(\lambda_{eff}\nabla T\right)\end{aligned} \quad (3)$$

$$v_{w,g}=-\frac{Kk_{w,g}}{\eta_{w,g}(T)}\cdot\nabla P_{w,g} \quad (4) \qquad y_i=\frac{\rho_i}{\rho_g} \quad (8)$$

$$P_w=P_g-P_c \quad (5) \qquad h_i=c_{pi}\left(T-T_{ref}\right) \quad (9)$$

$$P_i=\frac{\rho_i\tilde{R}T}{\tilde{M}_i} \quad (6) \qquad \dot{m}_v=\beta\frac{P_\infty\tilde{M}_v}{\tilde{R}T}\cdot\ln\left(\frac{P_\infty-P_{v,surf}}{P_\infty-P_{v,\infty}}\right) \quad (10)$$

$$\tilde{M}_g=\tilde{M}_a+(\tilde{M}_v-\tilde{M}_a)\frac{P_v}{P_g} \quad (7) \qquad \dot{q}=\alpha\left(T_\infty-T_{surf}\right)+\Delta h_v\cdot\dot{m}_v \quad (11)$$

filled r_{max} is essential. Here, λ denotes thermal conductivity, and σ is surface tension. Additional to free capillary water, water can be adsorbed on the pore walls so that total saturation is $S = S_{fw} + S_{sorb}$. The sorption isotherm (18) describes the relationship between S_{sorb} and P_v where P_v^* is equilibrium vapour pressure. Sorption is assumed to depend on surface properties alone and the effect of capillary condensation is neglected.

In the following, drying kinetics are presented for two examples of pore volume distributions, a mono-modal one with radius 100 ± 10nm, and another with two modes of equal volume with radii 100 ± 10nm and 1.0 ± 0.1μm (see Fig. 2a). Sample half-thickness is $L = 0.1$m, porosity $\psi = 0.5$ and sorption limit $S_{sorb} = 0.1$; the solid has thermal conductivity λ_s = 1W/m/K and volumetric heat capacity $(\rho c)_s = 2\cdot10^6$J/m^3/K. Initial saturation and temperature are S_0 = 0.9 and T_0 = 20°C. Gas pressure is P_∞ = 1bar; drying air has zero moisture and temperature T_∞ = 80°C. Transfer coefficients for the boundary layer are α = 95W/m^2/K for heat and β = 0.1m/s for mass.

The above set of model equations is solved using finite volumes and a Newton Raphson scheme with variable time step (Nasrallah and Perré 1988; Turner and Perré 1996) and non-uniform grid (41 elements) of geometric progression to better describe steep gradients near the product surface. Figure 2b shows the resulting drying rate curves as evaporation mass flux over liquid saturation; local evolutions of saturation, temperature and gas pressure are shown in Figs. 3 and 4; and in Fig. 5 saturation profiles are given.

Table 2. Effective parameters for transport and equilibrium in a bundle of capillaries

Effective diffusivity	$D_{eff}(S,T,P_g)=(1-S)\psi\delta_{va}(T,P_g)$	(12)
Effective thermal conductivity	$\lambda_{eff}(S,T)=(1-\psi)\lambda_s(T)+S\psi\lambda_w(T)$	(13)
Absolute permeability	$K=\frac{1}{8}\int r^2\frac{dS_{fw}}{dr}dr$	(14)
Relative liquid permeability	$k_w(S_{fw})=\frac{1}{8K}\int^{r_{max}} r^2\frac{dS_{fw}}{dr}dr$	(15)
Relative gas permeability	$k_g(S_{fw})=1-k_w(S_{fw})$	(16)
Capillary pressure	$P_c(S_{fw},T)=\frac{2\sigma(T)}{r_{max}(S_{fw})}$	(17)
Sorption isotherm	$\frac{P_v}{P_v^*(T)}=\begin{cases}\frac{S}{S_{sorb}}\left(2-\frac{S}{S_{sorb}}\right), & S<S_{sorb}\\ 1, & S\geq S_{sorb}\end{cases}$	(18)

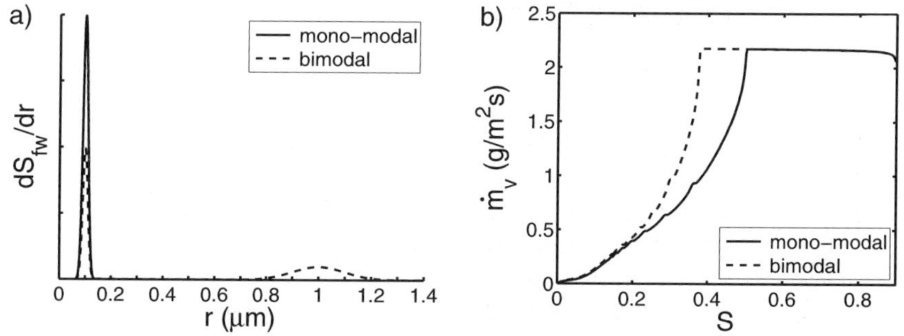

Fig. 2. Pore volume distributions for bundle of capillaries (a) and resulting drying rate curves (b)

It can be stated that a steep moisture gradient develops for the narrow mono-modal pore size distribution (Fig. 5a); as a consequence, all free surface water is removed at a total saturation of 0.5 (after 2.6 hours); this goes along with a drop in evaporation rate and marks the end of the first drying period. In the bimodal case, the broadly distributed macro pores first dry out completely with a small gradient, and a steeper gradient only occurs when the narrow mode of small pores empties (Fig. 5b). As a result, the first drying period extends down to a saturation of 0.38. The final stages of drying, including slow removal of adsorbed water, are similar for both cases.

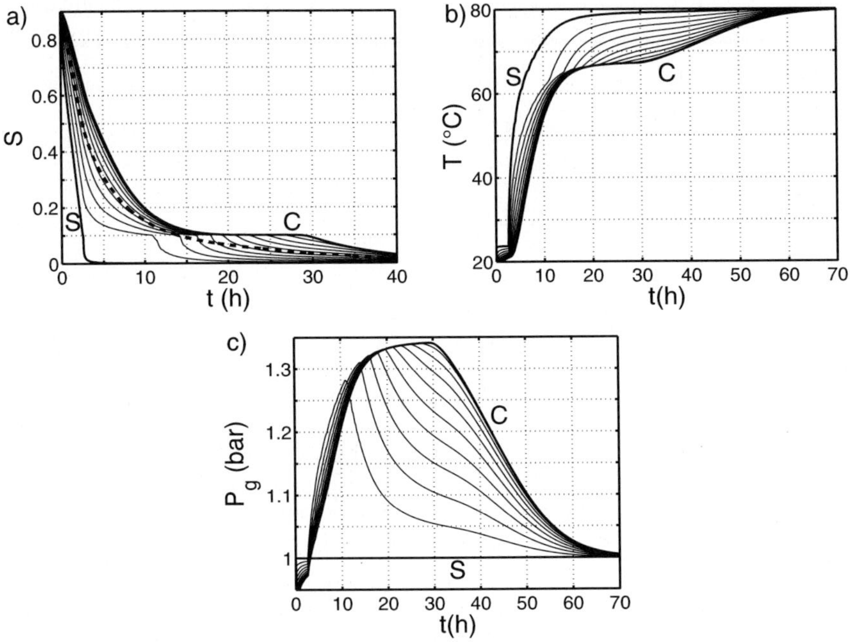

Fig. 3. Evolution of (a) saturation, (b) temperature and (c) gas pressure during drying of bundle of capillaries with mono-modal radius distribution. Solid curves show local values at equidistant positions between surface (S) and centre (C) of the porous medium; the dashed line in (a) indicates average saturation.

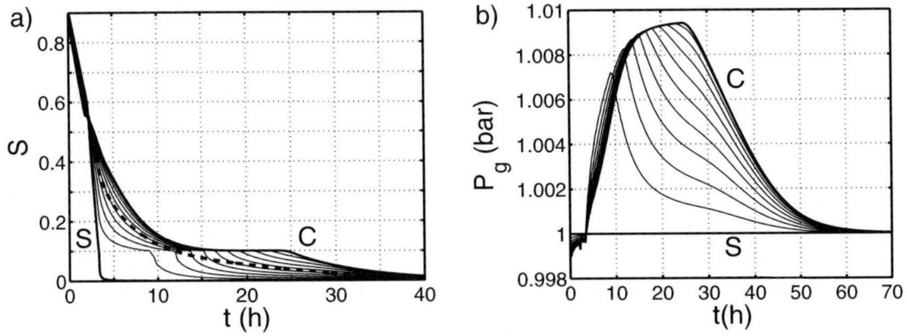

Fig. 4. Evolution of (a) saturation and (b) gas pressure during drying of the bundle of capillaries with bimodal radius distribution. Solid curves are local values at equidistant positions between surface (S) and centre (C) of the porous medium; the dashed line in (a) is average saturation.

Temperature evolution (shown in Fig. 3b for the mono-modal case and similar to bimodal results) can be divided into three stages. During the first drying period, the product approaches wet-bulb temperature; then, temperature of the drying front

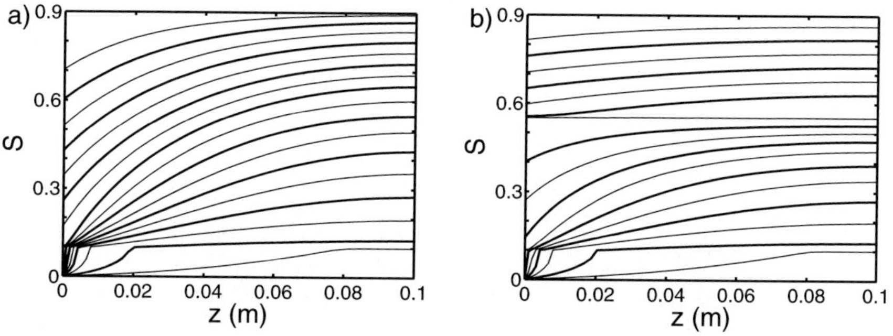

Fig. 5. Saturation profiles during drying of a bundle of capillaries with (a) mono-modal and (b) bimodal radius distribution, plotted for steps of 0.05 in network saturation (bold: steps of 0.1)

increases as it recedes into the material; finally, when all free water has been removed after 27 hours, the material gradually heats up to drying air temperature. Concerning gas pressures, an overpressure is observed during the second drying period when most of the water evaporates from inside the material; the corresponding convective flow is due to the Stefan effect in the vicinity of an evaporating interface that is impermeable to air. The magnitude of overpressure may be significant, as in the mono-modal case (Fig. 3c), and depends on permeability. For the mono-modal distribution, absolute permeability K is $1.26 \cdot 10^{-15}$ m^2, whereas the macro pores in the bimodal case increase the value to $6.33 \cdot 10^{-14}$ m^2, thereby reducing overpressure drastically (Fig. 4b).

The continuous drying model with a bundle-of-capillaries geometry can help studying the influence of pore size distribution on drying behaviour. It is worth mentioning that not only drying rate curves can be assessed but that also local information on moisture content, temperature and pressure is available.

3 Account of Pore Structure in Discrete Pore Network Model

If more structural parameters of the porous material shall be accounted for, two- or three-dimensional representations of the pore space are needed. To keep the numerical cost limited, a network of geometrically similar pores is used to approximate real pore space. The proposed pore network model is based on a regular lattice of volumeless pores which are connected by cylindrical throats of randomly distributed radii r_{ij} (Irawan 2006). Figure 6 shows an example of a partially dried pore network that is extended by a discretised boundary layer to capture the first drying period (Irawan et al. 2005). Transport phenomena are described at the discrete level of pores and throats. The basic drying model is isothermal and considers non-viscous capillary flow in liquid-filled regions and vapour diffusion in gas-filled network regions and boundary layer. Model equations for vapour transport are given in the first section of Table 3 (cf. p.209). Evaporation takes place at all menisci close to the surface. Vapour flow rates in throats connecting pores i and j are computed from Eq. (19) by solving the quasi-stationary diffusion problem (20) in the whole gas region; saturation vapour

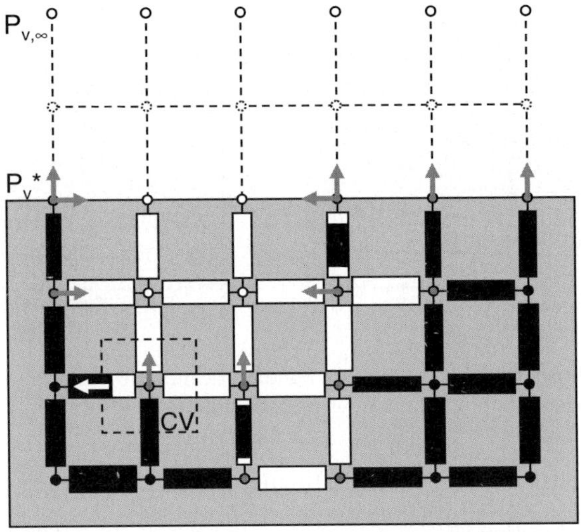

Fig. 6. Partially saturated pore network with diffusive boundary layer. Liquid pores and throats are in black; gas in white and solid in grey. Vapour diffusion is shown as grey arrows; its boundary conditions are indicated. One control volume for heat balance is illustrated.

pressure at menisci (21) and vapour pressure of drying air $P_{v,\infty}$ are imposed as boundary conditions (see Fig. 6).

Not all menisci will recede as a result of local evaporation, since liquid may be pumped to small menisci by capillary action. Equation (26) gives the lowest liquid pressure that a meniscus can produce (for perfect wetting). If viscosity is neglected, only the largest of all communicating meniscus throats will become empty – at a rate corresponding to the total evaporation rate of this liquid cluster. Time-stepping is imposed by complete emptying of a meniscus throat. Then, a new quasi-stationary vapour diffusion problem must be solved. Due to a random spatial distribution of the throat radii, the emptying order is also random, and many liquid clusters can form during drying. For liquid flow, it is important to track connectivity. A variant of the Hoshen-Kopelman algorithm (Al-Futaisi and Patzek 2003; Metzger et al. 2006) is used to label liquid clusters. More details on the pore network drying algorithm can be found in (Metzger et al. 2007d).

Drying of 48×51 networks (periodic in horizontal direction) with throat length L = 500μm has been simulated for a mass transfer coefficient β = 0.514mm/s. Figure 7 shows phase distributions for a mono-modal throat radius distribution with r_{ij} = 50 ± 2.5μm. One can see that inner throats may empty while liquid is pumped to surface throats where it evaporates (Fig. 7a). However, this capillary flow is interrupted when the liquid phase splits up into clusters (Fig. 7b). Such clusters have a screening effect and must evaporate before the drying front can penetrate further. In three dimensions, such trapping of liquid is much less pronounced (Le Bray and Prat 1999; Metzger and Tsotsas 2008). Here, we present two-dimensional simulations which can be better visualized.

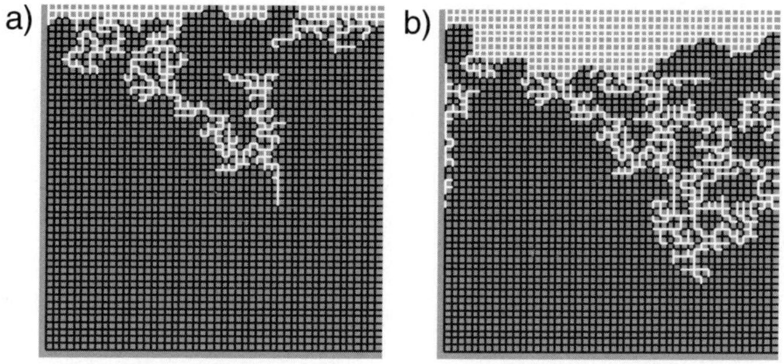

Fig. 7. Phase distributions of periodic mono-modal network at saturations (a) 0.9 and (b) 0.75

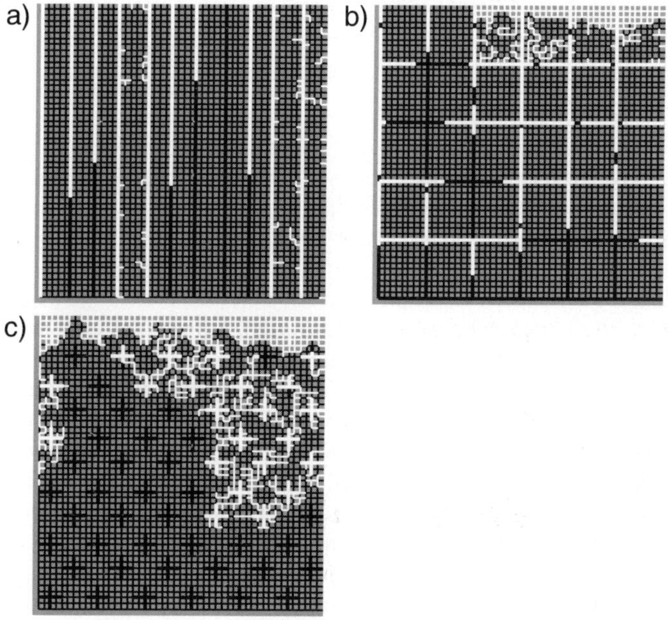

Fig. 8. Phase distributions at liquid saturation 0.75 for bimodal networks with different spatial correlation of macro throats (see text)

The influence of macro throats in networks with a bimodal radius distribution is illustrated in Figure 8. To this purpose, part of the throats has been enlarged to radii $r_{ij} = 100 \pm 5\mu m$ that make up for approximately 44% of total pore volume in all three cases. If macro throats are aligned to long channels perpendicular to the network surface (Fig. 8a), they become empty first while most of the surface throats stay saturated. In this case, micro- and macro-porous regions are continuous (strictly, this is only possible in three dimensions). Figure 8b shows a network with a super-lattice of macro throats. These are emptied favourably during drying, and the micro-porous regions are separated from each other, and then dry out layer by layer. Here, the

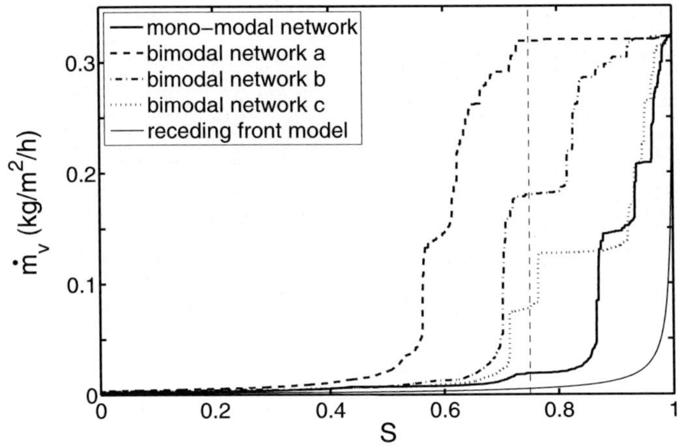

Fig. 9. Drying rate curves for mono- and bimodal networks (as named in Fig. 8) with limit of immobile liquid (saturation 0.75 is indicated)

micro-porous phase is discontinuous. Finally, if macro throats occur as separate clusters (Fig. 8c), the phase pattern resembles that of the mono-modal case, with the only difference that a macro-cluster always empties completely before gas invasion continues in micro throats.

For the mono-modal and all bimodal cases, drying rate curves are plotted in Fig. 9. They result from the phase patterns since the liquid-filled throats closest to the network surface determine the current drying rate.

As expected, the mono-modal network (Fig. 7b) and the bimodal one with disperse macro-pores (Fig. 8c) exhibit similarly unfavourable drying curves, whereas the rates are increased if the macro-porous phase is continuous (Fig. 8b), and further for a bi-continuous spatial distribution of pore radii (Fig. 8a) which has the longest first drying period. Capillary pumping plays a role for all networks, but with a different degree, as a comparison with the result for immobile water shows.

A Monte Carlo study with the above random networks (to eliminate sampling effects) as well as simulation results for regular networks with different coordination number can be found in (Metzger et al. 2007b).

4 Extension of Transport Phenomena in Pore Network Model

From studying structural effects, we now turn to the modelling of additional transport phenomena, namely viscous liquid flow and heat transfer. These will influence phase distributions and drying curves; and pore structure decides on the importance of the observed effects.

4.1 Viscous Effect on Capillary Flow

Liquid viscosity cannot be neglected if the porous material is dried rapidly, if it is of large size and if the pores are small with a narrow size distribution. The second

Table 3. Pore network model (index *i* denotes pore, *ij* denotes throat connecting pores *i* and *j*)

Vapour diffusion in gas pores	$\dot{M}_{v,ij} = \pi r_{ij}^2 \frac{\delta}{L} \frac{P_g \tilde{M}_v}{\tilde{R}T} \cdot \ln\left(\frac{P_g - P_{v,i}}{P_g - P_{v,j}}\right)$	(19)
Mass balance in gas pores	$\sum_j \dot{M}_{v,ij} = 0$	(20)
Gas-liquid interface	$P_{v,i} = P_v^*(T_i)$	(21)
Viscous **liquid** flow	$\dot{M}_{w,ij} = \frac{\rho_w \pi r_{ij}^4}{8\eta_w L S_{ij}} \left(P_{w,i} - P_{w,j}\right)$	(22)
Mass balance in liquid pores	$\sum_j \dot{M}_{w,ij} = 0$	(23)
Water balance at menisci	$\dot{M}_{w,ij} - \dot{M}_{ev,ij} = \rho_w \pi r_{ij}^2 L \frac{dS_{ij}}{dt}$	(24)
Stationary meniscus	$\dot{M}_{w,ij} = \dot{M}_{ev,ij}$	(25)
Moving meniscus	$P_{w,ij} = P_g - \frac{2\sigma(T_{ij})}{r_{ij}}$	(26)
Heat conduction	$\dot{Q}_{ij} = \frac{(\lambda A)_{ij}}{L}\left(T_i - T_j\right)$	(27)
Thermal conductivity	$(\lambda A)_{ij} = \lambda_s\left(L^2 - \pi r_{ij}^2\right) + \lambda_w \pi r_{ij}^2 S_{ij}$	(28)
Enthalpy balance	$(\rho c V)_i \frac{dT_i}{dt} = -\sum_j \dot{Q}_{ij} - \Delta h_v(T_i) \sum_j \dot{M}_{ev,ij}$	(29)
Heat capacity	$(\rho c V)_i = \rho_s c_s \left(L^3 - \frac{L}{2}\sum_j \pi r_{ij}^2\right) + \rho_w c_w \sum_j \pi r_{ij}^2 L \frac{S_{ij}}{2}$	(30)

section of Table 3 gives the equations for modelling the viscous effect. Liquid flow in each throat is described by Poiseuille's law (22), incompressible flow obeys the mass balances (23), and coupling to vapour diffusion is achieved by (24).

A meniscus can be *stationary*, i.e. capillary pressure can provide liquid water at the local evaporation rate, or *moving*, i.e. evaporation takes place at a higher rate than capillary flow. The procedure to find the state of the meniscus is iterative, since boundary conditions on the flow problem depend on meniscus states themselves,

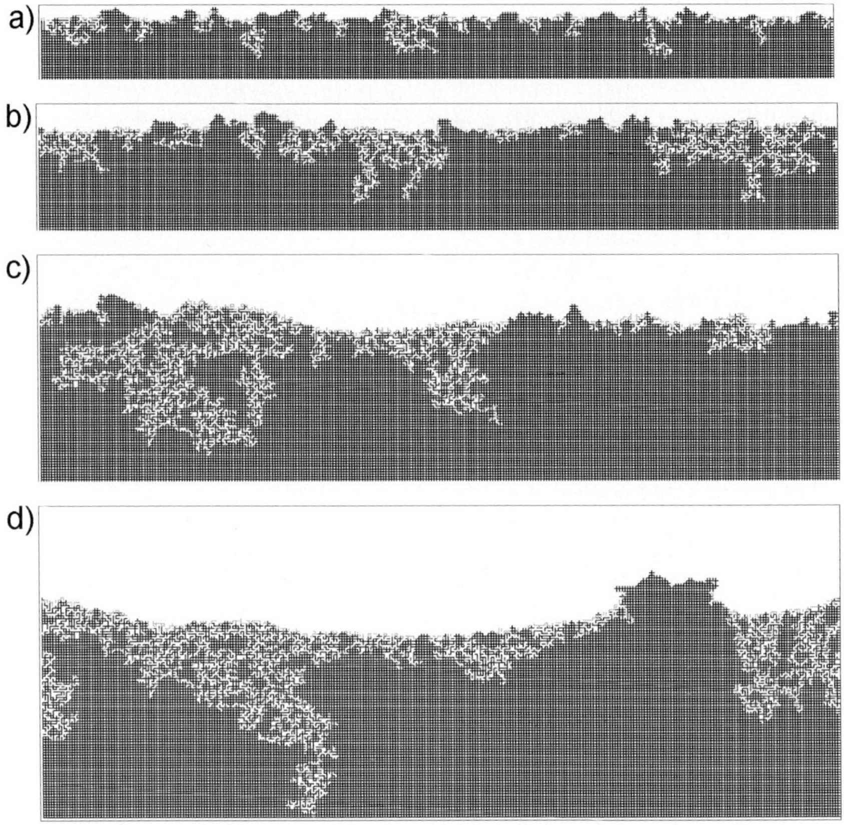

Fig. 10. Evolving drying front for viscous drying at network saturations (a) 0.95, (b) 0.9, (c) 0.75 and breakthrough saturation 0.583 (d); in (a) to (c) only top of network is shown

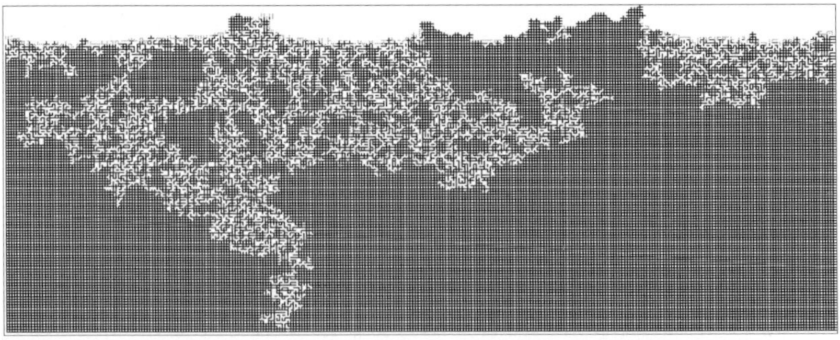

Fig. 11. Phase pattern for non-viscous drying, at breakthrough occurring at $S = 0.805$

namely Eqs. (25) and (26). For the correct boundary conditions, liquid flow rates at menisci are computed and meniscus velocities are obtained by Eq. (24). Time-stepping is imposed by the complete emptying or (re)filling of a meniscus throat; then, vapour and liquid flow problems must be solved for new boundary conditions. The algorithm is described in detail in (Metzger et al., 2007d). As a consequence of liquid viscosity, several menisci per cluster may move, and the distance over which capillary flow may take place at a given rate is limited.

Figure 10 shows simulated liquid phase distributions for a 300×120 network with a throat length L = 500nm, throat radii r_{ij} = 50 ± 1nm and a mass transfer coefficient β = 0.514m/s. The drying front is stabilized, i.e. of finite width, and gradually widens up (Fig. 10a-d) as it recedes into the porous material. The reason is the reduction of evaporation rate with increased mass transfer resistance. The result of a non-viscous drying simulation with the same network is depicted in Fig. 11. Here, the drying front is not stabilized and gas reaches the network bottom (breakthrough) before the first drying period ends due to a complete drying out of network surface. Drying rate curves are not shown here, but it is clear that viscous stabilization drastically reduces the duration of the first drying period. Viscosity counteracts capillary pumping, just as discontinuity of the liquid phase does in the previous section (see Fig. 9).

It has to be stated that the pore size distribution decides on the kind of drying behaviour: for wider pore size distributions or for bimodal networks, the stabilizing effect of liquid viscosity is reduced or negligible (Metzger, 2007a) since the available capillary pressure gradients are larger.

4.2 Thermal Effect on Capillary Flow and Vapour Diffusion

Convective drying is not an isothermal process: the porous medium is heated from the surface by the drying air and cooled where evaporation takes place. This leads to temperature gradients in the porous medium and to a rise of temperature as drying rates decrease during the process. Both phenomena can affect drying behaviour. Furthermore, evaporation in warm regions may coexist with condensation in cold regions. The resulting coupled mass and heat transfer from warm to cold phase boundaries is known as the heat pipe effect.

The non-isothermal model neglects liquid viscosity and describes heat transfer and its coupling to mass transfer by the equations in the third section of Table 3 (Surasani et al. 2008a). Inside the pore network, heat transfer is assumed to occur only by conduction, as in Eq. (27). Convective contributions will be considered in a future model version with liquid viscous flow. Effective thermal conductivities of throats and surrounding solid phase depend on throat saturation and are approximated by Eq. (28). Temperature evolution of each control volume (see Fig. 6) is obtained from dynamic enthalpy balances (29) that combine heat conduction with heat sinks/ sources at places of evaporation/ condensation. Effective heat capacity of the control volume is a function of saturation and approximated by Eq. (30).

The non-viscous liquid flow is described as in the basic model version (see Section 3), i.e. only one meniscus per liquid cluster is moving. The net phase change of the cluster determines whether the corresponding throat is emptying or filling. Net evaporation is modelled as above. Net condensation is accounted for as long as the moving meniscus throat is not fully saturated, or if it is neighbour to an evaporating liquid

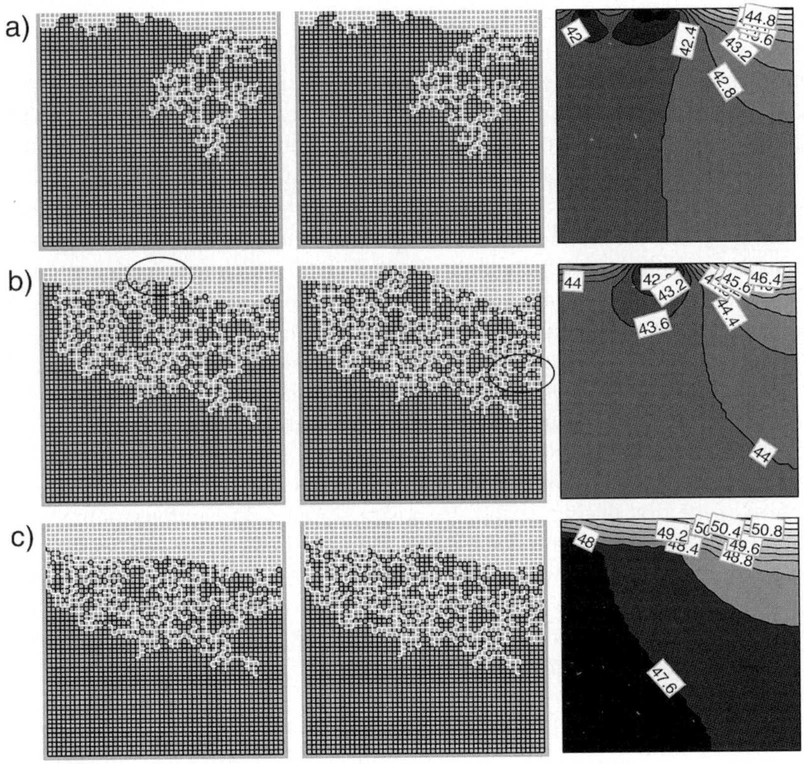

Fig. 12. Isothermal (left) and non-isothermal (middle) phase distributions and temperature fields (right, isotherms in °C) for network saturations (a) 0.88, (b) 0.75 and (c) 0.7

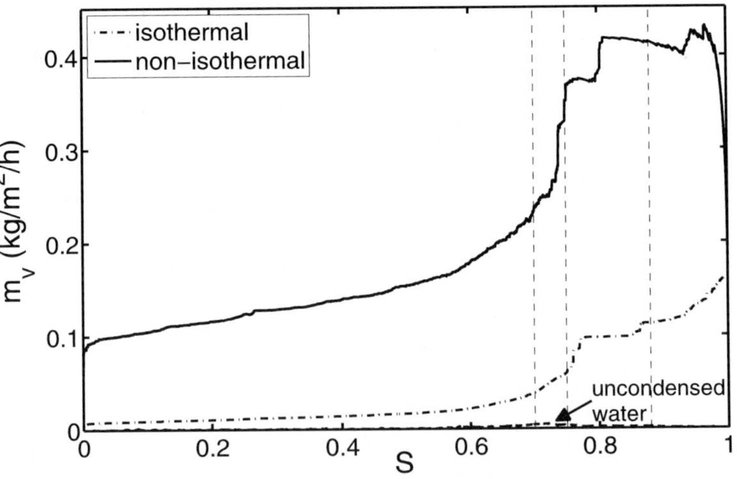

Fig. 13. Isothermal and non-isothermal drying rate curves, with error due to neglected condensation (saturations of Fig. 12 are indicated)

cluster. Further condensation would require new (imbibition) rules for filling of completely empty throats; this has not yet been addressed.

Non-isothermal drying has been simulated for a (non-periodic) 51×51 network with a throat length $L = 500\mu m$, radius distribution $r_{ij} = 40 \pm 2\mu m$ and a mass transfer coefficient $\beta = 0.257 mm/s$; drying air has zero moisture and 80°C. Results are shown as phase distributions (in comparison with isothermal ones) and temperature fields in Fig. 12 and as drying curves in Fig. 13. The error due to neglected condensation is also plotted.

Several coupling effects between heat and mass transfer can be observed. First, the surface temperature of the network is non-uniform. Wet spots are cooler than dry spots as a consequence of local differences in evaporative cooling. Second, network temperature gradually rises as drying rates are reduced. During warming up, the temperature gradients in the dry zone are more pronounced than in the wet zone beyond the evaporation front (see Fig. 12c, and also confer to Fig. 3b); with rising temperature, the saturation vapour pressure, Eq. (21), is increased, so that vapour diffusion is enhanced and drying rates can be much higher than in the isothermal case where network temperature remains uniformly at 20°C (see Fig. 13).

And third, phase distributions are influenced by temperature gradients. According to temperature dependence of Eq. (26), the liquid pressure in a warm pore throat is higher than in a cold one of same radius. Therefore, the randomness of throat emptying (caused by the radius distribution) is partially overruled by temperature gradients. The effect only seems to play a role for extremely narrow pore size distributions. For this condition, it is expected to stabilize the drying front, as viscosity does in the above simulation. For illustrative reasons, we have chosen an example with a pronounced but rather untypical behaviour. In the isothermal simulation, a fraction of the surface stays wet (Fig. 12a) for a while before it empties (Fig. 12b) according to the random spatial distribution of throat radii; in the non-isothermal simulation, however, the cold surface region stays wet for longer while warmer inner-network throats continue to dry out (see marked areas in Fig. 12b).

Further non-isothermal simulations with a bimodal network may be found in (Surasani et al. 2008a); a comparison between convective and contact heating is presented and discussed in (Surasani et al. 2008b). The influence of gravity has been addressed in (Surasani et al. 2007).

5 Conclusion and Outlook

Two approaches have been presented to investigate the influence of pore-scale structure and phenomena on macroscopic drying behaviour of porous materials. The traditional continuous model is numerically efficient, but it can not easily account for structural effects. Therefore, we have chosen a simple bundle-of-capillaries pore geometry to study the influence of pore size distribution.

Pore networks are a much more powerful tool for linking micro- and macro-scale. This link has been established by simulating the drying process for a network that represents the whole porous sample; in this way, the role of macro pores and their spatial distribution in the porous medium has been studied and viscous stabilization and non-isothermal effects have been investigated. In future work, representative pore networks shall be used to parameterise the more efficient continuous model. Additionally, the pore network model – with its strong assumptions on the geometry of individual pores – shall be assessed by more fundamental modelling techniques that can describe liquid-gas phase boundaries in real pore geometries, e.g. the volume-of-fluid method. Furthermore, experiments shall prove the validity of viscous and non-isothermal model extensions.

Besides investigating heat and mass transfer, there is considerable interest in describing mechanical effects that are induced by capillary forces during drying, such as shrinkage and product damage by cracks. To this purpose, the pore network model has been combined with the discrete element method in an ongoing GK project; first results indicate that the evolution of liquid phase distributions plays a crucial role for the number of micro cracks (Kharaghani et al. 2008).

These modelling efforts are undertaken in the belief that new discrete approaches to classical problems can help for a fundamental understanding (Metzger et al. 2007c). Since microscopic effects are explicitly described in such models, their influence on macroscopic behaviour can be directly investigated; and – unlike in continuous models – truly local information is available from the simulation.

Acknowledgments. The authors would like to express their thanks to the French researchers Patrick Perré (LERMAB, Nancy) and Marc Prat (INP Toulouse) who accompanied this work by fruitful discussions.

References

Al-Futaisi, A., Patzek, T.W.: Extension of Hoshen-Kopelman algorithm to non-lattice environments. Physica A 331, 665–678 (2003)

Huinink, H.P., Pel, L., Michels, M.A.J., Prat, M.: Drying processes in the presence of temperature gradients. Pore-scale modelling. Eur Physical J. E. 9, 487–498 (2002)

Irawan, A.: Isothermal drying of pore networks: Influence of pore structure on drying kinetics. PhD Thesis, Otto-von-Guericke-University Magdeburg, Germany (2006)

Irawan, A., Metzger, T., Tsotsas, E.: Pore network modelling of drying: combination with a boundary layer model to capture the first drying period. In: Proceedings of 7th World Congress of Chemical Engineering, Glasgow, Scotland, pp. 33–42 (2005)

Kharaghani, A., Metzger, T., Tsotsas, E.: Mechanical effects during isothermal drying: a new discrete modelling approach. In: 16th Int Drying Symposium, Hyderabad, India (submitted to 2008)

Laurindo, J.B., Prat, M.: Numerical and experimental network study of evaporation in capillary porous media. Phase distributions. Chem. Eng. Sci. 51, 5171–5185 (1996)

Laurindo, J.B., Prat, M.: Numerical and experimental network study of evaporation in capillary porous media. Drying rates. Chem. Eng. Sci. 53, 2257–2269 (1998)

Le Bray, Y., Prat, M.: Three dimensional pore network simulation of drying in capillary porous media. Int. J. Heat Mass Transfer 42, 4207–4224 (1999)

Metzger, T., Tsotsas, E.: Influence of pore size distribution on drying kinetics: a simple capillary model. Drying Technology 23, 1797–1809 (2005)

Metzger, T., Tsotsas, E.: Viscous stabilization of drying front: three-dimensional pore network simulations. Chem Eng Research Design, (in press, 2008)

Metzger, T., Irawan, A., Tsotsas, E.: Discrete modelling of drying kinetics of porous media. In: Eikevik, T.M., Alves-Filho, O., Strommen, I. (eds.) Proceedings of 3rd Nordic Drying Conference (NDC 2005), Karlstad, Schweden (2005)

Metzger, T., Irawan, A., Tsotsas, E.: Remarks on the paper "Extension of Hoshen-Kopelman algorithm to non-lattice environments. by Al-Futaisi, A., Patzek, T.W. Physica A 321, 665–678 (2006); Physica A 363, 558–560

Metzger, T., Irawan, A., Tsotsas, E.: Isothermal drying of pore networks: influence of friction for different pore structures. Drying Technology 25, 49–57 (2007a)

Metzger, T., Irawan, A., Tsotsas, E.: Influence of pore structure on drying kinetics: a pore network study. AIChE J. 53, 3029–3041 (2007b)

Metzger, T., Kwapinska, M., Peglow, M., Saage, G., Tsotsas, E.: Modern modelling methods in drying. Transport in Porous Media 66, 103–120 (2007c)

Metzger, T., Tsotsas, E., Prat, M.: Pore-network models: A powerful tool to study drying at the pore level and understand the influence of structure on drying kinetics. In: Tsotsas, E., Mujumdar, A.S. (eds.) Modern drying technology. Computational tools at different scales, vol. 1, Wiley-VCH, Weinheim (2007d)

Nasrallah, S.B., Perré, P.: Detailed study of a model of heat and mass transfer during convective drying of porous media. Int. J. Heat Mass Transfer 31, 957–967 (1988)

Nowicki, S.C., Davis, H.T., Scriven, L.E.: Microscopic determination of transport parameters in drying porous media. Drying Technology 10, 925–946 (1992)

Perré, P., Turner, I.W.: A 3-D version of TransPore: a comprehensive heat and mass transfer computational model for simulating the drying of porous media. Int. J. Heat Mass Transfer 42, 4501–4521 (1999)

Plourde, F., Prat, M.: Pore network simulations of drying of capillary media. Influence of thermal gradients. Int. J. Heat Mass Transfer 46, 1293–1307 (2003)

Prat, M.: Percolation model of drying under isothermal conditions in porous media. Int. J. Multiphase Flow 19, 691–704 (1993)

Prat, M.: On the influence of pore shape, contact angle and film flows on drying of capillary porous media. Int. J. Heat Mass Transfer 50, 1455–1468 (2007)

Segura, L.A., Toledo, P.G.: Pore-level modeling of isothermal drying of pore networks. Effects of gravity and pore shape and size distributions on saturation and transport parameters. Chem. Eng. J. 111, 237–252 (2005)

Surasani, V.K., Metzger, T., Tsotsas, E.: A non-isothermal pore network drying model: Influence of gravity. In: Proceedings of 6th European Congress of Chemical Engineering (ECCE-6), Copenhagen, No. 2131 (2007)

Surasani, V.K., Metzger, T., Tsotsas, E.: Consideration of heat transfer in pore network modelling of convective drying. Int. J. Heat Mass Transfer 51, 2506–2518 (2008a)

Surasani, V.K., Metzger, T., Tsotsas, E.: Influence of heating mode on drying behaviour of capillary porous media: pore scale modelling. Submitted to Chemical Engineering Science (2008b)

Turner, I.W., Perré, P.: A synopsis of the strategies and efficient resolution techniques used for modelling and numerically simulating the drying process. In: Turner, I., Mujumdar, A.S. (eds.) Mathematical modeling and numerical techniques in drying technology, Marcel Dekker, New York (1996)

Vu, T.H.: Influence of pore size distribution on drying behaviour of porous media by a continuous model, PhD Thesis, Otto-von-Guericke-University Magdeburg, Germany (2006a)

Vu, T.H., Metzger, T., Tsotsas, E.: Influence of pore size distribution via effective parameters in a continuous drying model. Proceedings of 15th International Drying Symposium, Budapest A, 554–560 (2006b)

Yiotis, A.G., Stubos, A.K., Boudouvis, A.G., Yortsos, Y.C.: A 2-D pore-network model of the drying of single-component liquids in porous media. Adv. Water Resour. 24, 439–460 (2001)
Yiotis, A.G., Boudouvis, A.G., Stubos, A.K., Tsimpanogiannis, I.N., Yortsos, Y.C.: The effect of liquid films on the drying of porous media. AIChE J. 50, 2721–2737 (2004)
Yiotis, A.G., Tsimpanogiannis, I.N., Stubos, A.K., Yortsos, Y.C.: Pore-network study of the characteristic periods in the drying of porous materials. J. Colloid Interface Science 297, 738–748 (2006)

Part IV

Dynamics of Particles and Particle Systems

Micro and Macro Aspects of the Elastoplastic Behaviour of Sand Piles

P. Roul[1], A. Schinner[2], and K. Kassner[1]

[1] Institut für Theoretische Physik, Otto-von-Guericke-Universität Magdeburg
[2] GeNUA mbH, Kirchheim

Abstract. We use a discrete element method to simulate the dynamics of granulates made up from arbitrarily shaped particles. Static and dynamic friction are accounted for in our force laws, which enables us to simulate the relaxation of (two-dimensional) sand piles to their final static state. Depending on the growth history, a dip in the pressure under a heap may or may not appear. Properties of the relaxed state are measured and averaged numerically to obtain the values of field quantities pertinent for a continuum description. In particular, we show that it is possible to obtain not only stresses but also displacements in the heap, by judicious use of an adiabatic relaxation experiment, in which gravity is slowly changed. Hence the full set of variables of the theory of elasticity is available, allowing comparison with elastoplastic models for granular aggregates. A surprising finding is the behaviour of the material density in a heap with dip, which increases where the pressure is minimum.

1 Introduction

In spite of their importance in applications, it is fair to say that there is as yet no fundamental understanding of granular materials. Such an understanding might manifest itself in a general continuum theory, applicable to the majority of granular assemblies, without the need of ad hoc assumptions for each new system considered. Even though continuum descriptions have been applied extensively to model granular materials, especially in the engineering community [1,2], neither are these based on a microscopic theory nor is their predictive power for new experiments on granulates impressive. In the physics community, continuum descriptions are based either on balance equations [3] or on symmetry considerations [4], i.e., on general principles that are not specific to the granular state, so these ideas may yield important constraints for a microscopic theory but cannot stand in its place. For static assemblies, phenomenological closure relations [5] as well as elastoplastic models [6] have been used in macroscale calculations of the stress tensor, leading to different stress distributions in a sand pile.

The pressure distribution under a sand pile is not independent of the conditions of its creation. Rather, in some cases the pressure exhibits a minimum below the tip of the sand pile whereas in others, it does not. Which behaviour is observed depends strongly on the characteristics of the granulate, especially the size and shape distribution of the particles. Moreover, it depends on the construction history of the sand pile, so two piles consisting of the same material may have different stress distributions. If grains are dropped from a point source, there usually is a pressure minimum; if they are dropped layerwise, then there is no minimum. This phenomenon has been observed both in experimental sand heaps [7] and in numerical simulations [8].

The counterintuitive behaviour of the stress distribution under a sand pile may be traced back to the fact that the aggregate consists of particles that can be considered rigid to a good approximation and that do not stick together, i.e., the material is non-cohesive. The pile will nevertheless be able to show elastic or plastic responses to external loads, as the particles can rearrange under pressure to fill voids more completely, so there will be a finite macroscopic deformation resulting from a finite load. Since the only effects that hold the pile together near its surface are friction and geometric constraints, the free surface of the heap has a tendency to flow, which means that in its vicinity *plastic* behaviour should be anticipated.

On the other hand, deep inside the pile, *elastic* behaviour is not necessarily to be expected, if mechanical aspects suggested by analogies from the field of structural rigidity are considered [9]. A network of rotatable bars is flexible (= hypostatic), isostatic or over-constrained (= hyperstatic), depending on whether the number of bars connecting vertices is smaller than, equal to, or larger than, the number needed to maintain stable equilibrium. If the links between touching grains in a sand pile are considered as the "bars" of a network, then the non-cohesive nature of the granular constituents allows only bars under compression, which rules out the possibility of an over-constrained network, leaving the sand pile to be either hypostatic or isostatic. Arguments based on the different scaling behaviour of self-stresses and imposed stresses [9] seemed to imply isostaticity for granular matter loaded only by its own weight. Then the average coordination number z of grains would have to correspond exactly to a critical value z^{crit} (6 in two dimensions for frictionless non-circular particles and 3 with friction). The mechanical equilibrium conditions of isostatic structures lead to hyperbolic field equations, whereas static elasticity is described by elliptic equations.

However, it has been pointed out that load and geometry are not independent [10] in sand piles, and the distinction should be between isostatic and non-isostatic *problems* rather than *structures* [10,11]. Solutions of isostatic problems with prescribed load may lead to hypostatic structures, describable by elliptic equations, hence the introduction of effective elastic coefficients may be meaningful [10].

In order to investigate the matter, we perform numerical simulations, in which a sand pile is constructed from several thousand convex polygonal particles with varying shapes, sizes and edge numbers. The particles are poured from either a point source, which regularly leads to a pressure minimum under the pile, or a line source. We use a discrete-element method with soft but shape-invariant particles: two particles in contact with each other are allowed to interpenetrate partially. On the one hand, it would be inefficient to solve the elastic equations for each collision between pairs of non-rigid particles, on the other, to implement an event-driven code allowing the (desirable) solution of the equations of motion for rigid particles would be too cumbersome with polygonal particles.

2 Simulation Method

We solve the equations of motion following from the balances of momenta and angular momenta of the particles, using a fifth-order Gear predictor-corrector method [12].

Colliding particles overlap. Forces are then calculated from the geometric characteristics overlap area and contact length (defined as the distance between the two points of intersection of the overlapping polygons) using the relative velocities of the two particles. The calculation involves phenomenological elastic constants as well as model parameters for friction and viscous damping. Details are given in [8].

In two dimensions, the momentum balance provides two equations per particle, the angular momentum balance one:

$$m_i \ddot{\mathbf{r}}_i = \sum_{j=1}^{n} \mathbf{F}_{ij} + \mathbf{G}_i , \qquad I_i \ddot{\varphi} = \sum_{j=1}^{n} L_{ij} . \tag{1}$$

Here, the subscript i runs over all the particles, the subscript j over all the contacts of particle i with other particles. That is, forces and torques are exchanged between particles only if they touch. Hence we have short-range forces, viz. contact forces. $\mathbf{G}_i$ is the force acting on particle i due to external fields, in our case just gravitation, $\mathbf{F}_{ij}$ the force created by the particle touching particle i in contact j.

The force calculation is the most time-consuming part of the algorithm. Of course, advantage is taken of the short-range nature of the forces by calculating only non-vanishing forces, i.e., forces between particles that are really in contact with each other. To achieve fast contact determination in a time that is proportional to the number of particles (not to its square), independent of the complexity, i.e., number of edges of the particles, algorithms from virtual reality and computational geometry were adapted. These use bounding boxes and Voronoi regions to determine overlaps of particles [8].

2.1 Stress Calculation

Once we have the forces, we can compute stresses. It is easy to derive a formula for the average stress obtained in a homogeneous polygonal particle [13], assuming that the forces given in the contact points act on the corresponding edge of the polygon:

$$\sigma_{ij}^{p} = \frac{1}{V^{p}} \sum_{c=1}^{m} x_i^c f_j^c , \tag{2}$$

where x_i^c is i-th component of the branch vector pointing from the centre of mass of the particle to the contact point c, and f_j^c is the j-th component of the total force in that contact point. V^p is the volume of particle p (actually an area, since we are in 2D) .

Expression (2) may be interpreted as the stress tensor associated with a single particle. This microscopic stress would not be a convenient means to describe the macroscopic sandpile, as it fluctuates wildly within a volume containing a few sand grains. In fact, it is undefined in the voids between the grains. Hence, for a continuum description, we need to average microscopic stresses. A representative volume element (RVE) is introduced via the requirement that the average becomes size independent, if the volume is taken equal to this value or larger. We find that box sizes containing 100-200 particles are sufficient to serve as RVE.

The averaged stress tensor was evaluated throughout the sand pile; typically, we represent it via a plot of tensor components as a function of the lateral coordinate x of the pile for layers of given heights $y_1, y_2, \ldots y_n$.

2.2 Determining Strains

While the calculation of stresses is rather straightforward, this is not true for strains. In fact, even the definition of strain is problematic after assuming particles to be essentially rigid. For this reason, most macroscopic descriptions proposed in the last few years try to get by without using strain at all. Whether this approach can be successful in the long run remains to be seen. In any case, even if it may be difficult or impossible to determine strains in *experiments* on sand piles, this is not so in a *simulation*.

We define strains with respect to a hypothetical reference state of zero gravity and a sand pile identical to the one at ambient gravity, except for slightly displaced particle centres; i.e., in the reference state, no particle rearrangements that modify neighbourhood relationships should be present in comparison with the actual state. We obtain the reference state from the ambient one by slowly changing gravity. In principle, it is not necessary to go down to zero gravity, as long as the strains increase linearly with the gravity level – one may then extrapolate to zero from the knowledge of the positions of the particle centres of mass at two arbitrary different gravity levels. But it is necessary to let the sand pile approach a rest state after a reduction of gravity. Moreover, linearity has to be checked by looking at different gravity levels.

Figure 1 shows effective elastic constants determined using the stress tensor evaluated according to the prescription of the last subsection and the strain tensor obtained by variation of gravity and measurement of the ensuing displacements.

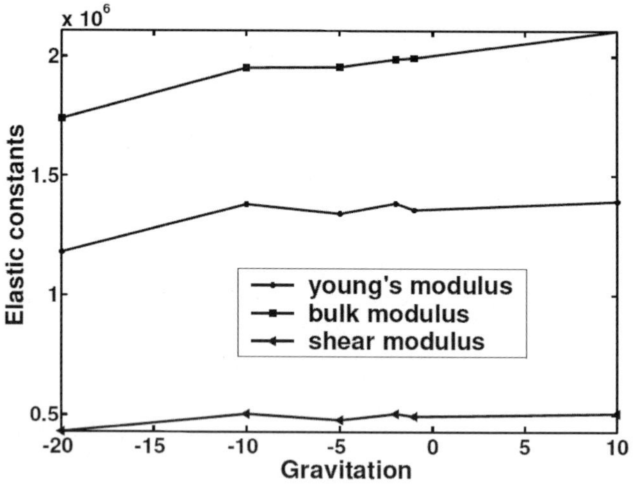

Fig. 1. Elastic moduli evaluated using displacement vectors as obtained from different changes of the gravitational acceleration. Zero corresponds to the ambient value of g (9.81m/s^2), the figures give the change in percent, used in the calculation of displacements. The Young modulus for a single particle was taken to be 10^7N/m.

It appears that too large a change of gravity leads to topological rearrangements of particles and plastic deformations of the sandpile. Nevertheless, there is a range of gravity levels g of about 10% about the ambient level, in which strains change linearly, meaning that essentially no rearrangements of this type have taken place and the ideal zero-gravity limit can be defined by extrapolation. Hence, our strain measurements are obtained by a meaningful procedure. The appropriate RVE for strain averaging turns out to have the same size as the one for stress averaging.

3 Analytic Descriptions

Next we would like to briefly describe two macroscopic approaches based on analytic descriptions [5,6] the quality of which we have checked with our simulations. All analytic approaches for sand pile physics have to respect the basic law of mechanical equilibrium, which in two dimensions reads

$$\begin{aligned} \partial_x \sigma_{xx} + \partial_y \sigma_{xy} &= 0 \\ \partial_x \sigma_{xy} + \partial_y \sigma_{yy} &= -\rho g \end{aligned} \tag{3}$$

where ρ is the density of the sand pile, taken constant in these theories. In our simulation, we have to evaluate this density as the product of the particle density, which is fixed, and the local volume fraction of the sand pile.

Moreover, it is generally agreed that the *surface* of a sand pile is in a state of incipient failure, i.e., it corresponds to a slip plane. Using this assumption, one can show that the normal-component free-interface condition $\sigma_{nn} = 0$ leads to the vanishing of *all* stress components (i.e. $\sigma_{nt} = 0$ and $\sigma_{tt} = 0$). This follows directly from the Mohr-Coulomb yield criterion

$$\left(\sigma_{xx} - \sigma_{yy}\right)^2 + 4\sigma_{xy}^2 - \left(\sigma_{xx} + \sigma_{yy}\right)^2 \sin^2 \varphi = 0\,, \tag{4}$$

applied in a coordinate system with x parallel to the surface (i.e., replace $x \rightarrow t$, $y \rightarrow n$ in (4)). Herein, φ is the internal friction angle (related to the friction coefficient μ via $\tan\varphi = \mu$). The assumption of incipient failure provides stress boundary conditions at the surface of the sand pile.

Because the two field equations are insufficient to determine the three stress components σ_{xx}, σ_{xy}, and σ_{yy}, a third equation, a so-called closure relation, is needed. In elasticity, this is a constitutive relation connecting stresses and strains.

Usually, it is then stated in the literature that for sand piles displacement fields are not available, which is true experimentally and also for the macroscopic analysis, as it does not have access to the microscopic particle displacements. Moreover, it is argued that for rigid particles these displacements are not meaningful. Both rigidity and Coulomb friction contribute to static indeterminacy of the pile.

A closure relation between the stress components is then sought for and postulated, to remove this static indeterminacy. Different approaches differ in their postulates concerning this "constitutive" relation. A common assumption of several theories is *radial stress field scaling* (RSF), which seems to be verified in experiments and is essentially based on the idea, that the stress fields of geometricalla similar piles should be the same up to a scale factor. Mathematically, this reads

$$\sigma_{ij} = \rho g y \, s_{ij}\left(\frac{x}{y\cot\varphi}\right). \tag{5}$$

One can then reduce the equilibrium equations to ordinary differential equations, once a closure relation has been found. (In three dimensions, several closure relations are needed – the expression for the divergence of the stress tensor yields only three equations, whereas the stress tensor has six independent components.)

A theory that created a lot of stir in the 90s is due to Wittmer et al. [5]. They considered a continuous family of closure relations of the form

$$\sigma_{ww} = K\sigma_{uu}, \tag{6}$$

where K is a constant, and σ_{ww} and σ_{uu} are the principal stress components along two orthogonal directions w and u, oriented at a prescribed fixed angle (the parameter of the family) with respect to the basic xy coordinate system. These models are called OSL models; OSL stands for "oriented stress linearity". The most interesting of these models, as it seemed to be justifiable more easily as a natural form incorporating the construction history of the sand pile, is the so-called *fixed-principal axis* model (FPA). It is given by $K = 1$ and the angle of the coordinate axis, along which σ_{uu} is to be measured, being equal to $\tau = (\pi\text{-}\varphi)/2$. This model can be derived more intuitively by assuming that the principal axes of the stress tensor take the fixed directions $\pm\psi$ on both sides of the central axis of the sand pile, where $\psi = (\pi\text{-}2\varphi)/4$, hence the name FPA. The FPA model leads to a pronounced dip in the pressure distribution under the tip of the sand pile.

Part of the debate about the model came from the fact, that with a closure relation such as (6), the field equations for the stress tensor became hyperbolic throughout the volume of the whole sand pile (corresponding to isostaticity).

A much more conventional approach is the elastoplastic theory by Didwania et al. [6]. They note that near the surface of the pile, plastic behaviour is to be expected, and the closure relation is simply given by Mohr's yield criterion, Eq. (4).

Near the centre of the pile, they assume that there is linearly elastic behaviour. The absence of measurable displacements is not a problem, as one can derive within linear elasticity stress compatibility relations, from which the elastic moduli scale out, so the limit $E \to \infty$ can be easily taken. In two dimensions, there is just one such relationship. It takes the form

$$\sigma_{xx,yy} + \sigma_{yy,xx} - 2\sigma_{xy,xy} = 0, \tag{7}$$

and if it is imposed, rigid-body indeterminacy is removed. Whenever a plastic region touches an elastic one, there are boundary conditions, requiring continuity of stresses but allowing discontinuous derivatives. When two elastic regions touch each other with non-matching stress derivatives, an infinitely thin layer of a yield region is assumed between them, along which equation (4) holds.

Cantelaube et al. assume RSF scaling as well. They obtain solutions which in the outer plastic domain obey the field equations (3) and (4), which FPA does near the sand pile surface, too, but strongly differ from FPA behaviour in the elastic core.

For symmetric sand wedges, the shape of the inner domain is that of an isosceles triangle with a steeper base angle ($\hat{\beta}$) or a smaller tip angle. They find three discrete solutions, of which one has a pressure minimum. Once the angle of repose φ of the pile is fixed, the theory contains no free parameters. For later reference, we write their solution here. The expressions for the elastic domain are

$$\sigma_{xx} = (a_2 - 1)y, \qquad \sigma_{yy} = (a_1 - 1)y + b_1|x|, \qquad \sigma_{xy} = -a_1 x, \tag{8}$$

those for the plastic domain ($\beta = \pi/2 - \varphi$)

$$\sigma_{xx} = a_{11}\left[\frac{|x|}{\tan\beta} - y\right], \quad \sigma_{yy} = a_{22}\left[\frac{|x|}{\tan\beta} - y\right], \quad \sigma_{xy} = a_{12}\left[\frac{|x|}{\tan\beta} - y\right]. \tag{9}$$

4 Simulation Results

To convey an impression of a typical result for a numerical aggregate obtained in a sand pile simulation, we show a final state of a computation comprising a few thousand particles dropped from a point source (this is the hopper above the pile).

Fig. 2. Simulated sand pile. The walls and the hopper are made of immobile specially shaped particles. Different grey levels correspond to particles dropped at different times. The number of polygon edges varies between 6 and 8.

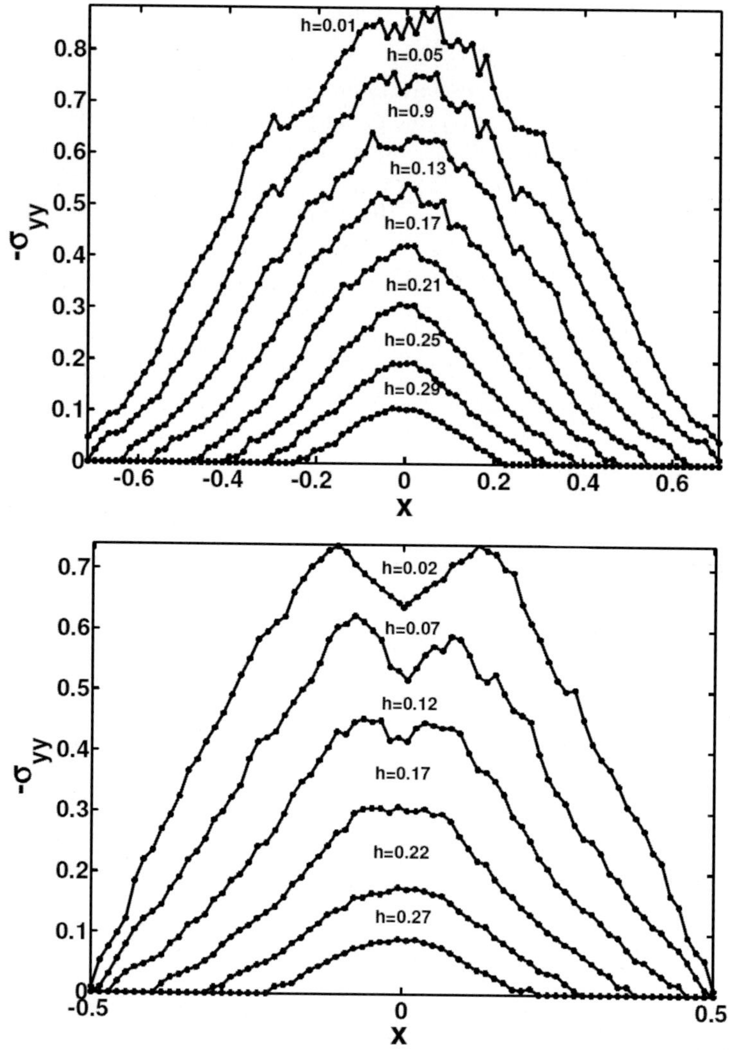

a)

b)

Fig. 3. Distribution of "pressure" on horizontal cuts at different heights through a simulated sandpile. a) sand piles constructed from a line source (average over 6 piles of 6600 particles each), b) sand piles deposited from a point source (average over 7 piles of 8000 particles each). The topmost curves correspond to the lowest cuts and vice versa.

Next we display the distributions of vertical stress components obtained in layer-wise deposition (line source) and in deposition from a central position (point source). Both results are averages over a number of simulations.

The next figure shows the effect of particle shape. Here, roughly elliptic particles with a ratio of major and minor axis of 2 were used, whereas the particles leading to Fig. 3 were inscribed into circles. The "dip" in the pressure distribution becomes

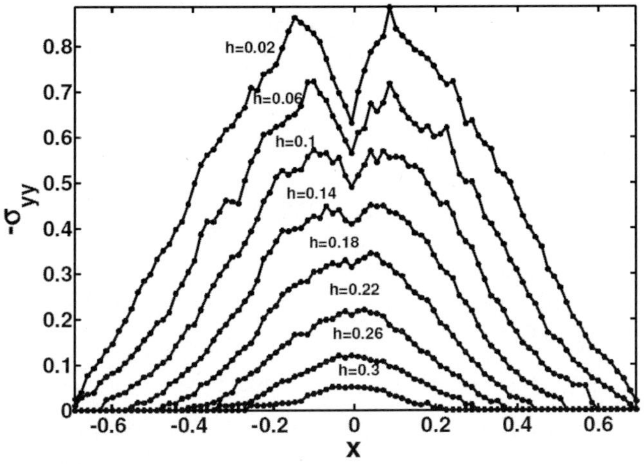

Fig. 4. Pressure distribution under sand pile constructed from elliptic particles. Average over 7 piles with 8000 particles each.

significantly more pronounced for elliptic particles. We also determined the orientational distribution of the elliptic particles. Their alignment is mostly horizontal, as one would expect.

5 Comparison with Theory

Our first observation in attempting to confront our data to theoretical results is that the orientation distribution of the principal axes of the stress tensor is varying smoothly throughout the sand pile. This rules out the FPA model as a quantitative predictor.

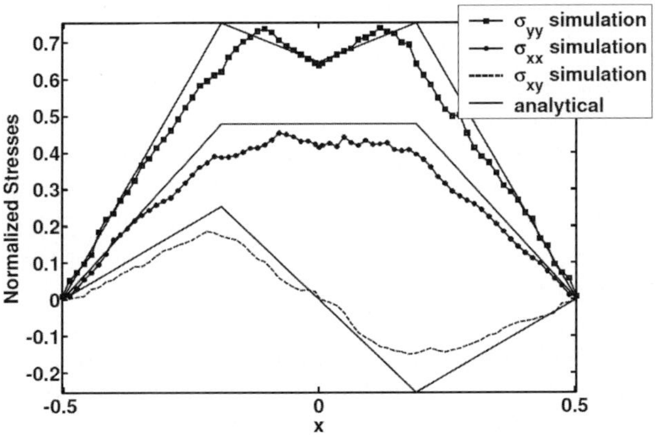

Fig. 5. Fit of the components of the (negative) stress tensor predicted by the OSL model to the point source simulations. Parameter values obtained: K=1.4, $\tau = 85°$.

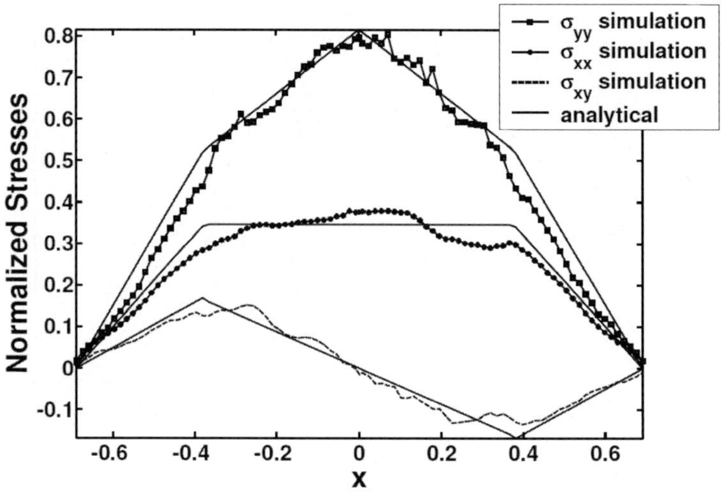

a)

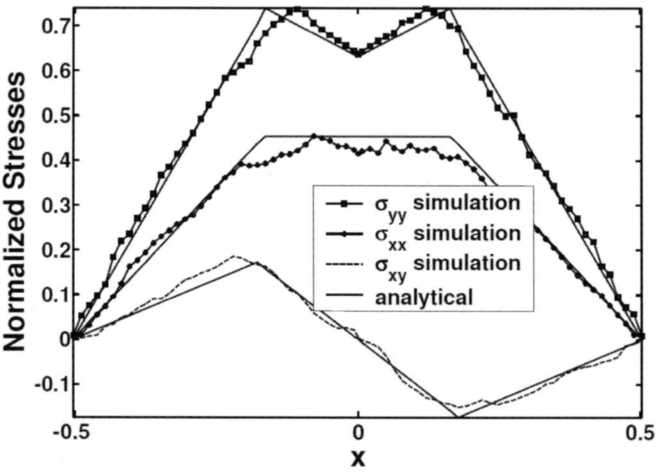

b)

Fig. 6. Comparison of simulation data with predictions from elastoplastic theory [6]. Shown components of the (negative) stress tensor are evaluated at the bottom of the pile. a) sand piles from line source, b) point source (compare with corresponding curves from Fig. 3).

Some of the less plausible OSL models give a better fit with our data. An example is shown in Fig. 5. For a given angle of repose, there is a relationship between the two parameters K and τ, so this is essentially a one-parameter fit.

As soon as we need to fit, however, we get comparable or better quality from fits to the elastoplastic model by Cantelaube et al. [6], to which we will turn now.

The theory predicts that the sum of the parameters a_{11} and a_{22} from Eq. (9) must be equal to 2, a relationship that may serve as a consistency check. For the piles on the bottom panel of Fig. 6, we find $a_{11} = 1.23$ and $a_{22} = 0.78$, so the relationship is satisfied to better than 1%. If we consider the elastoplastic approach as a theory with a fit parameter, the agreement with the simulations is quite satisfactory. However, for

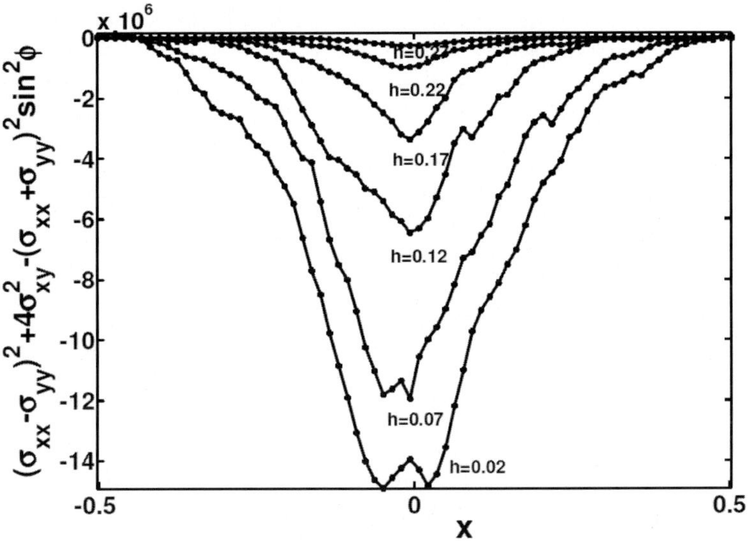

Fig. 7. The Coulomb-Mohr expression (4) evaluated for the average over sand piles from the bottom panel of Fig. 6

symmetric sand piles the theory does not contain any free parameters. In particular, it predicts the angle $\hat{\beta}$, which for $\varphi = 28°$, the angle of repose in our simulations from a point source, should be 22° for the solution producing a dip, but is obtained as 35° from the fit. For the case of the solution producing a plateau, appropriately describing sand piles constructed from a line source, the agreement is surprisingly good: both the theoretical and fitted angle are 49°.

A reason why the theory does not work as well for the solution that is discontinuous at the centre of the pile is that its assumption of a yield line along the axis of the pile is not really satisfied. This can be seen from Fig. 7, where we evaluated the expression on the left-hand side of Eq. (4), which should become zero in the plastic regions. Clearly, it approaches zero far from the centre of the pile ($x = 0$), so the existence of plastic regions near the surface of the pile can be confirmed (though not their triangular shape), but there is little indication of singular behaviour of the expression near the centre of the pile. For the plateau solution, there is no such singular behaviour even in the theory, which may explain why it works so well.

6 Conclusions

To conclude, we have performed simulations of two-dimensional granular aggregates consisting of convex polygons and measured microscopic force distributions of the resulting "sand piles". Via averaging over representative volume elements, for which a sufficient size was determined to contain 100-200 particles, we have determined stress and strain distributions. To obtain a measure for strain, the sandpile was allowed to relax under reduction or increase of gravity.

For a point source, we find, not unexpectedly, that the pressure is not only minimum at the bottom layer, but also in higher layers of the pile. However, it disappears in layers near the tip of the pile. The density profile of sand piles was also measured; we observe it to have a maximum where the pressure is minimum, a somewhat unexpected result, as it suggests the presence of a mechanical instability.

A similar pressure minimum was not obtained in piles poured from a line source, which demonstrates that the simulation reproduces pressure distributions corresponding to different experimental protocols. Dynamically, the two cases differ by the appearance of avalanches during the build-up of a pile from a point source, and their absence for layer-by-layer deposition.

While it may be difficult or impossible to determine the *strain* tensor in an experimental sand pile, it is feasible to obtain a reasonable approximation to it from simulations. We define the strain with respect to a hypothetical reference state of zero gravity. This reference state may be generated from the static pile obtained in a simulation, by slowly changing gravity and following the particle trajectories during the ensuing load change. Then, it is easy to compute the macroscopic strain tensor by averaging over an RVE. It turns out that the size of the RVE we need for converged strain tensors is the same as for stress tensors.

Comparison with simple analytic theories [5,6] for the macroscopic mechanical behaviour of a sand pile shows that these theories have certain deficiencies. Radical departures from conventional approaches such as the introduction of almost ad hoc closure relations [5] seem unnecessary, as an equally good or better fit of the data is obtained by a simple elastoplastic model [6]. Nevertheless, reality is not as simple as these models. One ingredient missing in all the models that use stresses only, is possible density variations in the sand pile.

As an outlook, it may be said that the consideration of varying density naturally leads to the idea that the internal texture of the pile is important and that a macroscopic description therefore probably has to go beyond a simple description in terms of stresses and must introduce additional variables such as fabric tensors. There have been some recent developments in this regard [14]. The question is then of course, how to calculate a macroscopic fabric tensor to close the theoretical description.

References

[1] Lade, P.V., Prabucki, M.-J.: Softening and preshearing effects in sand. Soils and Found., Jap. Geotechn. Soc. 35, 93–104 (1995)

[2] Gudehus, G.: A comprehensive constitutive equation for granular materials. Soils and Found., Jap. Geotech. Soc. 36, 1–12 (1996)

[3] Eggers, J., Riecke, H.: Continuum description of vibrated sand. Phys. Rev. E 59, 4476–4483 (1999)

[4] Csahok, Z., Misbah, C., Rioual, F., Valance, A.: Dynamics of aeolian sand ripples. Eur. Phys. J. E 3, 71–86 (2000)

[5] Wittmer, J.P., Cates, M.E., Claudin, P.: Stress propagation and Arching in Static Sand piles. J. Phys. I France 7, 39–80 (1997)

[6] Didwania, A.K., Cantelaube, F., Goddard, J.D.: Static multiplicity of stress states in granular heaps. Proc. R. Soc. Lond. A 456, 2569–2588 (2000)

[7] Vanel, L., Howell, D., Clark, D., Behringer, R.P., Clement, E.: Memories in sand: Experimental tests of construction history on stress distributions under sand piles. Phys. Rev. E 60, R5040–R5043 (1999)
[8] Schinner, A.: Ein Simulationssystem für granulare Aufschüttungen aus Teilchen variabler Form. PhD thesis, Univ. Magdeburg (2001)
[9] Moukarzel, C.F.: Isostatic Phase Transition and Instability in Stiff Granular Materials. Phys. Rev. Lett. 81, 1634–1637 (1998)
[10] Roux, J.N.: Geometric origin of mechanical properties of granular materials. Phys. Rev. E 61, 6802–6836 (2000)
[11] Unger, T.: Characterization of static and dynamic structures in granular materials. PhD thesis, Budapest Univ. (2004)
[12] Gear, C.W.: Numerical Initial Value Problems in Ordinary Differential Equations. Prentice-Hall, Englewood Cliffs (1971)
[13] Cundall, P.A., Strack, O.D.L.: A discrete numerical model for granular assemblies. Géotechnique 29, 47–65 (1979)
[14] Ball, R.C., Blumenfeld, R.: Stress Field in Granular Systems: Loop Forces and Potential Formulation. Phys. Rev. Lett. 88, 115505-1-115505-4 (2002)

Micro-Macro Deformation and Breakage Behaviour of Spherical Granules

S. Antonyuk[1], J. Tomas[2], and S. Heinrich[1]

[1] Institut für Feststoffverfahrenstechnik und Partikeltechnologie, TU Hamburg-Harburg
[2] Institut für Verfahrenstechnik, Otto-von-Guericke-Universität Magdeburg

Abstract. Compression and impact tests have been used to study the deformation and breakage behaviour of spherical granules (γ-Al_2O_3, zeolites 13X and 4A, sodium benzoate). The elastic compression behaviour of granules is described by means of force-displacement curves and by application of Hertz' contact theory. An elastic-plastic contact model is proposed to describe the deformation behaviour of elastic-plastic granules. The breakage probability is determined by using particle size distributions after the impacts at different velocities and described by a modified Weibull distribution. DEM is applied to analyse the dynamics and mechanisms of microscopic breakage behaviour of the experimentally investigated zeolite granules. The arrangement of the primary particles inside a granule, their coordination number and binder strength have been varied to obtain the breakage probability depending on the impact velocity.

1 Introduction

Many products, like e.g., catalysts, adsorbents, fertilizers, tablets etc., are often produced as granules. Obvious advantages of the granules in comparison to powders are the higher packing density, a better flow behaviour as well as less dust formation. Due to granulation some technological problems like long time consolidation or segregation of the bulk materials in bunkers and transport containers can be avoided. Moreover, desirable properties such as a regular shape, chemical composition, narrow particle size distribution, porosity, and internal surface can be obtained.

During the processing sequence, transportation and handling the granules are exposed to a lot of mechanical stressing due to granule-granule and granule-apparatus wall contacts (Fig. 1). The high contact forces often lead to the breakage of granules, when the granule is crushed into several fragments [2].

In general, the granules should not form dust and fragments during transportation, storage and handling. The maximum stressing conditions during these operations define the lower limit of the strength which all granules should have in order to be able to resist the stressing.

In order to optimise the existing production processes and minimize the product quality losses during transportation and handling, the deformation and breakage behaviour of granulates during mechanical loading must be investigated by experiments and physical modelling.

Compared to crystalline solids, the granules are particle compounds and tend to show plastic force-displacement behaviour during loading. A granule consists of primary

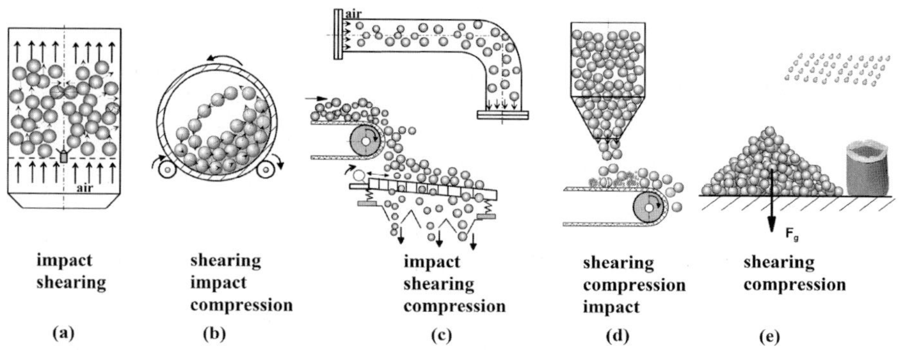

Fig. 1. Typical stressing of particles during granulation in fluidised bed (a), agglomeration or drying in a rotated drum (b), transportation (c), discharge (d) and storage (e)

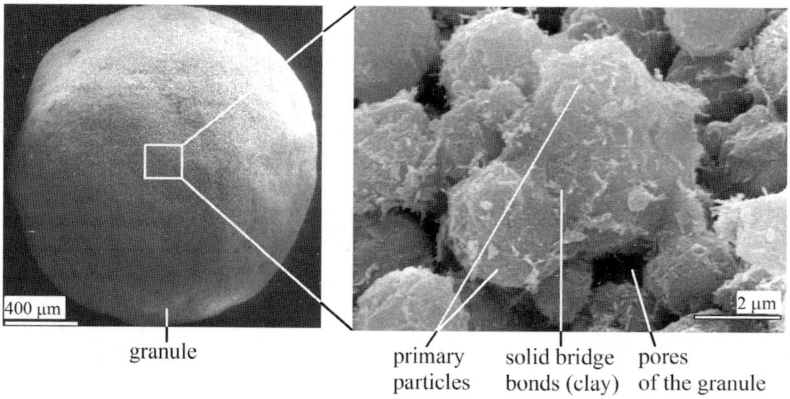

Fig. 2. A zeolite 13X granule with the microstructure

particles, which stick together by the adhesion force at their contacts. Depending on the granulation process, the internal adhesion force is influenced by the superposition of van der Waals interactions between fine primary particles, capillary or solid bridges, high-viscous binder, organic macromolecules, sintering or interlocking of granules. The mechanical breakage behaviour of granules is strongly determined by these micromechanisms.

SEM images (Fig. 2) show a granule of the synthetic zeolite 13X with its microstructure. In this case, the primary particles of zeolite are connected through solid bridge bonds from clay.

2 Experimental Methods and Materials

The macro-properties of granules, their mechanical behaviour and breakage mechanisms can be studied by different testing methods [1]. One of the most important tests is performed by simple compression of a granule up to the primary breakage, which determines the minimum energy requirement for the breakage.

For compression tests performed in this work a strength measuring system produced by Etewe GmbH is used. Fig. 3 shows the principle of a granulate compression test. During the movement of the punch towards the upper fixed plate, the stressing of a granule follows. During this period the displacement s and the force F are measured. To increase the statistical significance of results, the compression tests of the model granules have been repeated 100 times for each sample. All tests are performed at a constant stressing velocity of $v_B = 0.02$mm/s (with strain control).

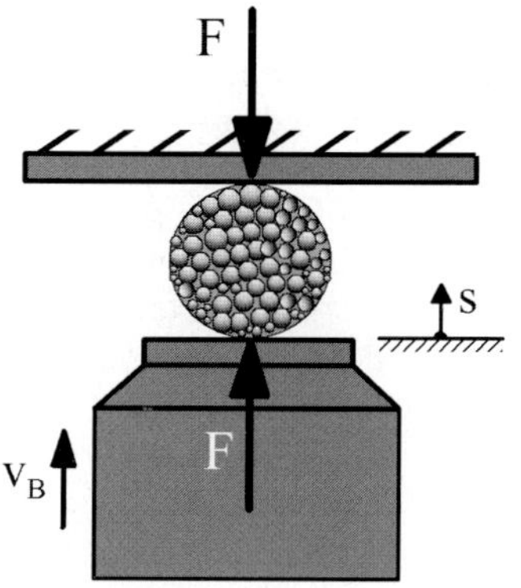

Fig. 3. Principle of single particle compression test [6]

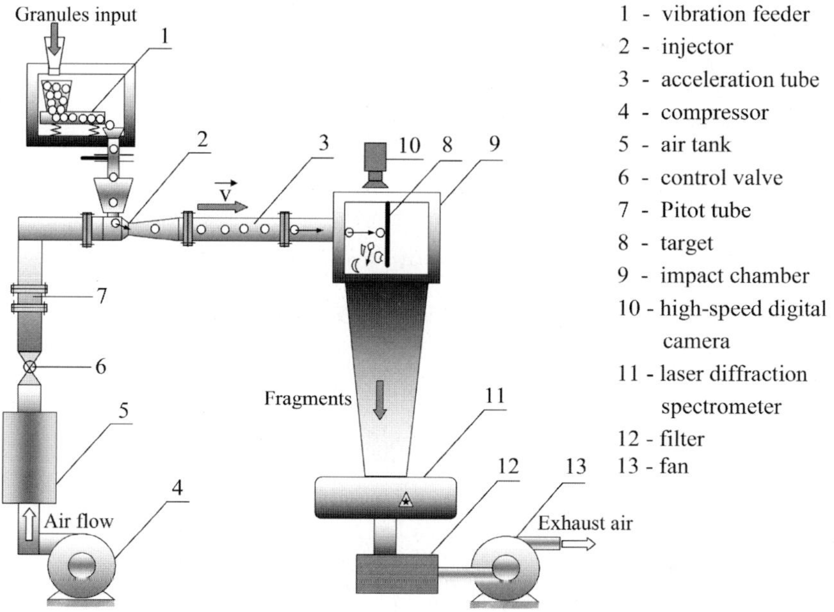

Fig. 4. Impact test rig [2]

The impact of granules has been investigated using an impact tester (Fig. 4). The granules are fed with the help of a vibrational feeder *1* into the hopper of the injector *2*. The granules are fed against the injector hopper acceleration tube *3* into the air stream at a defined velocity. The air pressure is generated by the compressor *4* in the compressed air tank *5*. The control valve *6* is installed between two nozzles, which are connected by means of two check valves, and, thus, the amount of air velocity in the acceleration tube can be adjusted. The air velocity can be measured with the installed Pitot tube *7* by the micro-manometer.

The granules collide horizontally with a hardened steel target *8* inside the impact chamber *9*. After impact, fragments and unbroken granules fall into a filter *12*. The falling velocity is accelerated by two parallel fans *13* at the outlet. The particle size distribution of the fragments is measured online with a laser diffraction spectrometer *11* (of the company Sympatec) before entering into the filter.

The granules have been examined in the impact velocity range of 10-50m/s. The impact test of each sample is repeated eight times.

The different spherical granules, γ-Al_2O_3, the synthetic zeolites 13X and 4A as well as sodium benzoate ($C_6H_5CO_2Na$), are used as test materials to examine the mechanical behaviour from elastic to plastic. A summary of the material properties of these granules is given in Table 1.

Table 1. Characteristics of the examined granules

Characteristics	γ-Al_2O_3	Zeolite 13X	Zeolite 4A	Sodium benzoate
Granule size in mm				
for compression tests	1.6 - 1.9	1.2 - 1.7	2.1 - 2.5	0.86 - 1.60; 1.24 - 1.60
for impact tests	1.6 - 1.9	1.0 - 1.6	-	0.7 - 1.1; 1.0 - 1.6
Sphericity	0.95	0.94	0.91	0.92
Granule density in kg/m^3	1040	1150	1140	1440

3 Experimental Results

3.1 Force-Displacement Behaviour during Compression

The typical force-displacement curve of a γ-Al_2O_3 granule is shown in Fig. 5. At the beginning of the punch-particle contact, the elastic contact deformation of the granule takes place. The elastic contact force is described by Hertz' theory [3] as

$$F_{el} = \frac{1}{6} E^* \sqrt{d\, s^3} \tag{1}$$

where s is the displacement in normal direction, and d is the granule diameter.

The effective modulus of elasticity E^* of both granule (without index) and punch (index w) is given by

$$E^* = 2\left(\frac{1-\upsilon^2}{E} + \frac{1-\upsilon_w^2}{E_w}\right)^{-1} \tag{2}$$

where υ and υ_w are the Poisson's ratios of the granule and the punch (wall), respectively.

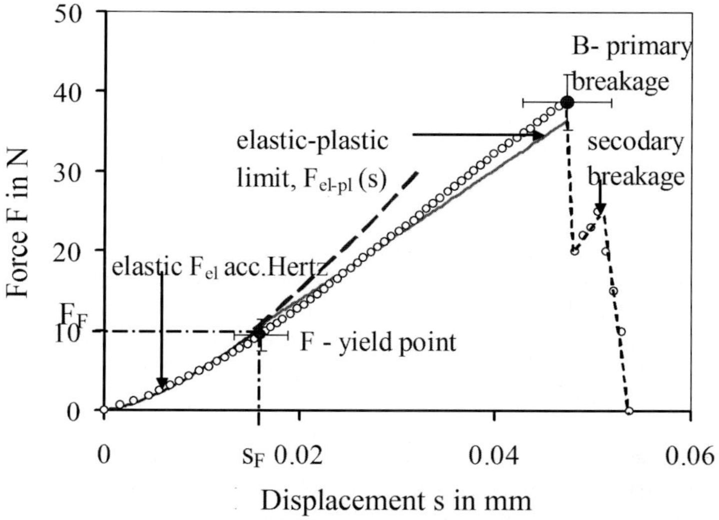

Fig. 5. Typical force-displacement curve of a γ-Al_2O_3 granule (d = 1.62-1.76 mm) during compression

Table 2. Mechanical characteristics of the examined granules by compression

Granules	Diameter	Yield point		Modulus of elasticity
	d in mm	F_F in N	s_F in µm	E in GPa
γ-Al_2O_3	1.6 - 1.9	8.5 ± 2.0	12.5 ± 3.8	14.5 ± 0.31
zeolite 4A	2.1 - 2.5	16.4 ± 0.9	50 ± 5.0	2.7 ± 0.21
zeolite 13X	1.2 - 1.7	1.0 ± 0.08	10 ± 0.9	2.3 ± 0.16
sodium benzoate	0.80 - 1.0 1.24 - 1.60	< 0.1	< 1	-

continued	Stiffness in normal direction in N/mm		
	k_{el} at $F = F_F$	k_{el-pl} at $F = 0.98 \cdot F_B^*$	k_{pl}
γ-Al_2O_3	2120 ± 330	1819 ± 174	-
zeolite 4A	983 ± 76	893 ± 49	-
zeolite 13X	294 ± 34	256 ± 26	-
sodium benzoate	-	-	288 ± 61 491 ± 98

* F_B is the breakage limit.

Due to the parabolic curvature $F_{el}(s)$, the average contact stiffness in normal direction increases with the displacement and the size of the granule [4]

$$k_{el}^{*} = \frac{dF_{el}}{ds} = \frac{1}{4} E^{*} \sqrt{d\,s} = \left(\frac{2}{k_{el}} + \frac{2}{k_{el,w}} \right)^{-1} . \tag{3}$$

Based on Eqs. (2) and (3), the modulus of elasticity and the stiffness of the granule during the elastic deformation are obtained.

Table 2 shows the values of these properties in the yield point F (at the force of $F = F_F$ and the displacement of $s = s_F$, Fig. 5). When this elastic limit is reached the plastic deformation begins. This is confirmed by the increasing deviation of the experimental curve F-B from the theoretical Hertz curve. The plastic deformation is near to the contact centres of the nearly circular contact areas of the granule, and the elastic deformation is generated at the limit of the contacts where the pressure is smaller than the yield limit. Therefore, the plastic deformation is additionally affected by a continuous elastic deformation. The force-displacement relationship for elastic-plastic deformation of the granule without adhesion can be given as

$$F_{el\text{-}pl} = \frac{1}{4} \pi\, d\, p_F \left(1 - \frac{1}{3} \sqrt[3]{\frac{s_F}{s}} \right) s \tag{4}$$

where p_F is the micro-yield strength or yield pressure of the granule, and s_F is the displacement up to the yield point ($s = s_F$).

The slope of the curve F-B is a measure of the average elastic-plastic stiffness, which is proportional to the granule diameter and the yield pressure

$$k_{el-pl}^{*} = \frac{1}{4} \pi\, d\, p_F \left(1 - \frac{2}{9} \sqrt[3]{\frac{s_F}{s}} \right) . \tag{5}$$

The sodium benzoate granules deform perfectly plastic during compression [6]. In this case the repulsive force can be described as a linear function of the plastic displacement

$$F_{pl} = \frac{1}{2} \pi\, d\, p_F\, s \tag{6}$$

with an average constant plastic stiffness

$$k_{pl}^{*} = \frac{dF_{pl}}{ds} = \frac{1}{2} \pi\, d\, p_F . \tag{7}$$

The average values of the measured mechanical properties of the examined granules are summarised in the Table 2.

3.2 Breakage Probability of Granules during Normal Impact

The breakage probability has been determined by using the particle size distributions of fragments measured at the impact tests at different velocities (Fig. 4). The breakage probability is described by using a modified Weibull distribution, Eq. (8) [2]. Fig. 6 shows the breakage probability as function of the mass-related impact energy. The mass-related impact energy W_m is calculated from the kinetic energy of the granule having impact velocity v, as $W_m = v^2/2$.

$$P = 1 - \exp\left[-\left(\frac{d_{50,3}}{d_0}\right)^2 \cdot \left(\frac{W_m}{W_{m,0}}\right)^z \right] \tag{8}$$

For Eq. (8), the mean value $d_{50,3}$ of an initial granule size distribution is used. The exponent z determines the slope of the breakage probability distribution. The respective energy barrier ($W_{m,min}$ - $W_{m,max}$) becomes narrower for the larger z value of the granules. Therefore the constant z in Eq. (8) characterizes the material properties and the defect distribution in the granules.

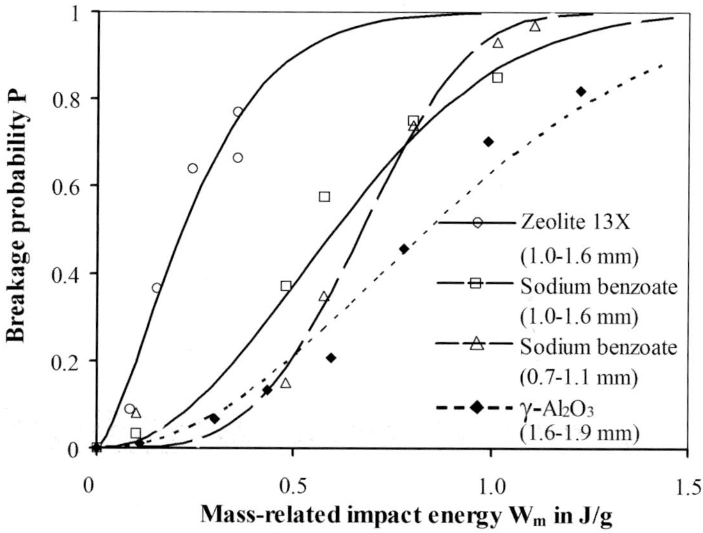

Fig. 6. Breakage probability P of examined granules versus mass-related impact energy W_m

Table 3. Weibull distribution parameters for the breakage probability of granules

Granules	$d_{50,3}$ in mm	z	$W_{m,0}$ in J/g
zeolite 13X	1.38	1.44	0.44
γ-Al_2O_3	1.80	3.63	1.03
sodium	1.31	2.00	0.87
Benzoate	0.93	2.10	0.93

The second parameter $W_{m,0}$ is a characteristic value of mass-related breakage energy of the granules. This parameter and the particle size $d_{50,3}$ affect the position of the breakage probability distribution function. The related parameter d_0 (1 mm) in Eq. (8) is used for dimensionless representation of probability function.

Table 3 shows the values for the parameters $d_{50,3}$, z and $W_{m,0}$ for three examined granules. The breakage probability of sodium benzoate is represented for small and large fractions. As expected, the curve shifts towards the left with increasing granule size. This shows that small granules require more mass-related kinetic energy to break than large granules for the same amount of breakage.

4 Simulation of the Granules Breakage with Discrete Element Method

In the DEM all the constituents are considered as distinct elements where Euler's laws of motion and selected contact models are applied to each element. The DE model is developed with the Particle Flow Code (from Itasca company) [5]. A detailed description of the DE model of the granules is given in [2].

4.1 Simulation Data

The chosen model is in analogy with the experimentally studied zeolite 13X granule. The macro-properties of these granules (Table 4) are used during the calibration of the DE model for the fitting of the micro-characteristics of primary particles and bonding agent.

Table 4. Macro-properties of the modelled granules and micro-properties of their primary particles and solid bridge bonds

Macro-properties	Values	Micro-properties	Values
Granule diameter in mm	1.2	Normal and tangential stiffness of particles in N/m	$6 \cdot 10^4$
Granule density in kg/m^3	1150	Friction coefficient of particles	0.5
Normal stiffness in N/m	$1.5 \cdot 10^5$	Coordination number of particles	4.4
Compressive strength in N/mm^2	8.8	Stiffness of solid bridge bonds in N/m^3	$1.5 \cdot 10^{10}$
Elastic modulus in kN/mm^2	2.3	Strength of solid bridge bonds in N/m^2	$1.8 \cdot 10^4$
		Bond radius ratio R_i/R_b	0.7

The particle size distribution of the primary particles of the modelled granule corresponds to the normal distribution. The primary particles of zeolite granule are connected by the solid bridge bonds (Fig. 2). This bond can transform normal and shear forces F_b and bending moments M_b between the particles, which can be reduced to the normal and shear stresses acting within the cross-section of the bond (Fig. 7). If these stresses reach their corresponding bond strength limit, the solid bridge bond breaks.

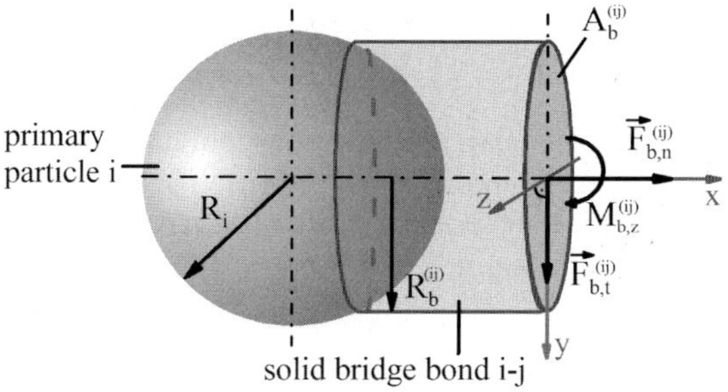

Fig. 7. Solid bridge bond model

The characteristics of the modelled solid bridge bonds are given in Table 4. The bond radius ratio connects the radius of the solid bridge bond between a particle pair to the radius of the smaller particle. The wall properties have been selected according to the typical steel with the stiffness of 10^8N/m and the friction coefficient of 0.5. The value of time steps by simulation is in the range of 0.15μs to 0.22μs.

4.2 Simulation Results

A two dimensional discrete element simulation of the impact tests of granules at different velocities has been performed. The impact events of the specimen at an impact velocity of 30m/s are shown in Fig. 8.

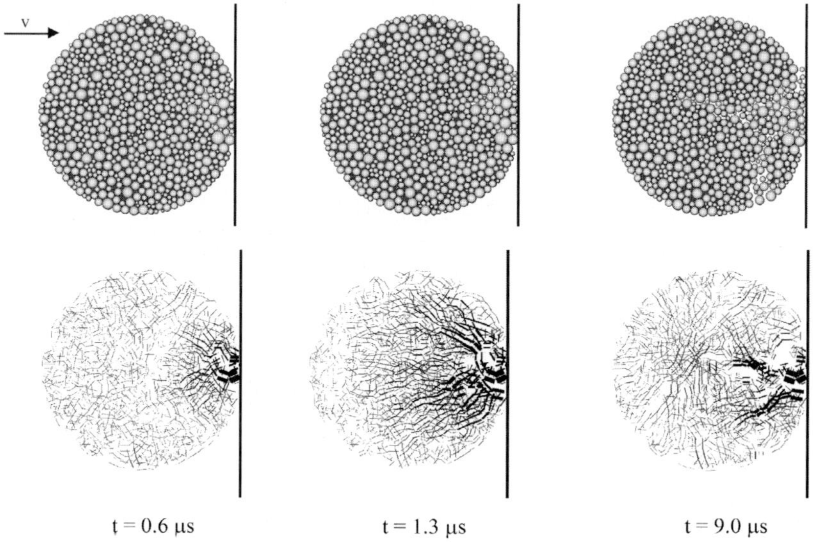

Fig. 8. Fracture events (top) and corresponding normal force network in solid bridge bonds (bottom) with line thickness proportional to force value, impact velocity v = 30m/s, t is impact time

The granule undergoes a contact deformation as it touches the target wall. The solid bridges bonds have to bear the maximum stresses. Reaching their limits, the bonds fail, and cracks between bonded particles are generated. The primary cracks are formed under the maximum shear stress at the contact and propagate between the primary particles under the maximum tensile stress. These cracks can be seen during the decrease of the contacts in the corresponding normal force network in solid bridge bonds during impact (Fig. 8 bottom).

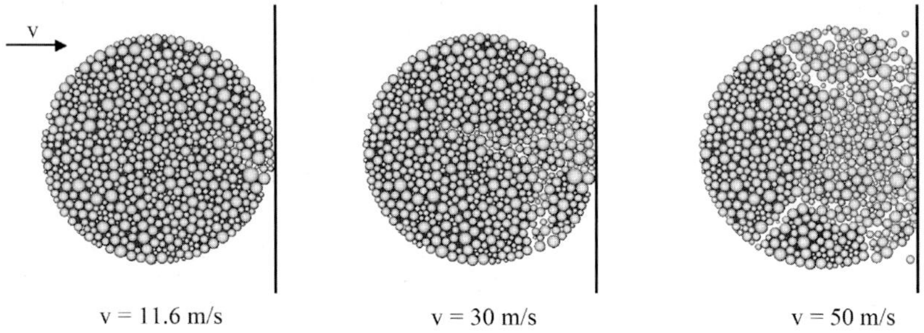

Fig. 9. Fractured specimens at different impact velocities v, impact time of t = 9μs

The fractured specimens at three different impact velocities (Fig. 9) clearly demonstrate that the increase in impact velocity causes an increase of contact deformation and an increasing zone of broken bonds. This zone is similar to the cone of fines in impact experiments. The impact stressing produces a large number of fragments at larger velocities. At 50m/s, the majority of solid bridge bonds between primary particles are broken.

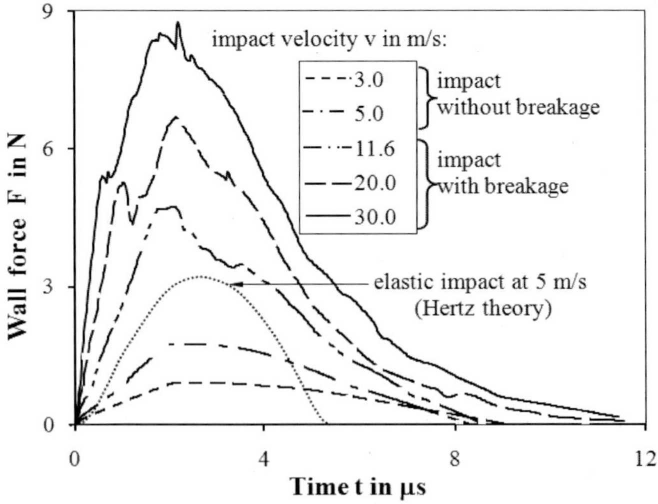

Fig. 10. Effect of impact velocity of the granule on the force-time behaviour during central impact with the wall

Fig. 10 illustrates the effect of the impact velocity on the force and time of the impact. The minimum velocity, which leads to the breakage of the granule, is obtained by v_{min} = 11.6m/s. This value has a standard deviation of 0.7m/s which has been obtained from simulation of impacts for the same model granule at the different impact points. Thus, the model reproduces the typical anisotropy of the breakage behaviour of real granules. Moreover, the breakage force of the granule from the simulation results in 4.7 ± 0.2N, which is approximately 20% smaller than the breakage force during the compression test of these granules, see in [6].

Fig. 10 shows a force-time curve during the elastic impact of a particle at 5m/s calculated according to the Hertz' theory. Compared to the elastic impact, the impact force of a viscoelastic granule (DE model) is smaller and the impact time is larger. The energy losses during impact are resulting from the viscous damping and the friction within the structure of granules.

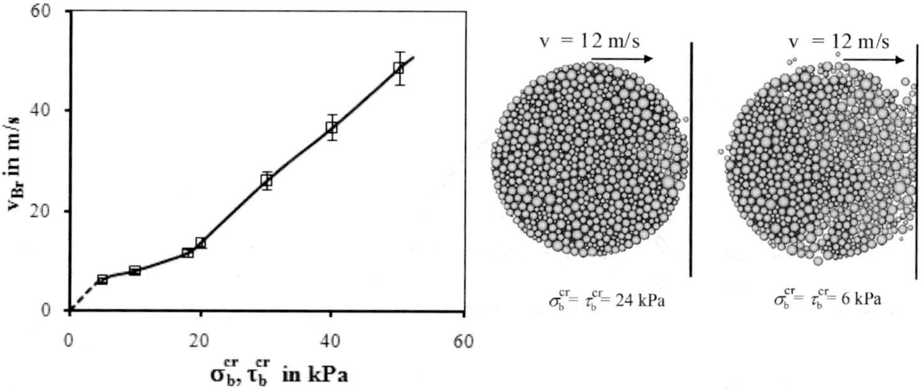

Fig. 11. Effect of the solid bridges stresses ($\sigma_b^{cr} = \tau_b^{cr}$) on: the critical impact velocity v_B(left)$_r$, the breakage phenomena (right)

To examine the influence of the solid bridges strength on the minimum impact velocity, i.e., the critical velocity which leads to the breakage of the granule, the normal σ_b^{cr} and shear τ_b^{cr} stresses of the solid bridge bonds have been varied in the model in the range from 5kN/m^2 to 52kN/m^2. As expected, the decrease of the solid bridge strength leads to the decreasing the critical velocity (Fig. 11 left) and to an increasing the number of fragments. The simulations at the same impact velocity but different strength clearly demonstrate this influence, Fig. 11 (right).

The microproperties in the DE model concerning the arrangement of the primary particles inside a granule and their coordination number (from 2.0 to 4.4) as well as the impact point on the granule have been varied to obtain the breakage probability depending on the impact velocity. A good correspondence between simulation and experiment results of the breakage probability is observed (Fig. 12).

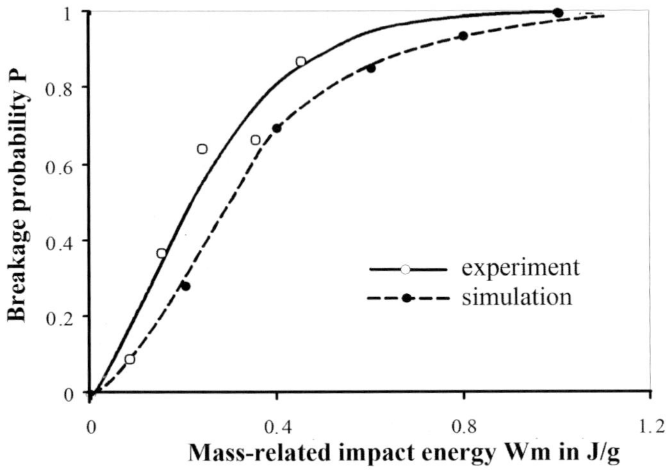

Fig. 12. Comparison of breakage probability functions of zeolite granules from experiment and simulations

5 Conclusions

In the case of compression, γ-Al_2O_3, zeolites 13X and 4A granules show both elastic and elastic-plastic behaviour. The Hertz theory predicts satisfactorily the elastic force-displacement behaviour of these granules. To describe the repulsive force during the elastic-plastic deformation beyond the yield point, a new force-displacement model has been developed. The characteristic parameters are the contact pressure and the deformation at the yield point. These are determined using measured force-displacement curves. The sodium benzoate granules show dominantly plastic behaviour. Their linear force-displacement curves have been described using perfectly plastic contact model.

The breakage probability of granules measured in the impact tests has been described with modified Weibull distribution function. The performed DEM simulations of the impact tests of granules show the similar cone type disturbed region as shown by experiments for brittle granules. The large impact velocities lead to the complete disintegration (break up) of the granule into predefined primary particles. The breakage probability obtained from the simulation shows a good agreement with the experiments.

References

[1] Hintz, W., Antonyuk, S., Schubert, W., Ebenau, B., Haack, A., Tomas, J.: Advanced determination of physical properties of fine particles, nanoparticles and particle beds. In: Tsotsas, E. (ed.) Modern drying technology. Elsevier, Amsterdam (2008)

[2] Antonyuk, S., Khanal, M., Tomas, J., Heinrich, S., Mörl, L.: Impact breakage of spherical granules: Experimental study and DEM simulation. Chem. Eng. and Processing 45, 838–856 (2006)

[3] Hertz, H.: Über die Berührung fester elastischer Körper. J. reine u. angew. Math. 92, 156–171 (1882)
[4] Tomas, J.: Assessment of mechanical properties of cohesive particulate solids – part 1: particle contact constitutive model. Particulate Science & Technology 19(2), 95–110 (2001)
[5] Particle Flow Code in 2 Dimensions, Manual, Vers. 3.0, Itasca Consulting Group Inc., Minneapolis (2003)
[6] Antonyuk, S., Tomas, J., Heinrich, S., Mörl, L.: Breakage behaviour of spherical granulates by compression. Chem. Eng. Sci. 60, 4031–4044 (2005)

Investigations of the Restitution Coefficient of Granules

P. Müller[1], S. Antonyuk[2], J. Tomas[1], and S. Heinrich[2]

[1] Institut für Verfahrenstechnik, Otto-von-Guericke-Universität Magdeburg
[2] Institut für Feststoffverfahrenstechnik und Partikeltechnologie, TU Hamburg-Harburg

Abstract. The energy adsorption during the normal and oblique impacts of different spherical granules (γ-Al_2O_3, zeolite 4A, sodium benzoate) has been experimentally studied. Free-fall tests have been carried out to obtain the normal and tangential restitution coefficients that characterise the energy losses durizng loading and unloading. The effects of the granule impact velocity in the range of 0.5 to 4.5m/s and the impact angle from 0° to 80° on the restitution coefficient have been examined. Free-fall experiments with wet granules have been shown that the moisture content has non-negligible influence on the damping behaviour of plastic granules.

1 Introduction

Many granules exhibit elastic-plastic or dominantly plastic behaviour during their collisions [1]. The absorption (loss) of the kinetic impact energy is the consequence of plastic deformation, adhesion and friction within the granule contacts or granule apparatus wall contacts. The restitution coefficient is a very important material parameter that is needed to describe the energy absorption and damping force in discrete numerical modelling of particles and agglomerates [2].

The restitution coefficient is defined as the square root of the ratio of elastic energy $W_{kin,R}$ released during the restitution to the impact energy, i.e., initial kinetic impact energy W_{kin}

$$e = \sqrt{\frac{W_{kin,R}}{W_{kin}}} = \sqrt{1 - \frac{W_{abs}}{W_{kin}}} = \frac{|v_R|}{v}, \quad (1)$$

where W_{abs} is the energy absorption during the impact, v and v_R are velocities of the impact and the rebound, respectively.

In case of a ideal elastic impact the impact energy absorbed during compression of the impacted bodies is fully restored during the rebound, and so the relative velocity of contact partners before impact is equal to that after the impact, e = 1 in Eq. (1). In the case of full absorption of initial kinetic energy the impact bodies are not separated after the unloading, e = 0 in Eq. (1). For an elastic-plastic impact, the restitution coefficient is in the range of $0 < e < 1$, see examples [3-5].

From the ratio of energies in Eq. (1), it follows that the restitution coefficient is a ratio of relative rebound velocity v_R to that before the impact v. The normal and oblique impacts are described by normal and tangential coefficients of restitution

$$e_n = |v_{R,n}| / v_n \ , \ e_t = |v_{R,t}| / v_t \ . \tag{2}$$

The coefficient of restitution depends on many factors, like impact velocity, material behaviour of impacted bodies, particle size, shape, roughness, moisture content, adhesion properties, and process conditions, like temperature [6]. Many interesting results of investigations of these effects for rigid bodies have been reported in the literature [7-9].

The granules are particle compounds bonded by binding agents like clay, hardened cement paste or cellulose. During the wet granulation process, the granules are containing different of moisture contents. As a consequence of wetting the mechanical properties of dry granules can be changed.

The objectives of our work are to obtain the restitution coefficients of granules with different microstructure, deformation behaviour and moisture content as well as to reveal the most important factors affecting their dynamic response.

2 Testing Method and Material

The most frequently used method to measure the restitution coefficient is the free-fall test, see examples in [6-8]. Fig. 1 schematically shows the experimental setup that was used for the free-fall experiments presented in this work. Before a granule is

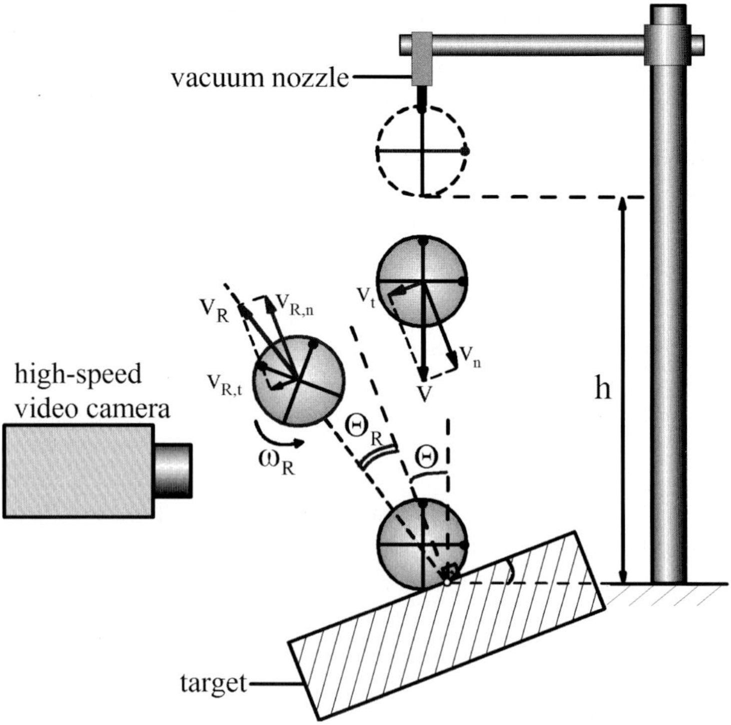

Fig. 1. Free-fall apparatus

dropped, it is held at a predetermined height h above the target (steel flat plane with thickness of 30mm) with the aid of vacuum nozzle that releases the granule with zero initial velocity and rotation. The granule falls freely onto the target. The movement of the granule near the contact point before and after the impact is recorded by a high-speed video camera with a frequency of 4.000 frames per second. These velocities were determined from captured impacts using an own software procedure on the basis of MATLAB.

Three different industrial spherical granules – γ-Al_2O_3, synthetic zeolite 4A, and sodium benzoate ($C_6H_5CO_2Na$) have been used as test materials. Previously, we have tested many other granules to find some model materials and have selected these granules for varying the deformation behaviour from elastic-plastic (γ-Al_2O_3, zeolite) to plastic (sodium benzoate). The properties of these granules are given by Antonyuk et al. [1].

In order to examine the influence of the impact velocity on the restitution coefficient of granules, the drop height has been changed from 12mm to 1.5m, i.e., the impact velocity has been adjusted in the range of 0.5-4.5m/s. To describe the oblique impact the angle of incidence (Θ in Fig. 1) has been varied from 0° to 80°. Before the testing the granules have been marked to obtain the angular velocity ω_R during image processing.

To vary the moisture content of the granules X_W (Eq. (3)) an air conditioned chamber at the temperature of 45° and the relative air humidity of 95% has been used. The moisture content is determined from the mass increase during wetting. To obtain larger moisture contents the granules have been temporary placed in water

$$X_W = \frac{m_{tot} - m_{TS}}{m_{TS}}. \tag{3}$$

With m_{tot} total mass of the wet granule and m_{TS} mass of the dry granule. The pore saturation is calculated by

$$S = \frac{1-\varepsilon}{\varepsilon}\frac{\rho_s}{\rho_l}X_W \tag{4}$$

with ρ_s solid density and ρ_l liquid density. The measured values of the moisture content and pore saturation are shown in Table 1.

Table 1. Maximum generated moisture content of granules using different wetting methods

Materials	Wetting method	Moisture content X_W in kg_{Water}/kg_{TS} wet air	water bath	Pore saturation S wet air	water bath
γ-Al_2O_3		0.48	0.69	0.72	1.06
Zeolite 4A		0.23	0.5	0.38	0.82
Sodium benzoate		0.26	soluble	soluble	soluble

3 Experimental Results and Discussion

a. Restitution coefficient of dry granules

The normal restitution coefficients of the granules have been obtained from the free-fall tests at different impact velocities (Fig. 2). All granules show an elastic-plastic impact in the examined velocity range. Furthermore, the following sequence from elastic-plastic to dominantly plastic behaviour (based on the average coefficient of restitution) can be observed: γ-Al_2O_3 ($e_n = 0.735$), zeolite 4A ($e_n = 0.653$) and sodium benzoate ($e_n = 0.532$).

The standard deviations of the restitution coefficients measured by 50 repeated tests are relatively large. This was expected because of inhomogeneity of granules. For a given particle size, the contact stiffness and the position of the yield point are not constant, since the mechanical characteristics of the primary particles and the bonding agents are randomly distributed within a fraction of granules. The individual granules differ in surface roughness and distribution of pore sizes.

For all examined granules increasing the impact velocity does not change the mean normal coefficient of restitution. In other words, the granules exhibit elastic-plastic behaviour without a viscous effect during the impact in this velocity range. The measured values of the restitution coefficient have been approximated by a model of Walton and Braun [10] in Eq. (5), shown in Fig. 2 as horizontal lines.

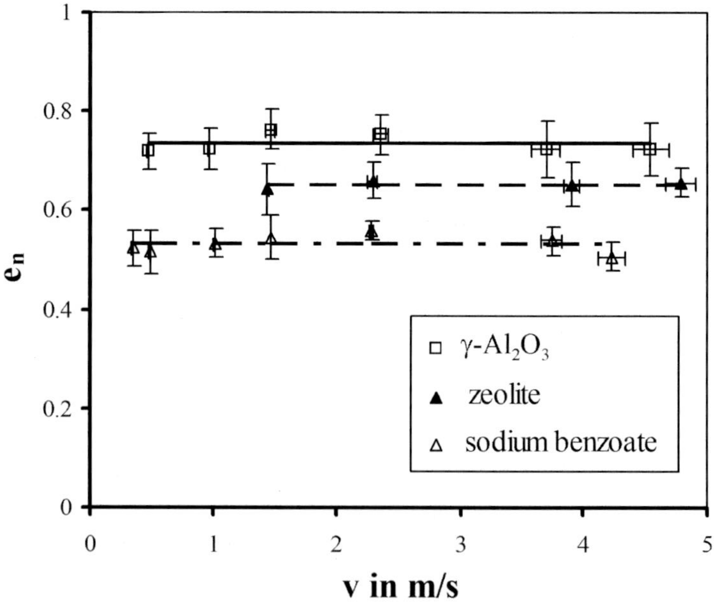

Fig. 2. Normal coefficient of restitution of dry granules e_n versus impact velocity v

$$e = \sqrt{\frac{W_{kin,R}}{W_{kin}}} = \sqrt{\frac{k_L}{k_U}}, \quad (5)$$

where k_L and k_U are constant contact stiffness of load and of unload, respectively.

The coefficient of restitution in the normal direction does not depend on the impact angle (Fig. 3 (a)). However, the coefficient of tangential restitution decreases with target inclination and it reaches a minimum at an impact angle close to 30°. By the impact below this angle the granule rolls. At angles larger than about 55° the coefficient of tangential restitution exceeds the normal coefficient of restitution and reaches a value of 1 at angles above 80° by glancing incidence.

Fig. 3 (b) summarizes the measurements of rebound angles and angular velocities of granules versus the impact angle. When the impact takes place at an angle below 55°, the rebound angle of γ-Al_2O_3 granules is slightly lower than the impact angle. The dotted line shows an ideal case of rolling at $\Theta = \Theta_R$ and the coefficient of friction: $\mu \approx 0$. The angular velocity after rebound increases with increasing the impact angle, reaches a maximum at an angle of about 50° that shows the transition from rolling to sliding.

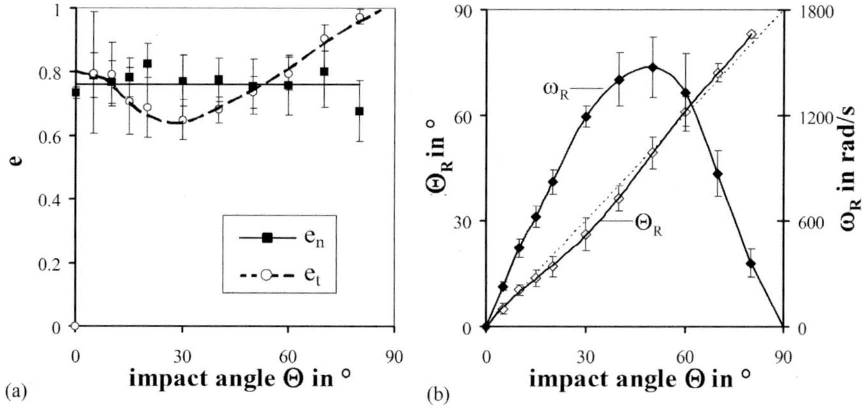

Fig. 3. Effect of the impact angle Θ on: (a) the coefficient of restitution in normal e_n and tangential e_t direction, (b) the rebound angle Θ_R and the angular velocity ω_R for γ-Al_2O_3 granules at an impact velocity of 2.3 m/s.

b. Restitution coefficient of wet granules

The granules have been wetted up to different moisture contents and then the free-fall tests by normal impact were performed. The normal restitution coefficients of wet granules impacted at two different velocities are shown in Fig. 4 as a function of the moisture content. For sodium benzoate granules, the restitution coefficient decreases with increasing moisture loading. These granules deforms more plastically during the impact with increasing moisture because of the granule solubility and softening. The restitution coefficient of zeolite 4A slowly decreases with increasing moisture. During

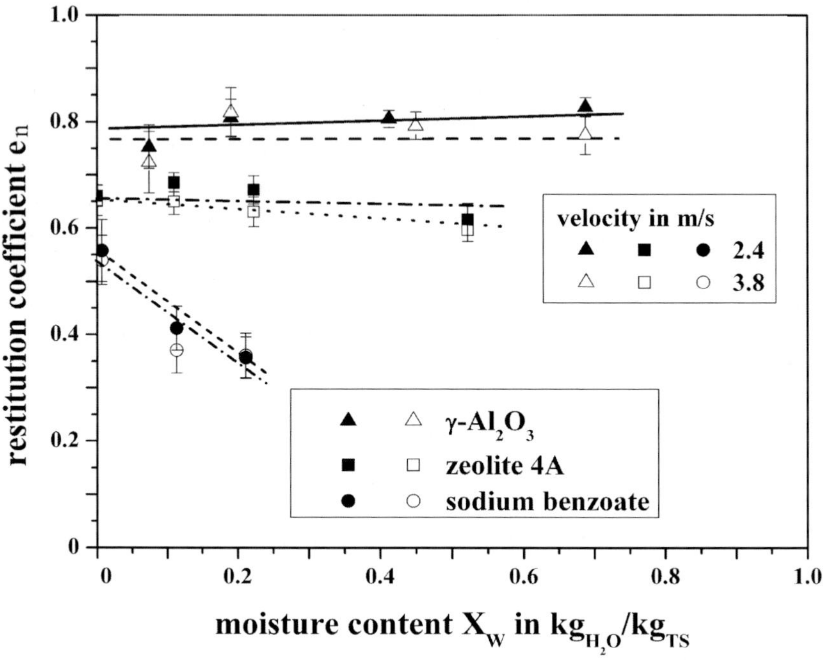

Fig. 4. Normal restitution coefficient of granules versus moisture and impact velocity

the wetting the liquid removes the solid bridge bonds of the granule. The moisture content has not shown a significant effect on the restitution coefficient of γ-Al_2O_3 granules consisting of stiff ceramic microstructure.

The restitution coefficient of wet granules decreases with increasing impact velocity. Higher impact velocities determine higher impact forces and the contribution of the energy absorption due to plastic deformation increases. This effect has been observed in the examined velocity range only for wet granules.

4 Conclusions

Using the free-fall tests and high-speed video recording of impacts, the restitution coefficients of granules (γ-Al_2O_3, zeolite 4A, and sodium benzoate) have been measured. The restitution coefficient of examined dry granules is independent of the impact velocity in the range of 0.5-4.5 m/s. Hence, if the granules are wet then their restitution coefficient increases with increasing impact velocity.

From the study of oblique impact of γ-Al_2O_3 granules we have found that the coefficient of normal restitution is independent of the impact angle. Hence, the coefficient of tangential restitution shows a minimum value at angles of 30°. The rebound angle of γ-Al_2O_3 granules is slightly lowers (at the impact angles in the range of 10°-55°) or equal to the impact angle. The angular velocity after rebound shows a maximum at an angle of about 50°.

Due to softening the restitution coefficients of sodium benzoate and zeolite 4A granules increase with increasing moisture content. Hence, the deformation behaviour of brittle γ-Al_2O_3 granules is independent of the moisture content.

Therefore, for DEM simulation of a fluidized bed with injection of a binder agent, where the particles are soft and wet, lower values of the restitution coefficient have to be considered than for dry particles.

References

[1] Antonyuk, S., Khanal, M., Tomas, J., Heinrich, S., Mörl, L.: Impact breakage of spherical granules: Experimental study and DEM simulation. Chem. Eng. and Processing 45, 838–856 (2006)

[2] Kruggel-Emden, H., Wirtz, S., Simsek, E., Scherer, E.: Modeling of granular flow and combined heat transfer in hoppers by the Discrete Element Method (DEM). Journal of Pressure Vessel Technology 128, 3 (2006)

[3] Walton, O.R., Braun, R.L.: Viscosity, granular-temperature, and stress calculations for shearing assemblies of inelastic, frictional disks. J. Rheol. 30, 949–980 (1986)

[4] Fu, J.S., Cheong, Y.S., Reynolds, G.K., Adams, M.J., Salman, A.D., Hounslow, M.J.: Impact deformation and rebound of wet granules, an experimental study of the variability in the properties and quality of wet granules. Powder Technol. 140, 209–216 (2004)

[5] Antonyuk, S.: Deformations- und Bruchverhalten von kugelförmigen Granulaten bei Druck- und Stoßbeanspruchung. Docupoint Publishers, Magdeburg (2006)

[6] Hintz, W., Antonyuk, S., Schubert, W., Ebenau, B., Haack, A., Tomas, J.: Advanced determination of physical properties of fine particles, nanoparticles and particle beds. In: Tsotsas, E. (ed.) Modern drying technology. Elsevier, Amsterdam (2008)

[7] Goldsmith, W.: Impact: The theory and physical behavior of colliding solids. Edward Arnold Ltd., London (1960)

[8] Louge, M.Y., Adams, M.E.: Anomalous behavior of normal kinematic restitution in the oblique impacts of a hard sphere on an elastoplastic plate. Physical review E 65 (2002)

[9] Labous, L., Rosato, A.D., Dave, R.N.: Measurements of collisional properties of spheres using high-speed video analysis. Physical review E 56(5B), 5717–5725 (1977)

[10] Walton, O.R., Braun, R.L.: Viscosity, granular-temperature, and stress calculations for shearing assemblies of inelastic, frictional disks. J. Rheol. 30, 949–980 (1986)

Oblique Impact Simulations of High Strength Agglomerates

M. Khanal[1], W. Schubert[2], and J. Tomas[3]

[1] JKMRC, University of Queensland
[2] Titania AS, Hauge i Dalane
[3] Institut für Verfahrenstechnik, Otto-von-Guericke-Universität Magdeburg

Abstract. Different type of particle compounds like concrete particles can be considered as a model material of high strength agglomerates. It is necessary to investigate and understand the fracture behaviour of these agglomerates in order to avoid breakage during storage, handling and transportation. The aim of the research is to examine the comminution behaviour of high strength agglomerates during oblique impact loadings. A two dimensional finite element analysis has been carried out to understand stress pattern distributions before crack initiation. Then a two dimensional discrete element method has been performed to study the fragmentation behaviour of the agglomerates. Concrete particles of B35 strength category have been chosen to represent the high strength agglomerates. The analysis is done with oblique impact loadings at different velocities from 7.7 m/s to 180 m/s. The stressing conditions comprise low flow rate transportation and handling to high speed impacts during fall down in bunker, stock piles, ship loading or stressing in crushers and mill operations. Particle size distributions and new surface generation have also been evaluated in the paper. It is shown that at higher velocities, particle size distributions are identical to each other regardless of the impact angle. Increasing impact velocity does not necessarily produce more new surfaces after a certain velocity limit.

1 Introduction

In many cases comminution is a desired activity. For rocks and coarse particles it is necessary to have suitable fragment size distributions. In agglomerates, it is desired to have stable particle dispersions. In recycling / mineral processing it is adopted to liberate valuable aggregates from cheaper matrixes. The material has to be fractured in all these cases. Heterogeneous and discontinuous materials show a complicated failure during fragmentation. There are different loading methods to comminute particles, like impact, double impact, compression etc. In this research, the loading conditions comprise low flow rate transportation and handling to high speed impacts during fall down in bunker, stock piles, ship loading or stressing in crushers and mill operations.

Different types of samples, like mixture of sand cement particles [1-4], PMMA particles [5,6] and other agglomerates [7-13] are used to study the comminution behaviour of particles under different loading conditions. Schubert [14] has shown that the impact or double impact loading has higher fracture probability than the compressive loading. During impact experiments of particles, a cone shaped fragment can be observed at the impact site [1-3,5]. As the velocity of impact increases, the size of the cone shaped fragment decreases, and at larger impact velocities the cone shaped fragment will be commuted to become fines [3]. Damage patterns and mechanism vary with impact velocity [5] and behaviour of particles [3]. The tensile stress generated under impact is found to be responsible for producing diametrical cracks [4], i.e., cracks propagating from the impact side to the other side of the specimens.

Due to the nature of the two dimensional simulation, one would expect a wedge shaped fragment in place of the cone shaped fragment observed in 3D simulations and experiments. However, according to Schönert [17], who investigated the breakage of spheres and discs, the particle fracture mechanism is the same in both samples. This demonstrates the relevancy of experimental and three dimensional analyses in relation to two dimensional simulations.

Both 2 dimensional [4,8,12,15,16] and 3 dimensional [9,13] discrete element method (DEM) simulations have been performed to investigate the failure behaviour of particles. The two dimensional simulations are relatively faster than the three dimensional simulations for a large number of particles. An increase in impact velocity results in a maximum wall force, shorter impact event (shorter time) and more broken bonds. Cracks initiate from the zone of contact and propagate towards the centre of the agglomerates. A localized damage zone has been observed adjacent to the contact [15,16]. A force-time curve can be used to describe failure mechanism of the agglomerates. Kienzler and Baudendistel [7] have performed continuum simulations of a sphere for the elasto-viscoplastic conditions and used stress patterns to relate the particle failure. Some authors [2,17] have used the photo-elasticity method to observe the stress pattern distributions during particle fracture. The impact simulations for spherical materials are in their majority carried out on engineering agglomerates. Thornton et al. [8] have reported that agglomerates either rebound, fracture or shatter depending upon the impact velocities. The velocities used to impact the agglomerates [8,9] are much less than the velocities required for a comminution of high strength agglomerates.

In this research, concrete particles are considered as an agglomerate with porosities of different sized primary particles having random properties. Though the material seems to be complex, it can be modelled with the principle of discrete element method [18]. The discrete element solution scheme assumes each of its constituents to be a separate entity, and applies force and momentum balances and constitutive models (force – displacement relations) at contact. The force and momentum balances are used to determine the motion of each particle arising from the contact and body forces acting upon it. The constitutive models are used to update the contact forces arising from the relative motion at each contact. The constituent particles are bonded by solid bridge bonds. The individual particle allows to delete the solid bridge bonds when it experiences the stress equivalent to the critical material strength. This process shows the fragmentation behaviour of the impact system.

2 Finite and Discrete Element Modelling

This paper uses 2D finite and discrete element models described in [4] to simulate the oblique impact loading of particles. The finite and discrete element simulations have been performed with ANSYS [18] and PFC [19], respectively.

Finite element analysis is performed to understand the stress distributions before fracture initiation. In the finite element simulation, a particle (model) of radius

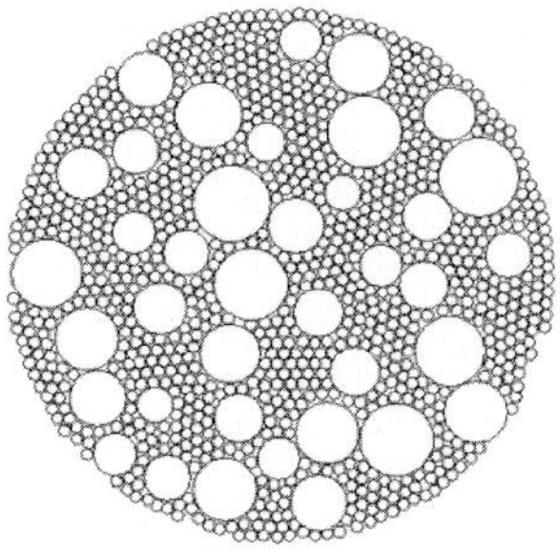

Fig. 1. Tentative arrangement of aggregates and hardened cement paste

150 mm is impacted on the target at a velocity of 20m/s. The specimen is constructed by 825 plane 182 type elements. The surface to surface contact is assigned between the specimen and the target plate. Elastic modulus, density and Poission's ratio are 15GPa, 2382kg/m^3 and 0.28, respectively. The model material is assumed to have the stress-strain relationship as characterized by linear elastic behaviour and perfect plastic deformation without hardening.

The DEM model (disc) consists of gravel as large particles and sand as small particles, and bonded with hardened cement paste. 962 small particles having 1 mm radius are mixed with 38 large particles, radii of 4 to 6 mm with a Gaussain distribution. The larger particles have a normal and shear stiffness of 1x10^{10}N/m and a density of 2870kg/m^3, whereas the smaller particles have the stiffness of 1x10^{7}N/m with density 1790kg/m^3. The normal strength and shear strength of the model are 4.1MPa. The parallel bond normal stiffness and shear stiffness are 1x10^{11}N/m^3 and 6.3x10^{10}N/m^3, respectively. The target wall has a stiffness of 1x10^{20}N/m. Gravity is also allowed to act on particles during the simulation. The default time step calculation used in PFC [19] is adopted. The linear spring mass contact model is chosen. The tentative arrangement of the hardened cement paste and aggregates for the DEM modelled specimen is shown in Fig. 1. The larger particles represent gravels and the smaller represent sand. The required number and size of particles are randomly generated and assigned bonding properties.

Under oblique impacts the target plate angle should have some influence on fragmentation (forming the wedge and on initiation of cracks). In this work, the comminution behaviour of the particles has been studied at 30, 45 and 60 degrees oblique impacts. The central (normal) impact can be considered as a part of the oblique impacts having 0 degree inclined target. The friction coefficient 0.3 is assigned between the specimen and the target wall because one of the aims of the research is to comminute the particle by generating asymmetrical stress fields.

3 Result and Discussions

3.1 Finite Element Simulation

Maximum principal stress distributions during 30 and 60 degree oblique impacts are shown in Fig. 2 and Fig. 3, respectively (negative sign shows compression and positive sign shows tension).

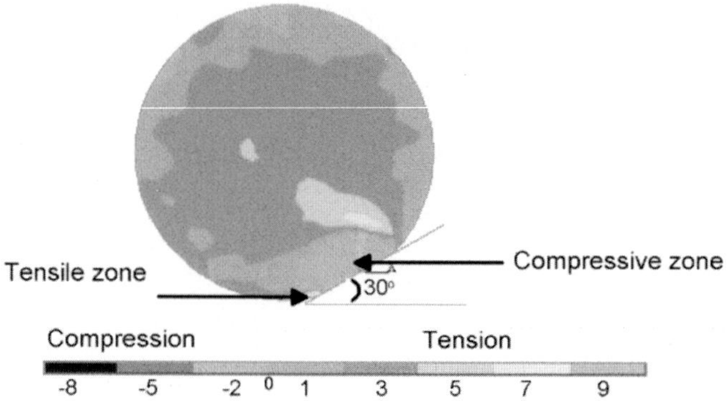

Fig. 2. Maximum principal stress distribution in MPa at 30 degree impact, v = 20m/s, t = 0.2ms

The direction of impact is vertically downward. The compressive stresses are generated at the contact region, and the surface nearer to the contact area generates tensile stresses. Due to the loading angle, a wedge like compressed zone at the contact also has some inclination. The maximum tensile stress observed at the periphery of the contact region can be termed as a lateral tensile stress in 2 dimensional simulations. If the generated tensile stress at the perimeter of the compressive zone is equal to the strength of the material then crack initiates at this point. The tensile waves are propagated on the periphery of the compressed region. This may suggest that the crack should follow this path. But in reality, after crack initiation existing stress fields may be completely modified. The direction of crack propagation may be changed [17].

The compressive zones observed in Fig. 2 and Fig. 3 are similar to the wedge which are obtained during central impact loadings [4], but with some angle. On the forward side of the contact deformation, a dynamic tensile zone can be observed. The generation of the tensile region can be explained in a practical case. When a particle is

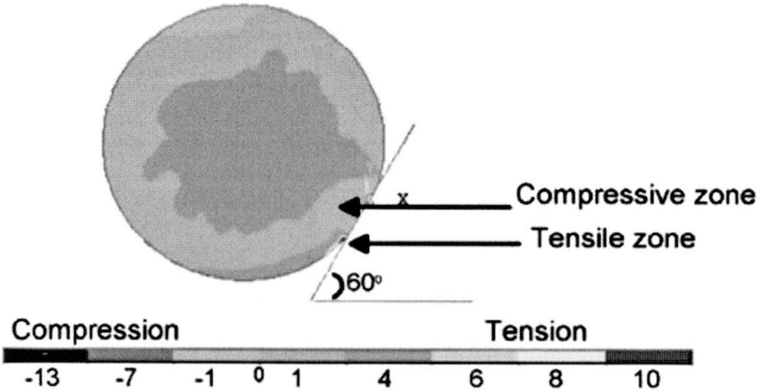

Fig. 3. Maximum principal stress distribution in MPa at 60 degree impact, v = 20m/s, t = 0.2ms

pressed on an inclined frictional surface, the leading side of the particle has a tensile zone on the boundary of the contact, whereas the trailing side has a compressive zone because of the friction. If the coefficient of friction is increased then this tensile zone on the leading side and the compressive zone on the trailing side will also be increased.

3.2 Discrete Element Simulation

3.2.1 Force Time Behaviour

Different fracture events during central impact loading at 20m/s are shown in Fig. 4. The first increase in the force curve shows that elastic strain energy is stored in a specimen due to the stressing input energy. The first crack (primary failure) is observed when there is a drop in the force curve. This drop shows that the energy is utilized to generate new surfaces, in other words, to initiate and propagate cracks. The further drop in the force curve shows that the energy is utilized to generate and propagate multiple cracks (simultaneously forming new surfaces). It is interesting to observe an increase in peak force even after the primary failure. The similar phenomenon has been observed by Unland and Szczelina [21] under compression crushing of brittle rocks. Here, the primary failure refers to the breaking of few primary particles from the parent particle (see the first agglomerate in the figure with the fragment at the contact zone). This is because of the force generated by inertia of aggregates present inside finer matrixes (see the location of aggregates and the target wall) and the newly generated fragments. After this stage the drop in the force curve infers propagation of cracks for generating new surfaces and fragments.

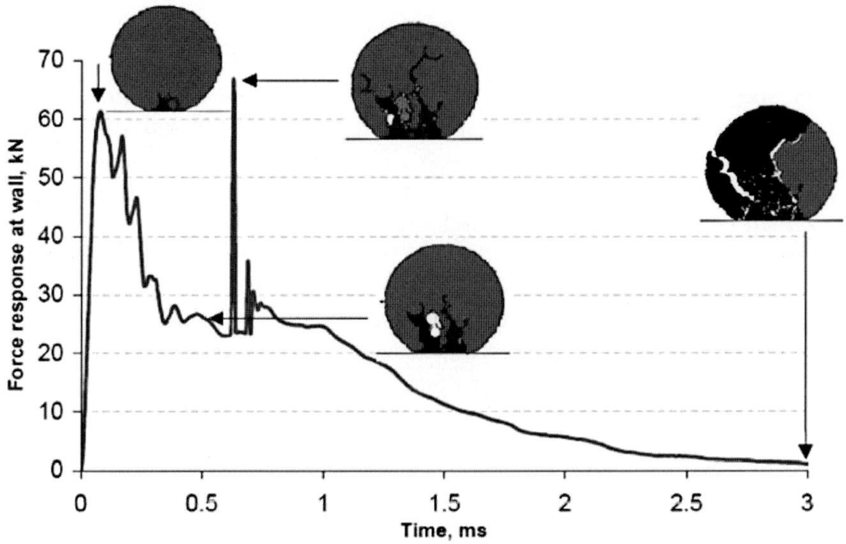

Fig. 4. Different fracture events at 20 m/s (central impact, coefficient of friction 0.3)

3.2.2 Fracture Pattern

Fig. 5 and Fig. 6 show the fractured specimen obtained from discrete element simulations at 30 and 60 degrees oblique impacts, respectively. The different separated parts in the figure are the generated fragments under impacts. These figures illustrate that a maximum damage occurs at the contact zone, and has a wedge like shape. This wedge like shape is clearly visible in the early stage of the impact simulation. The shown figures are taken at 5 ms, hence, the wedge like fragment further disintegrated into aggregates and matrixes. The wedge like shape can be related with the compressed region as observed in the finite element simulation. The respective figures also show the bond status of the fractured specimen.

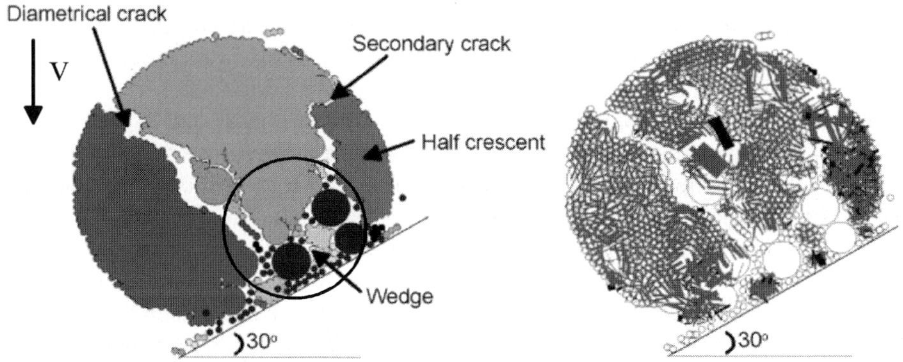

Fig. 5. The fractured specimen (left) and bond status (right, grey – tension, black – compression) at 30 degree oblique impact. v=20m/s, t=5ms. The fragmented wedge is shown in circle.

The large primary particles inside the wedge represent aggregates liberated out from the specimen during impact. The liberated aggregates, sand particles and fragments of the wedge try to escape in outward directions. As a result, the circumferential fragments are also pushed outwards. The similar observation is reported by Bergstrom and Sollenberger [20]. The circumferential fragments can be seen as half convex (half crescent) fragments, which are separated out on both sides of the fractured specimen. In oblique impact stressing, because of the orientation of the angle of the wedge, the size of the half crescent is also altered.

In central impacts these half crescent fragments are similar in nature [4], whereas in oblique impacts, they are not symmetrical. The orientation of the angle of the wedge in oblique impacts causes the half crescent fragment to be smaller in right hand side than that of the left hand side, and the disproportion increases with an increasing angle of the target. The dimension of the wedge depends upon the impact velocity of the specimen. The faster the impact speed the larger the impact zone, hence, the larger wedges. The variation in the impact angle causes the angle of the wedge to alter. From the figures, it is seen that maximum damage occurred at the rear part of the specimen, as observed by Samimi et al. [10].

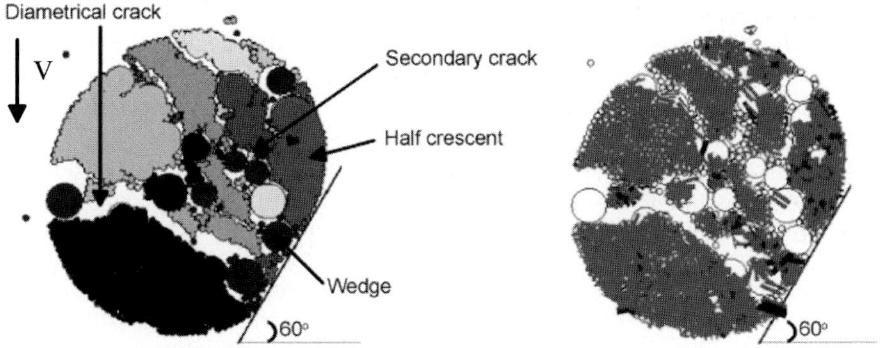

Fig. 6. The fractured specimen (left) and bond status (right, grey – tension, black – compression) at 60 degree oblique impact. v=20m/s, t=5ms

3.2.3 Particle Size Distribution

Particle size distribution provides information about the extent to which the particles of particular size intervals are generated after fracture. The particle size distribution is analysed by detecting the contacts between particles. The FISH program [18] has been written to evaluate the particle size distributions. Each particle is checked for contacts, and the particles in same fragments are stored in arrays. For each fragment, the searching procedure starts with the last scanned particle, as it is the first element in the linked list (inbuilt PFC [18]). After finding the particles those are in contact with the last particle, the search is continued in the similar way for all the particles those are in contact. Fig. 7 shows a flow chart to evaluate the fragment distributions in simulations.

The scanning method provides the particle size distributions on a number basis, which can be converted to a mass basis with a known density. The number basis is

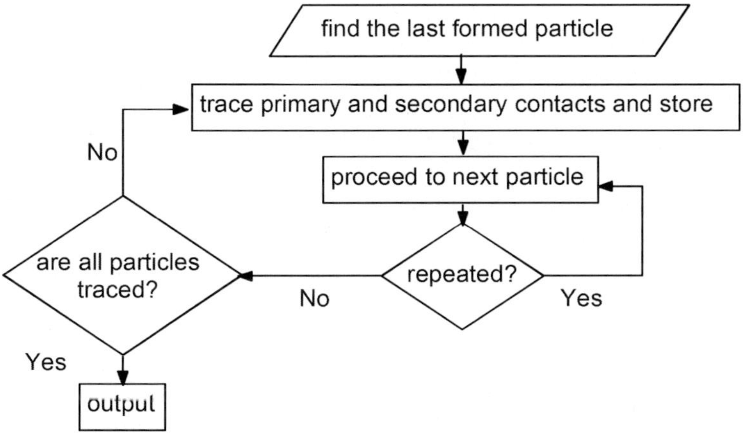

Fig. 7. Flow chart to evaluate particle size distributions

defined as the number of fragments obtained below a particular size. The third dimension is assumed to be a unit thickness, which is a reasonable assumption because we are interested to compare the mass distribution trend after fragmentation.

For every impact velocity, the particle size distributions are evaluated and compared at a predefined size interval. The particle size distributions are also compared with the central impact simulations to study the effect of impact angles on fragments. Fig. 8 shows the comparison of the particle size distributions at 20m/s and 50m/s. At higher velocities, the particle size distributions are identical to each other. This suggests that at higher impact velocities, the difference in the effect of the kinetic energy on fragmentation is minimal. Also at higher impact velocities, the effect of the impact angle in fragment generation is very minimal. It is seen from the figure that central impact produces finer fragments. This is noticeable at the impact velocity of 20m/s. At 20m/s, 45 degree oblique impact produces coarser fragments. It can be observed from the figure that the impact angle influences the fragment size distributions, which is also reported by Moreno et al. [15].

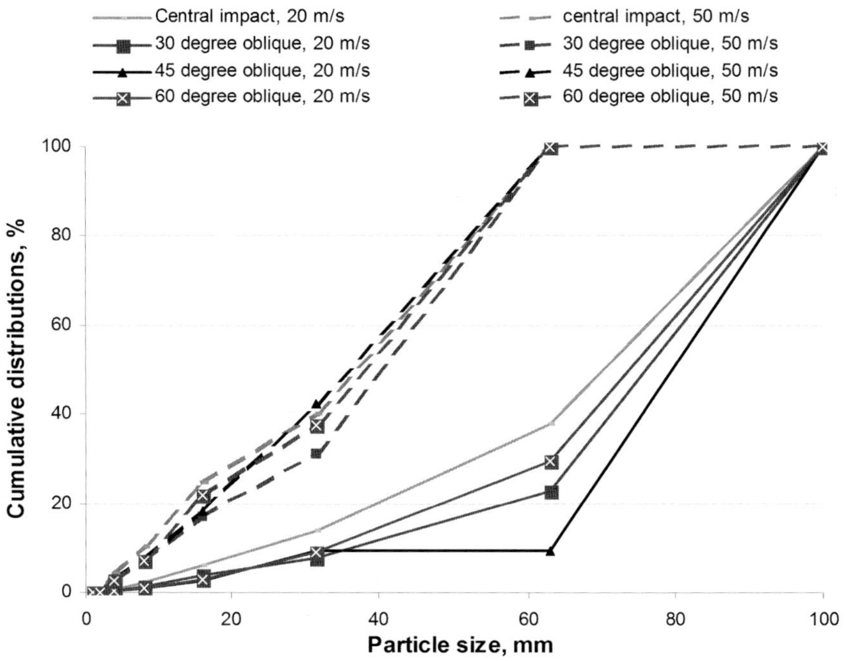

Fig. 8. Comparison of particle size distributions at 20m/s and 50m/s

3.2.4 New Surface Generation

When particles are fractured, they generate new surfaces. Primary particles should delete contacts among themselves to generate the new surfaces. Hence, the new surface generation is proportional to the broken bonds. The new surface generation has been evaluated the discrete element simulations and calculated by assuming a unit thickness in the depth direction, i.e., the length of new surfaces has been assumed as the area of new surfaces. A FISH program [18] has been written and coupled with the discrete element software to evaluate the new surface generation.

First, the maximum possible open surface of any particle (without bonding) is calculated. Then, the blocked surface which is in contact with other adjoining particles is calculated. The difference between the maximum possible open surface and the blocked surface at a particular instant gives the newly formed surface. 1 and 2 imply particle 1 and particle 2 respectively in Fig. 9. For example, the maximum possible new surface which can be generated from an aggregate having n number of particles (having unit thickness) is $\sum_{i=1}^{n} 2\pi r_i$.

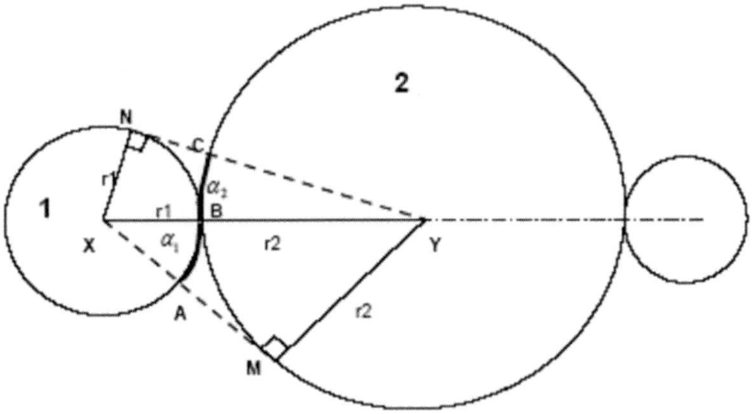

Fig. 9. Blocked surface for analysing the new surface generation

In the figure, particle 1 has $2\pi r_1$ (for simplification, assume a unit thickness) maximum possible open surface. The distance *CA* is a blocked surface at this moment. Hence, the new surface generation at this particular moment is the difference between $2\pi r_1$ and the blocked surface. The blocked surface is calculated as follows. Consider the triangle XNY, the surface S_1 blocked by disk 2 is,

$$\sin\alpha_1 = \frac{r_2}{r_1 + r_2} \tag{1}$$

$$S_1 = 2AB = 2r_1 \arcsin\frac{r_2}{r_1 + r_2} \tag{2}$$

Likewise, the surface S_2 blocked by disk 1 is calculated by considering triangle XMY. Similar trigonometrical calculations are performed for each pair of contacts, and the blocked surfaces are evaluated to find the new surface generation. For

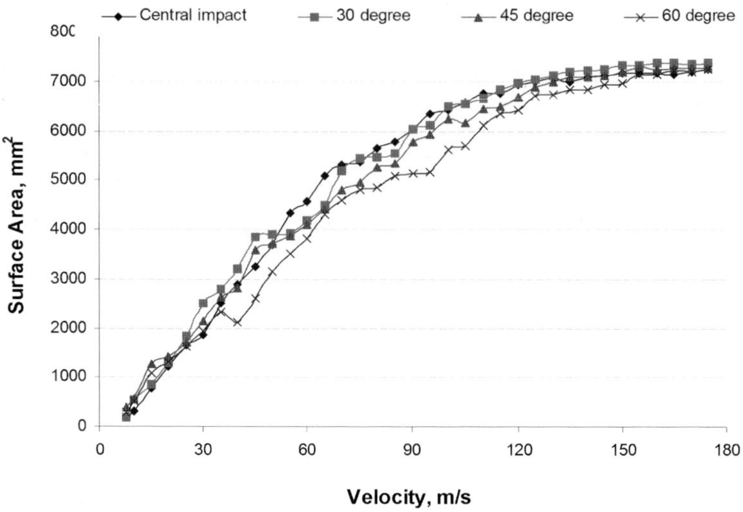

Fig. 10. New surface generation versus velocity at different impacts

example, in the shown figure, if particle 1 and particle 2 have equal radius, then the new surface created will be $2AC$.

Fig. 10 shows the new surface generation during different angular impacts at varying velocities. With an increasing velocity the new surface generation also increases up to an impact velocity of 150m/s for the considered model. After this limit even with increasing velocity there is no increment in the new surface generation. This suggests that after a certain limit of the velocity, the damage reaches the saturation level i.e. the particle cannot be comminuted further.

4 Conclusions

From the above observations, it has been found that oblique impacts also follow the same crack initiation and propagation mechanisms as those of central impacts. In oblique impacts either stress in the impinging direction or in its normal direction (in 2D) generates the maximum tensile stress. In oblique impacts the presence of the friction between the specimen and the target causes the tearing (sliding) effect. The tension at the forward side also adds-up to increase the tensile stress and helps to initiate the crack in oblique impacts. Therefore, it can be inferred that in oblique impacts, first crack initiates from the leading side. At higher velocities, the particle size distributions are identical to each other regardless of the impact angle. The increasing impact velocity does not necessarily produce more new surfaces after certain velocity limit.

The two dimensional simulations provide a convenient means of studying the cracking mechanisms (and crack propagation) with less computational time. However, it is recognized that three dimensional simulations would provide better comparison with experimental data.

References

[1] Kiss, L., Schnört, K.: Aufschlusszerkleinerung eines zweikomponentigen Modellstoffes unter Einzelkornbeanspruchung durch Druck- und Prallbeanspruchung. Aufbereit.-Tech. 30(5), 223–230 (1980)

[2] Arbiter, N., Harris, C.C., Stamboltzis, G.A.: Single fracture of brittle spheres. Soc. Min. Eng. AIME, Trans. 244, 118–133 (1969)

[3] Tomas, J., Schreiber, M., Gröger, T., Ehlers, S.: Impact crushing of concrete for liberation and recycling. Powder Technology 105, 39–51 (1999)

[4] Khanal, M., Schubert, W., Tomas, J.: Ball impact and crack propagation – simulations of particle compound material. Granular Matter 5(4), 177–184 (2004)

[5] Gorham, D.A., Salman, A.D.: The failure of spherical particles under impact. Wear 258, 580–587 (2005)

[6] Salman, A.D., Gorham, D.A., Verba, A.: A study of solid particle failure under normal and oblique impact. Wear 186-187, 92–98 (1995)

[7] Kienzler, R., Baudendistel, E.: Numerische Beanspruchungsanalyse einer definiert belasteten Kugel 2/4. Final research report DFG Schm 522(3-2), 522(3-4) (1985)

[8] Thornton, C., Yin, K.K., Adams, M.J.: Numerical simulations of impact fracture and fragmentation of agglomerates. Journal of Physics D: Applied Physics 29, 424–435 (1996)

[9] Mishra, B.K., Thornton, C.: Impact breakage of a particle agglomerates. International Journal of Mineral Processing 61, 225–239 (2001)

[10] Samimi, S., Moreno, R., Ghadiri, M.: Analysis of impact damage of agglomerates: effect of impact angle. Powder Technology 143-144, 97–109 (2004)

[11] Shipway, P.H., Hutchings, I.M.: Attrition of brittle spheres by fracture under compression and impact loading. Powder Technology 76, 23–30 (1993)

[12] Potapov, A.V., Campbell, C.S.: The two mechanisms of particle impact breakage and the velocity effect. Powder Technology 93, 13–21 (1997)

[13] Moreno, R., Ghadiri, M., Antony, S.J.: Effect of impact angle on the breakage of agglomerates: a numerical study using DEM. Powder Technology 130, 132–137 (2003)

[14] Schubert, H.: Zur Energieausnutzung bei Zerkleinerungsprozessen. Aufbereit.–Tech. 34(10), 495–505 (1993)

[15] Thornton, C., Ciomocos, M.T., Yin, K.K., Adams, M.J.: Fracture of particulate solids. Powders and Grains, Rotterdam (1997)

[16] Thornton, C., Ciomocos, M.T., Adams, M.J.: Numerical simulations of diametrical compression tests on agglomerates. European Congress on Comminution (2002)

[17] Schönert, K.: Breakage of spheres and circular discs. Powder Technology 143-144, 2–18 (2004)

[18] Particle Flow Code in 2 Dimensions Manual Vers. 3.0. Itasca Consulting Group Inc., Minneapolis, Minnesota, US (2002)

[19] FEM Manual ANSYS 6.1, ANSYS Inc., Southpointe, 275 Technology Drive, Canonsburg, PA 15317

[20] Bergstrom, B.H., Sollenberger, C.L.: Kinetic energy effect in single particle crushing. Trans. SME, AMIE 220, 373–379 (1961)

[21] Unland, G., Szczelina, P.: Coarse crushing of brittle rocks by compression. International Journal of Mineral Processing 74, 209–217 (2004)

Shear Dynamics of Ultrafine Cohesive Powders

R. Tykhoniuk[1] and J. Tomas[2]

[1] Institut für Computerphysik, Universität Stuttgart
[2] Institut für Verfahrenstechnik, Otto-von-Guericke-Universität Magdeburg

Abstract. Discrete element method simulations are presented with respect to modelling the standardised shear testers like Jenike shear cell. The main impact is made on modelling the flow behaviour of commercially widely used ultrafine cohesive powders (TiO_2, $CaCO_3$) and the influence of hysteretic contact models describing the microscopic particle-particle interaction behaviour with the load-dependent contact adhesion. Therefore, the macroscopic dynamic behaviour of cohesive powder flow can be microscopically investigated and understood. Quantitative agreement between the 3D-simulations and experiments is reached.

1 Introduction

The well-known flow problems of cohesive particulate solids in storage and transportation containers, conveyors or process apparatuses, include bridging, channelling, segregation, flooding, avalanching etc. In addition, the insufficient apparatus and system reliability of solid processing plants are also related to flow problems. Taking into account this list of selected technical problems and hazards, this motivates us to deal with the fundamentals of cohesive powder consolidation and flow behaviour, i.e. to develop a reasonable combination of particle and continuum mechanics. The goal of the present work is to build a numerical bridge between the microscopic particle properties based on the atomic force microscope (AFM) measurements, theoretical models of elastic-plastic contact behaviour and the macroscopic powder flow behaviour of cohesive powders.

Continuum mechanical models and appropriate measuring methods were successfully applied to describe the flow behaviour of cohesive powders, as well as for practical design of process apparatus, e.g., silos. However, the essential constitutive functions of the powder "continuum" can be better described and understood with the help of particle mechanics [1]. The Discrete Element Method (DEM) [2] is an alternative solution, which allows us to take into account the contact and, what is especially important in our case, adhesion forces, and introduce them into the equations of motion of the particles. By this method, the dynamics of cohesive powders can be studied and understood microscopically.

Combining the theoretical background and macroscopic shear tests of cohesive powders as well as the microscopic AFM measurements of the particle interaction forces [3], it becomes possible to develop an appropriate contact constitutive model to describe the deformation behaviour of ultrafine, cohesive frictional particles. In this context, to simulate efficiently the shear dynamics of cohesive powders, the implementation of an irreversible inelastic contact flattening, which is an essential element and physical reason of the load-dependent increase of the adhesion force, is of vital importance. There exists already a realistic and flexible microscopic model for contact laws with elastic, plastic, and adhesion forces, as based on macroscopic observations from bulk experiments [1,4]. The model in a simplified form is applied to the Jenike shear test [5], in order to find out the relationships between the mechanical parameters of a single particle on a microscopic level and flow parameters of the powder continuum on the macroscopic level.

2 Contact Forces between Single Particles and Contact Constitutive Models

In terms of particle technology, powder processing and handling, the consolidation and non-rapid flow of dry, ultrafine and cohesive powders (particle diameter $d < 10\mu m$) can be explained by load-history dependent adhesion forces at particle contacts. Here we intend to focus on a characteristic, soft contact of two isotropic, stiff, linear elastic, smooth spherical particles. Thus, this soft or compliant contact displacement is assumed to be small $h_K/d << 1$ compared to the diameter of the stiff particle. The contact area consists of a representative number of molecules. Hence, continuum approaches are only used here to describe the force-displacement behaviour in terms of nanomechanics.

We consider the contact of two isotropic, smooth spherical particles as the typical components of a particle packing under static load f_N. The realistic and rather complicated contact model presented by Tomas [1] is very machine time spending to be implemented in the simulation software at present. That is why we use the simplified contact law developed by Luding [6] (Fig. 1) for the simulations.

At the point-contact moment, i.e., when the contact deformation or particle-particle overlap is equal to 0, the interaction between the particles starts from a predefined

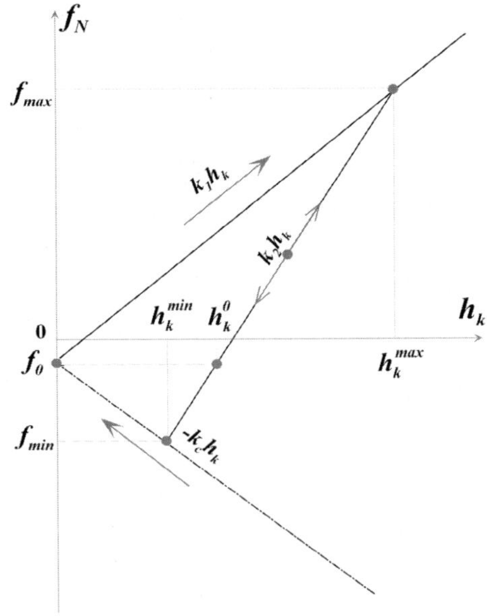

Fig. 1. Linearised normal force-displacement law for the DEM simulations [6]

adhesion level ($-f_0$, here attractive forces negative). The normal force then goes along the linear plastic yield limit with stiffness k_1, that takes care of a perfect plastic repulsion. The elastic deformation at the contact level is added by a linear spring, with a larger stiffness k_2 for unloading and reloading, so that the stiffness increases. This plastic loading and elastic unloading take care of rate-independent energy dissipation during contact; a linear dashpot is also considered that accounts for velocity proportional dissipation.

The variable adhesion force (or cohesion on a microlevel) between the particles comes into the model by a cohesive stiffness k_c, which allows for changing the attractive forces up to a maximal attractive force f_{min} (per absolute value). One should notice also the force equilibrium state at a non-zero contact deformation of h_k^0. Cast into an equation, the normal force on particle is

$$f_N = \begin{cases} k_1 h_k - f_0 & \text{for loading} \\ k_2\left(h_k - h_k^0\right) - f_0 & \text{for un-/reloading} \\ -k_c h_k - f_0 & \text{for unloading} \end{cases} \quad (1)$$

The tangential force involves dissipation due to Coulomb friction, but also some tangential elasticity that allows for stick-slip behaviour on the contact level.

3 Reference Shear Experiments

In order to quantitatively compare simulations with experiments (the stresses as well as the volumetric strain values), laser position sensors are mounted on a translational shear cell in order to measure the volumetric strains (Fig. 2). The three triangulating

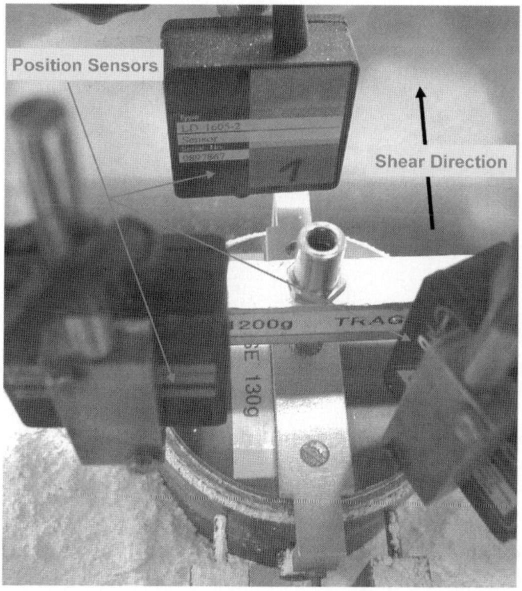

Fig. 2. Translational shear cell with shear lid height measurement by laser sensors

laser displacement sensors allow to measure the dynamic change (movement and tilt) of the shear lid during the shear test, and thus the volumetric strain versus shear displacement. A typical experiment consists of a pre-consolidation step (without tangential shear, but with twist or vibration to enhance the consolidation) at normal stress $\sigma_{N,0}$, a pre-shear step with normal stress σ_{pre}, and the shear steps with lower normal stresses, from which the yield stresses are eventually obtained. The consolidation is needed to properly consolidate the powder, so that the following shear tests can be performed within the very short strain path (6mm) available in this set-up. The preshear is used to get into the critical state flow regime, so that the subsequent shear tests are always starting from well defined initial conditions.

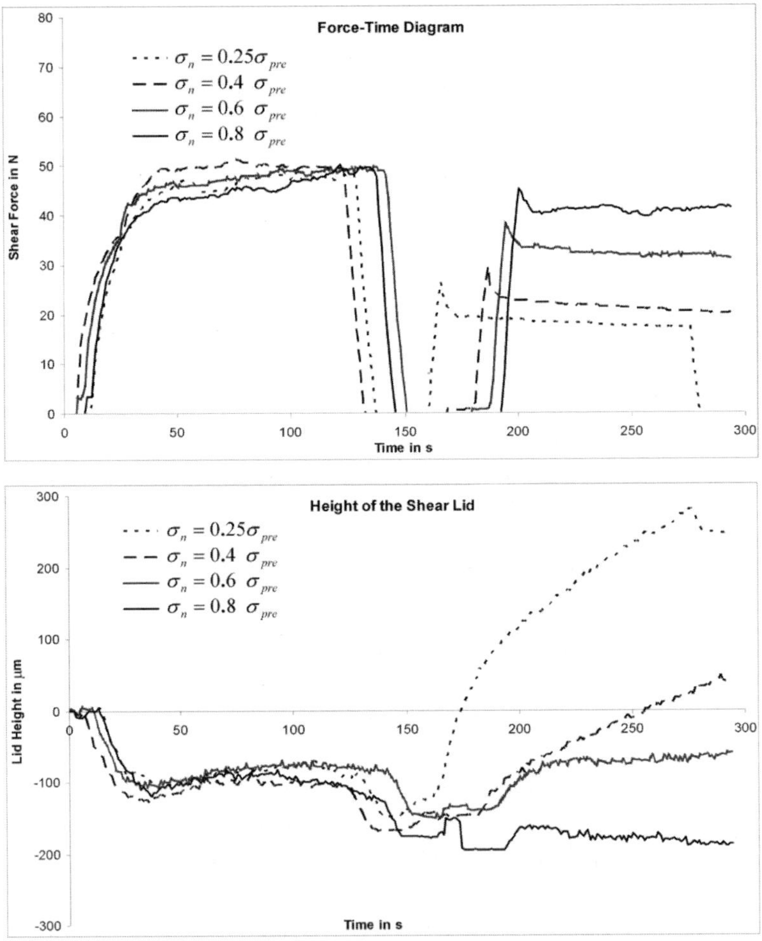

Fig. 3. Shear force measurements during shear tests (top) and volumetric strain measurements during the same shear test (bottom)

The tests have been accomplished for a limestone powder sample with average particle diameter $d_{50} = 1.2\mu m$, density $\rho_s = 2740 kg/m^3$, and a shear rate of $v_s = 2 mm/min$. Fig. 3 shows an example of the stress and volumetric strain measurements. The powder was preconsolidated with the normal stress $\sigma_{N,0} = 16 kPa$ and pre-sheared with the normal stress $\sigma_{pre} = 8 kPa$. The normal stress σ_n during the shear procedure was taken proportional to σ_{pre} with the factors 0.25, 0.4, 0.6 and 0.8. Some compression (about 0.1mm, i.e., about 80 particle diameters) is caused by the reorganization of initial powder packing at the beginning of the pre-shear process. After 40s (1.3mm), a maximum density is reached, followed by some dilatancy (height increase of about 20 particle diameters) before the steady-state is reached after 100s (2.6mm). The unloading (i.e. the shear drive is moved with the same rate in opposite direction, taking off in this way the force on the shear ring) allows the shear lid to compress the packing structure down to a lower porosity (higher density) state again. During the load change ("jumps" up and down on the figure), when the normal stress is reduced to the desired σ_n, the powder reacts again by expanding to higher porosity (lower density). This response as function of the normal stress can be also used to understand the elastic properties of a consolidated (and presheared) powder continuum. The lower is the normal stress σ_n, the higher "jumps" of the shear lid are.

4 Simulations

One possibility to gain insight upon the material behaviour of a granular packing is to perform elementary tests in the laboratory. Here, we chose as alternative the simulation with the discrete element model [7-9,13].

If the particles are assumed to be rigid, and the behaviour of the contacts is characterized using a soft contact approach, then the mechanical behaviour of such a system is described in terms of the movement of each particle and the inter-particle forces acting at each contact point. Newton's and Euler's laws of motion for the translational and rotational degrees of freedom give the fundamental relationship between particle motion and the forces that induce them

$$m_i \frac{d^2}{dt^2} \vec{r}_i = \vec{f}_i + m_i \vec{g} \quad \text{and} \quad I_i \frac{d}{dt} \vec{\omega}_i = \vec{t}_i \tag{2}$$

with the gravitational acceleration $\vec{g}$, mass m_i of the particle, its position $\vec{r}_i$, the total force $\vec{f}_i$ acting on it due to contacts with other particles or with the walls, its moment of inertia $\vec{I}_i$, its angular velocity $\vec{\omega}_i$, and the total torque $\vec{t}_i$.

The classical translational shear cell, developed by Jenike [5], is modelled (Fig. 4). Considering a suitable CPU-time for a certain number of particles one should note that we simulate here only a small three-dimensional (3D) element from the real shear cell. Fig. 4 shows the model for 2000 titanium dioxide particles with diameter of about $(3 \pm 0.5)\mu m$. The upper wall (shear lid) is stress controlled, i.e. when the reaction force F_N changes because of the particle reorganization, the height of the

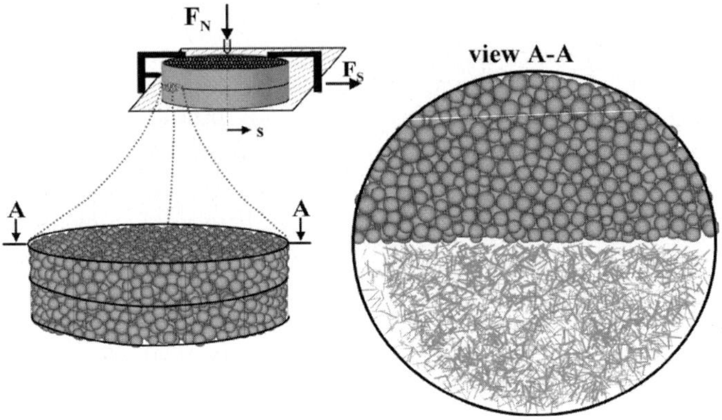

Fig. 4. The shear cell model system for the simulations (lines in the A-A view show the contact shear forces, with line thickness proportional to force value)

shear lid is changed as well. The horizontal shear rate of the upper part of the cell is preset, i.e. the upper ring is strain driven. As the direct response, the corresponding values of the reaction force are obtained, which acts on the lateral walls. Furthermore, the corresponding shear stresses $\tau = F_S/A$ are calculated enabling in this way to find the flow parameters of the simulated model powder. The shear rate applied here is about 1-4mm/min (similar to the one used in the Jenike shear cell in laboratory tests). During the simulation, the shear stress τ is recorded as function of the shear strain, for different normal stresses $\sigma = F_N/A$.

The translational shear cell studied in the following contains $N \approx 3500$ particles with radii randomly drawn from a homogeneous distribution with minimum 6μm and maximum 12μm. The material properties correspond to those of titanium dioxide powder. The friction coefficient used in the two-dimensional simulations is $\mu = 0.5$, shear rate $v_s = 2$mm/min. For convenience, a constant adhesion force of 1-10mN (0.1-1% of average contact forces of loading) is used to approximate the load-history dependent pull-off force [1]. Then a series of simulations is done applying the more general dissipative contact model for adhesive particles (Fig. 1) first in 2 dimensions, then in 3 dimensions, and the comparison is made.

Figure 5 shows the contact force network during the shearing with line thickness proportional to shear force. The force lines run mostly from the upper left wall, where the shear force acts, to the lower right wall, where the corresponding resisting force acts. Examining the shear strain, more contacts and stronger forces are parallel to the compressive direction of shear (top-left to bottom-right). This is correlated to the fact that the orientation of the major principal stress σ_1 is tilted counter-clockwise, acting against the shear. The lower left and the upper right walls experience much weaker forces, correlated to the tensile strain direction (bottom-left to top-right).

Additionally, Fig. 5 shows the force-displacement diagram (left) and the volumetric strain (right) at the constant normal stress of $\sigma_N = 3$kPa for three different values of porosity $\varepsilon = 1 - \rho_b/\rho_s$ (where ρ_b is the bulk material density, and ρ_s the solid density), i.e. three preshear tests. The upper curves in both graphs are obtained at a

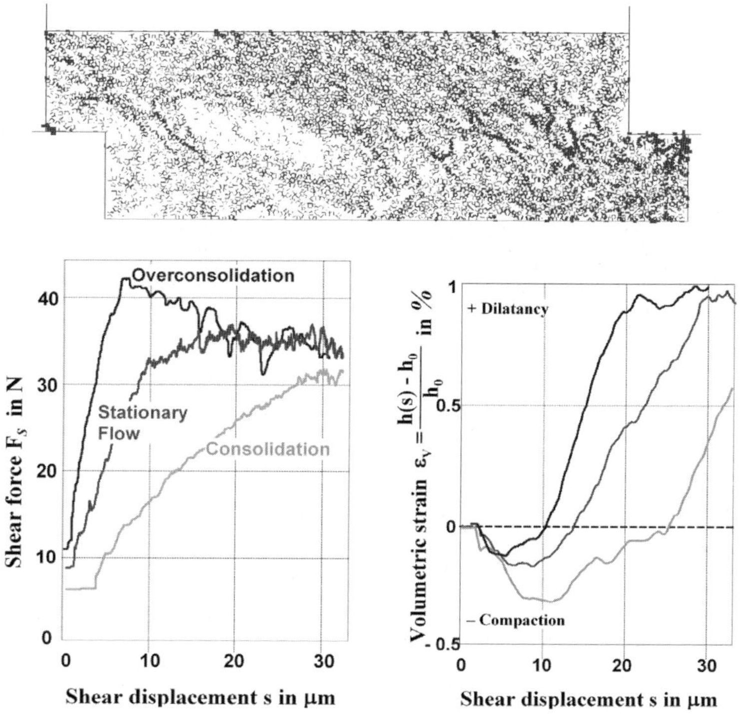

Fig. 5. Left: force-displacement diagram (steady-state flow), right: volumetric strain for three different initial porosity values

two-dimensional porosity of the particle system of $\varepsilon_{2D} = 0.16$ (it corresponds approx. $\varepsilon_{3D} = 0.46$ for three dimensions [10]). The typical behaviour of the over-consolidated powder is observed in this case achieving the peak force value along with the first-stage compaction at the beginning and tending to the steady state flow later on. The middle curves at $\varepsilon_{2D} = 0.18$ ($\varepsilon_{3D} \sim 0.50$) show almost ideal steady-state flow with the remark that the volumetric strain is coming to the relatively stable state only at the end of the shear process, which does not come into agreement with the theoretical expectations. However, new measurements of the vertical movement of the shear lid (volumetric strain) performed with the real Jenike shear cell show the same tendency. The fluctuations of the shear force can be explained by means of the temporary and local shear-thickening and shear-thinning processes. The lower curves correspond to $\varepsilon_{2D} = 0.20$ ($\varepsilon_{3D} \sim 0.54$) and show the tendency of an under-consolidated powder. Taking into account the shear testing experiences, the good qualitative agreement is reached between the simulation results and the laboratory tests.

Using the *steady-state shear stresses* for different applied normal stresses, as obtained from the Jenike cell experiments and simulations, the yield locus is defined. The slope of the yield locus gives the angle of internal friction of the material and the intersection with the vertical axis the macroscopic cohesion in stress units.

Fig. 6 shows the comparison of the yield locus calculated from the Jenike shear cell experiments with the simulated yield loci obtained by applying different contact laws with different parameters. The experimental line (upper line) shows the internal friction angle of the powder continuum to be equal to $\varphi_i = 26°$, and the cohesion $\tau_c = 3.09$kPa. The lowermost line is the result of the simulated shear test, where the simplest microscopic interaction between the particles is used, the linear-spring-dashpot contact model in normal direction, extended only by the constant adhesion force between the particles in contact $f_0 = 10$mN, see Fig. 1. The preshear normal stress was σ_{pre}=20kPa and, for the shear process, the adhesion force values f_0 have been taken proportional to the reduced macroscopic normal stresses. The average contact forces during preshear are in the range of 200-500mN. The centre two lines are obtained using the elastic-plastic contact model with variable adhesion, Fig. 1, with $k_1 = 10^8$N/m, $k_2 = 2k_1$, and the "microscopic adhesion stiffness" $k_c = k_2$ (lower curve) $k_c = 4k_2$ (higher curve), see Eq. (1), the latter leading to the macroscopic result for the internal friction angle $\varphi_i = 20°$, and the cohesion $\tau_c = 1.13$kPa.

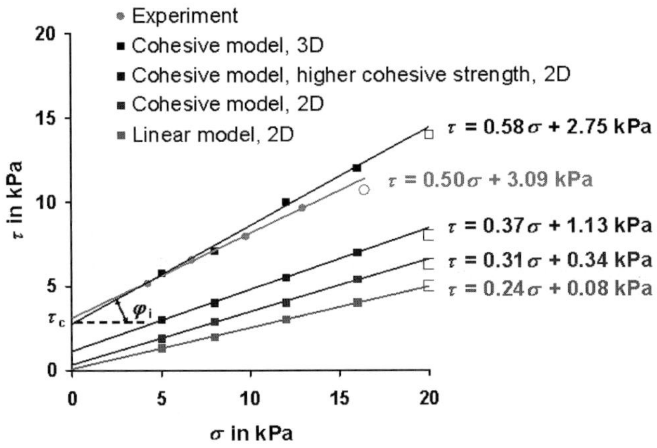

Fig. 6. Shear stress – normal stress diagram of yield locus 4 of model material (TiO_2, ρ=3870kg/m^3, v_s=2mm/min) for experiments and simulations

Thus, comparison of the simulated data in 2 dimensions with experiments did not yield satisfactory quantitative agreement. That is why a series of simulations with input data of the last 2D simulation has been performed in 3 dimensions (dashed line) and led to near to perfect quantitative agreement with the experiments. The same tendency of the increase of macroscopic adhesion in dependence on the contact model and contact model parameters was observed also in 3D simulations. Therefore, the implemented "adhesion" contact model shows a positive influence in order to mimic the macroscopic flow behaviour of a consolidated powder bed, when compared to the oversimplified linear contact model with load independent adhesion.

5 Conclusions

In summary, a set of DEM simulations based on different contact constitutive models is presented in 2 and in 3 dimensions, and several macroscopic material parameters like, e.g., the friction angle, are extracted from the simulation data with cohesion (no friction) and with friction (no cohesion). Altogether this is a first step of a micromodelling approach for cohesive frictional powders by means of going the whole long way from a measurement of ultra-fine single particles and implementing the complex microscopic contact constitutive laws for the contacts between the particles up to the shear dynamics of big particle systems finding out in this way the macroscopic flow parameters of bulk materials.

Microscopic insights as observed from the discrete element simulations (like the peculiar force/stress-patterns) are helpful to find out the main control parameters for the further calculations. Also the predicted particle redistribution in the shear zone was verified by the experiments. So far, starting from realistic contact properties, good qualitative agreement of the 2D simulations with experiments is reached, and the quantitative agreement with physically based input parameters could be reached only in 3 dimensions. The implemented cohesive contact model shows a positive influence on the macroscopic flow behaviour of a consolidated powder bed. The model shows to be capable of simulating the cohesive properties of a material with variable adhesion (pull-off) force depending among others on the preconsolidation history of every inelastic particle contact deformation. An important result is the fact that the macroscopic cohesion can be related to the maximal microscopic attractive force.

Further material parameters have to be identified, and also the role of particle rotations is an open issue, as related to micro-polar constitutive models. In both simulations and experiments, rotations are active in the shear band where the rotational degree of freedom is activated. The corresponding parameter identification and the micro-macro-transition for anisotropic micro-polar continuum models is a challenge for the future, like the implementation and simulation of experimentally determined force-laws [12] in more complex three-dimensional systems.

References

[1] Tomas, J.: Assessment of mechanical properties of cohesive particulate solids – part 1: particle contact constitutive model. Particulate Sci. & Tech. 19(2), 95–110 (2001)

[2] Cundall, P.A., Hart, R.D.: Numerical modelling of diskontinua. In: 1st US Conference on DEM. Eng. Comput., vol. 9, pp. 101–113 (1992)

[3] Tykhoniuk, R., Tomas, J., Luding, S., et al.: Ultrafine cohesive powders: From interparticle contacts to continuum behaviour. Chem. Eng. Sci. 67, 2843–2865 (2006)

[4] Tomas, J.: In: Mittal, K.L.: Particles on Surfaces 8: Detection, Adhesion and Removal, VSP Utrecht (2003)

[5] Jenike, A.W.: Storage and flow of solids. Eng. Exp. Station, Bull. No. 123, Univ. Utah (1964)

[6] Luding, S., Tykhoniuk, R., Tomas, J.: Anisotropic material behaviour in dense, cohesive powders. Chem. Eng. Tech. 26(12), 1229–1232 (2003)

[7] D'Addetta, G.A., Kun, F., Ramm, E.: On the application of a discrete model to the fracture process of cohesive granular materials. Gran. Matter 4(2), 77–90 (2002)

[8] Oda, M., Iwashita, K.: Study on couple stress and shear band development in granular media based on numerical simulation analyses. Int. J. of Eng. Sci. 38, 1713–1740 (2000)

[9] Luding, S., Lätzel, M., Volk, W., et al.: From discrete element simulations to a continuum model. Comp. Meth. Appl. Mech. Eng. 191, 21–28 (2001)

[10] Deresiewicz, H.: Mechanics of granular matter. Advances in Appl. Mech. 5, 233–306 (1958)
[11] Derjaguin, B.V., Abrikosova, I.I., Lifshitz, E.M.: Quart. Rev. Chem. Soc. 10, 295–329 (1956)
[12] Tejchman, J., Wu, W.: Numerical study on patterning of shear bands in a Cosserat continuum. Acta Mechanica 99, 61–74 (1993)
[13] Tykhoniuk, R., Luding, S., Tomas, J.: Simulation der Scherdynamik kohäsiver Pulver. Chemie Ingenieur Technik 76(1-2), 59–62 (2004)

CFD-Modelling of the Fluid Dynamics in Spouted Beds

O. Gryczka[1], S. Heinrich[2], and J. Tomas[1]

[1] Institut für Verfahrenstechnik, Otto-von-Guericke-Universität Magdeburg
[2] Institut für Feststoffverfahrenstechnik und Partikeltechnologie, TU Hamburg-Harburg

Abstract. Recently the importance of the spouted bed technology has significantly increased in the context of drying processes as well as granulation, agglomeration or coating processes. However, the understanding of the complex interactions within and between the single phases is still low and needs further improvement. Several research groups apply both continuum as well as discrete element simulations to understand the hydrodynamics of the spouting process. Within this work, the hydrodynamical behaviour of a novel spouted bed apparatus is modelled by applying the Euler/Euler continuum approach in two dimensions in the commercial CFD software-package FLUENT 6.2. The focus of this study is the influence of different gas-particle drag models on the flow structure in the spouted bed apparatus. Simulated spatial distributions of the granular solid phase in the apparatus are compared with images taken during experiments. Characteristic gas phase fluctuations over the entire bed are simultaneously measured by a high frequency pressure detector and will be compared with calculated fluctuations. Also a Fourier analysis on the measured and simulated pressure frequency spectra will be demonstrated.

1 Introduction

In food, pharmaceutical and chemical industry, fine and polydisperse particulate solids are treated and produced. Like the fluidised bed technology, the spouted bed technology can be applied for mixing of particulate systems, for heat and mass transfer processes like, e.g., cooling, drying, calcinations, combustion, gasification as well as for complex multiphase processes like spray granulation, agglomeration, particle layering and coating and also for chemical reactions. However, the understanding of the complex gas-particle and particle-particle interactions is still very low and needs further investigation. Due to a rapid development of computer performances in the last decade, modelling the hydrodynamics of spouted and fluidised beds by application of modern simulation tools, e.g., Lattice-Boltzmann simulations or Computational Fluid Dynamics (CFD) has established.

The focus of this work is to investigate experimentally and simulate the hydrodynamics of a novel spouted bed apparatus. The intention of the simulations is the calculation of the entire multi-phase processes including a scale-up procedure.

2 CFD-Multiphase Modelling

Two phases are considered in the simulations: a gas phase (air) and a particle (granular solid) phase (γ-Al_2O_3-particles). Properties of the experimental material and of the fluidising gas can be found in Gryczka et al. [5]. The particles of the granular solid phase are modelled as monodisperse ($d_s = d_{50,3} = 1.75$mm) smooth non-adhesive spheres. The Eulerian approach, where the contributing phases are considered as interpenetrating continua, is applied to the gas and solid phase. The governing equations of the continuum model are presented in the following.

The conservation equations solved by FLUENT 6.2 can be considered as generalization of the Navier-Stokes equations for interacting continua. The mass conservation equation (continuity equation) for a general phase q is

$$\frac{\partial}{\partial t}\left(\varepsilon_q \rho_q\right) + \nabla \cdot \left(\varepsilon_q \rho_q \vec{v}_q\right) = 0 . \tag{1}$$

The index q stands either for the solid (particle) phase $q = s$ or for the gas phase $q = g$. Here, ε_q is the phase volume fraction; ρ_q is the phase density, and $\vec{v}_q$ the velocity vector of phase q. The momentum balance equation for phase q is

$$\frac{\partial}{\partial t}\left(\varepsilon_q \rho_q \vec{v}_q\right) + \nabla\left(\varepsilon_q \rho_q \vec{v}_q \vec{v}_q\right) = -\varepsilon_q \nabla p + \nabla \bar{\bar{\tau}}_q + K_{qq} \Delta \vec{v}_q + \varepsilon_q \rho_q \vec{F}_g . \tag{2}$$

In this equation, $\Delta \vec{v}_q$ is the relative velocity between the two phases, K_{qq} is the drag coefficient, p is the pressure shared by all phases, and $\vec{F}_g$ is the force of gravity. $\bar{\bar{\tau}}_q$ stands for the qth phase stress tensor and is calculated as follows

$$\bar{\bar{\tau}}_q = \varepsilon_q \mu_q \left(\nabla \vec{v}_q + \nabla \vec{v}_q^T\right) + \varepsilon_q \left(\lambda_q - \frac{2}{3}\mu_q\right) \nabla \cdot \vec{v}_q \bar{\bar{I}} . \tag{3}$$

Here μ_q *is the shear viscosity and* λ_q *the bulk viscosity of the* q^{th} *phase.* For the description of the solid phase rheology, constitutive equations based on the kinetic theory of granular flow are incorporated in FLUENT 6.2. This theory uses an analogy to the classical theory of molecular gases [1] and describes the dependency of the rheological properties on the local particle concentration and on the particle fluctuating velocity due to particle-particle collisions.

The shear stress tensor of the solid contains shear and bulk viscosities *($\mu_{s,kin}$) arising from particle momentum exchange due to translation, rotation and collision. Furthermore, a dissipative and/or frictional* component of the viscosity is included to account for the visco-plastic transition that occurs when particles of a solid phase reach the maximum particle volume fraction. All three contributions are considered additively:

$$\mu_s = \mu_{s,col} + \mu_{s,kin} + \mu_{s,fr} . \tag{4}$$

The collisional part of the shear viscosity $\mu_{s,col}$ is modelled as following

$$\mu_{s,col} = \frac{4}{5}\varepsilon_s \rho_s d_s g_{0,ss}\left(1 + e_{ss}\right)\left(\frac{\theta_s}{\pi}\right)^{1/2} . \tag{5}$$

Here, e_{ss} is the coefficient of restitution which accounts for energy dissipation due to inelastic particle collisions, and $g_{0,ss}$ is the radial distribution function that contains a correction factor which modifies the probability of particle collisions when the particle phase becomes densely packed. The granular temperature θ_s characterizes the kinetic energy of the random particle motion due to particle collisions. The kinetic part $\mu_{s,kin}$ is calculated by the expression of Gidaspow et al. [3]

$$\mu_{s,kin} = \frac{10\rho_s d_s \sqrt{\theta_s \pi}}{96\varepsilon_s (1+e_{ss}) g_{0,ss}} \left[1 + \frac{5}{4} g_{0,ss} \varepsilon_s (1+e_{ss})\right]^2 . \quad (6)$$

Energy losses due to Coulomb friction between the non-adhesive particles are described by the frictional viscosity model proposed by Schaeffer [9]

$$\mu_{s,fr} = \frac{p_s \sin\phi}{2\sqrt{I_{2D}}} . \quad (7)$$

In this equation, ϕ *denotes the angle of internal friction,* p_s *is the averaged particle bed pressure, and* I_{2D} *is the second invariant of the deviatoric stress tensor, which characterizes the shear stress within the flowing particle bed.*

The granular bulk viscosity λ_s *accounts for the resistance of the granular particles by compression and expansion. The correlation of Lun et al. [8] is used to calculate this bulk viscosity*

$$\lambda_s = \frac{4}{3} \varepsilon_s \rho_s d_s g_{0,ss} (1+e_{ss}) \left(\frac{\theta_s}{\pi}\right)^{1/2} \quad (8)$$

3 Gas-Particle Drag Models

An appropriate description of the drag between the gas and the solid phase is of significant importance, since fluidisation is affected by the force exposure of the gas flow in the gap between the particles. Therefore, several gas-particle drag models are tested to obtain realistic flow structures. In the following, the definitions of the applied drag models and drag coefficients are presented.

- The oldest drag model is the Schiller and Naumann [10] model, where the gas-solid exchange coefficient K_{gs} is calculated by

$$K_{gs} = \frac{\varepsilon_s \rho_s}{\tau_s} \frac{C_D \, Re}{24} . \quad (9)$$

τ_s is the particulate relaxation time and the drag coefficient C_D has the form

$$C_D = \begin{cases} 24\left(1+0.15\,Re^{0.687}\right)/Re & Re \le 1000 \\ 0.44 & Re > 1000 \end{cases} . \quad (10)$$

The Reynolds number is defined as follows

$$\mathrm{Re} = \frac{\rho_g d_s \left|\vec{v}_s - \vec{v}_g\right|}{\mu_g}. \tag{11}$$

- The Wen and Yu [14] drag model is recommended for diluted particle-gas systems. The gas-solid exchange coefficient K_{gs} is defined by

$$K_{gs} = \frac{3}{4} C_D \frac{\varepsilon_s (1-\varepsilon_s) \rho_g \left|\vec{v}_s - \vec{v}_g\right|}{d_s}. \tag{12}$$

Here, C_D is calculated as follows

$$C_D = \frac{24}{(1-\varepsilon_s)\mathrm{Re}_s}\left[1 + 0.15\left((1-\varepsilon_s)\mathrm{Re}_s\right)^{0.687}\right]. \tag{13}$$

- The Syamlal-O`Brien [11] drag model is appropriate when the solid shear stresses are defined according to Syamlal et al. [12]. The gas-solid momentum exchange coefficient is calculated as

$$K_{gs} = \frac{3\varepsilon_s (1-\varepsilon_s) \rho_g}{4 v_{r,s}^2 d_s}\left(0.63 + \frac{4.8}{\sqrt{\mathrm{Re}_s / v_{r,s}}}\right)^2 \left(\frac{\mathrm{Re}}{v_{r,s}}\right)\left|\vec{v}_s - \vec{v}_g\right|. \tag{14}$$

- The Gidaspow [3] drag model is a combination of the Wen and Yu [14] model and the Ergun [2] equation. For very low particle concentrations (volume fraction $\varepsilon_s<0.2$), K_{gs} is defined as

$$K_{gs} = \frac{3}{4}\frac{\varepsilon_s (1-\varepsilon_s) \rho_g \left|\vec{v}_s - \vec{v}_g\right|}{d_s} \cdot \left(\frac{24}{(1-\varepsilon_s)\mathrm{Re}_s}\left[1 + 0.15\left((1-\varepsilon_s)\mathrm{Re}_s\right)^{0.687}\right]\right)(1-\varepsilon_s)^{-2.65} \tag{15}$$

For high particle concentrations $\varepsilon_s \geq 0.2$, the gas-solid exchange coefficient is calculated by the Ergun [2] equation

$$K_{gs} = 150 \frac{\varepsilon_s^2 \mu_g}{(1-\varepsilon_s) d_s^2} + 1.75 \frac{\varepsilon_s \rho_g \left|\vec{v}_s - \vec{v}_g\right|}{d_s}. \tag{16}$$

- The Koch and Hill [6] drag model has been established by Lattice-Boltzmann simulations.

$$\begin{aligned} K_{gs} &= \frac{18\mu_g (1-\varepsilon_s)^2 \varepsilon_s}{d_s^2}\left(F_0(\varepsilon_s) + \frac{1}{4}F_1(\varepsilon_s)\mathrm{Re}_s^2\right) && \mathrm{Re} < 40 \\ K_{gs} &= \frac{18\mu_g (1-\varepsilon_s)^2 \varepsilon_s}{d_s^2}\left(F_0(\varepsilon_s) + \frac{1}{2}F_3(\varepsilon_s)\mathrm{Re}_s\right) && \mathrm{Re} > 40 \end{aligned} \tag{17}$$

where F_0, F_1 and F_3 are dimensionless coefficients. The drag coefficient C_D is modelled by Eq. 10. The disadvantage of this drag model is that it is not valid for the intermediate Re regime which reflects in a strong discontinuity at Re = 40. The dependency of the local particle Reynolds number on the gas-particle drag coefficient K_{gs} of all applied drag models is depicted in Fig. 1 in dimensionless form.

- The van der Hoef et al. [13] model is the newest drag model and is valid for the entire range of Reynolds numbers ($0<Re<10^3$) without the discontinuity encountering in the Koch and Hill [5] drag model. It was also established by Lattice-Boltzmann simulations.

$$K_{gs} = \frac{18\mu_g (1-\varepsilon_s)^2 \varepsilon_s}{d_s^2} \left(F_0(\varepsilon_s) + F_1(\varepsilon_s) Re_s \right) \tag{18}$$

Fig. 1 reveals the differences between the gas-particle drag models in dependency of the particle Reynolds number. In Fig. 2 a simulated cumulative distribution of the particle Reynolds number Re_s for the entire process chamber and especially for the jet zone are depicted. One can see that the Reynolds numbers of the particles in the jet zone are very high and assume values between 2,000 and 10,000. Also in the entire process chamber, more than 80% of the particles have Reynolds numbers above 1,000. These perceptions can be used to evaluate the influence of the different drag models on the spouted bed flow structure.

4 Simulation Results

An entire two dimensional geometry model of the spouted bed apparatus has been created and meshed by means of the pre-processor GAMBIT 2.3.16 (Geometry And Mesh Building Intelligent Toolkit). The equations of the multiphase continuum model are solved by means of a finite volume technique provided by the commercial software FLUENT 6.2. The domain is discretised into a finite set of control cells and the general transport equations for mass, momentum, etc. are applied to each cell using the implicit segregated solver. The pressure-velocity coupling is accomplished by the SIMPLE-algorithm. The First-Order-Upwind-method is used for interpolating the cell-centre values to the cell boundaries. In the following, the influence of the different gas-particle drag models on the flow structure are investigated. The distribution of the simulated granular solid phase in the apparatus is compared with images taken during experiments. Fig. 3 shows representative images of the measured and simulated distributions of the granular solid phase in the apparatus using different gas-particle drag models after 2 seconds of simulation.

Fig. 3 shows that all applied drag models are able to simulate the typical flow structures encountering in the spouted bed apparatus. The jet zone, the fountain zone, and the backflow zone can be clearly identified. However, the flow structures calculated with the Wen and Yu [14] model (Fig. 3c), the Syamlal-O`Brien [11] model (Fig. 3d) and the Gidaspow [3] model (Fig. 3e) show differences compared to the experimental observed flow structures (Fig. 3a). The bed expansion is considerably over-predicted.

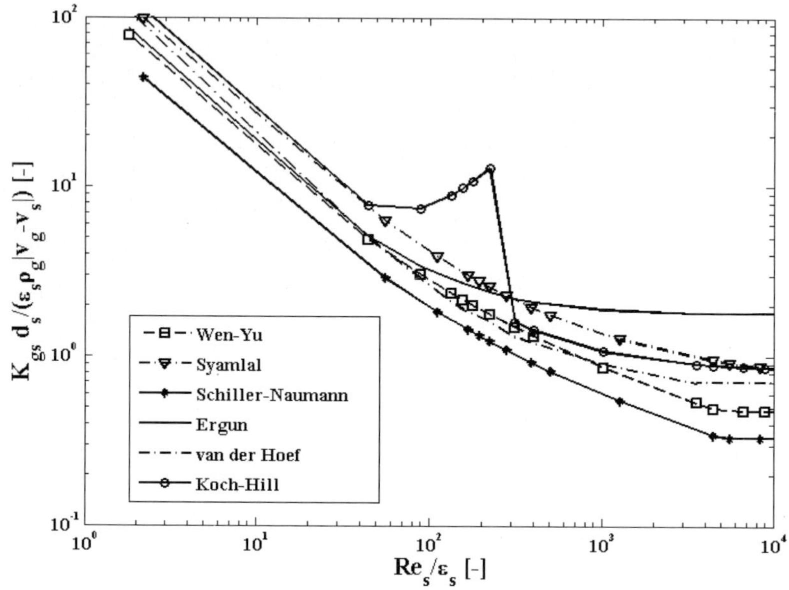

Fig. 1. Dependency of particle Reynolds number on the gas-particle drag coefficient (ε_s=0.2)

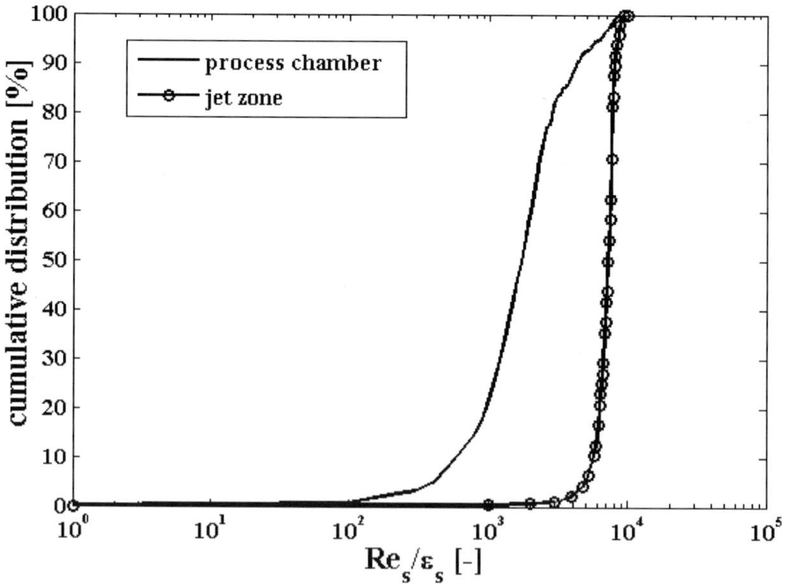

Fig. 2. Cumulative distribution of particle Reynolds numbers for 1 second of simulation

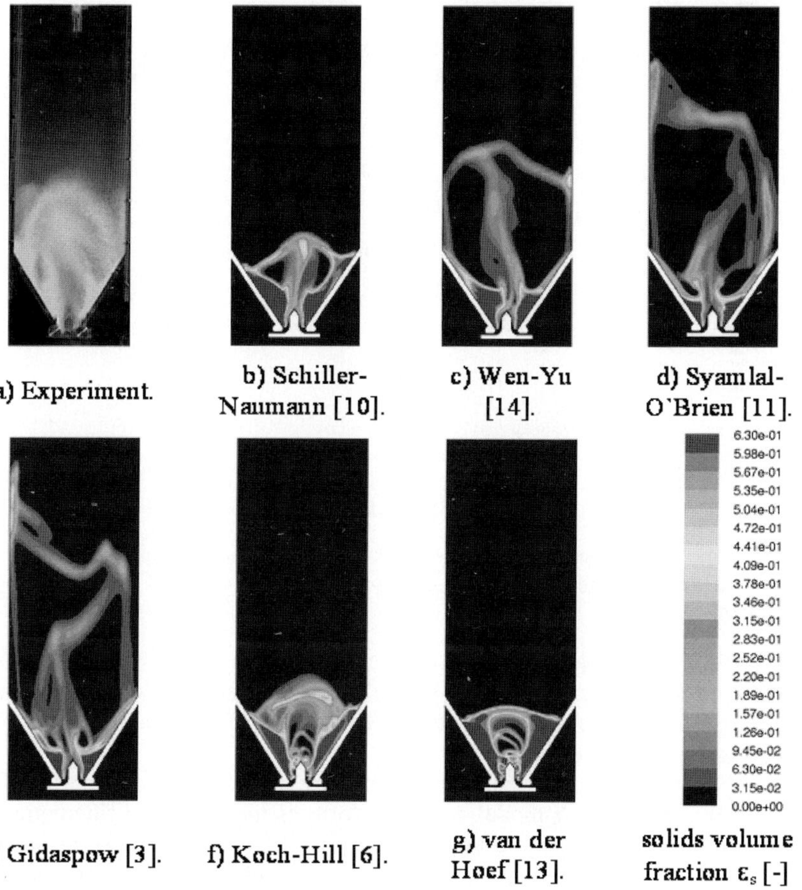

Fig. 3. Comparison of the spatial distribution of the granular solid phase between the experiment (a) and the simulations with different gas-particle drag models (b – g) after 2 seconds of simulation

The best resemblance with the experiment is obtained by applying the Schiller-Naumann [10] model (Fig. 3b), the Koch and Hill [5] model (Fig. 3f) and the van der Hoef et al. [13] model (Fig. 3g). Discrepancies regarding the under-prediction of the bed expansion may result from the frictional viscosity model by Schaeffer [9]. Investigations of dense gas-solid fluidised beds by Goldschmidt et al. [4] show that the application of the frictional viscosity model proposed by Laux [7] yield better resemblance with experiments which will be tested in near future.

Another characteristic of the investigated spouting process with an inlet air volume flow of $0.043 m^3/s$ is a periodic bursting of gas bubbles on the bed surface causing periodic bed pressure drop fluctuations (see Fig. 4 top) which have been detected by a high frequency pressure detector. In the following, the simulated and the measured bed pressure drop fluctuations as well as a Fourier analysis of these spectra are

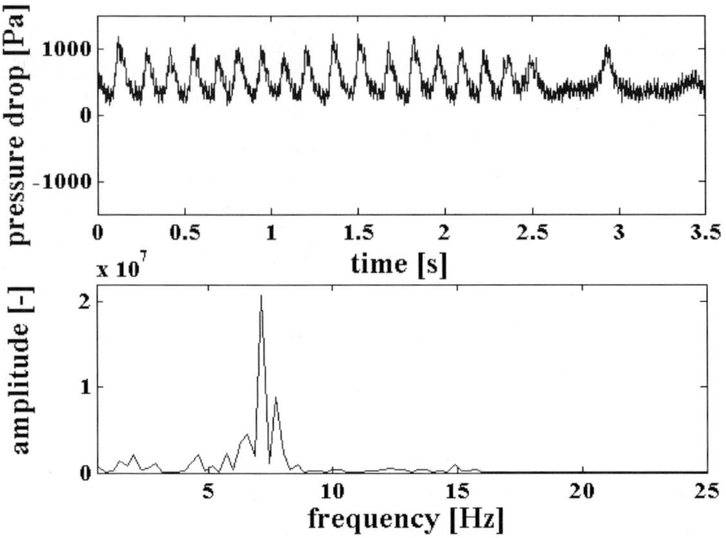

Fig. 4. Measured bed pressure drop fluctuations (top) and Fourier-transformation of the frequency spectra of these fluctuations (bottom). γ-Al_2O_3-particles, d_s = 1.75mm, m_{bed} = 1.0kg, $\dot{V}_{gas,in}$ = 0.043m^3/s.

contrasted to do even more accurate comparisons between the simulations and the experiment. In Fig. 5 the calculated bed pressure drop fluctuations of the different gas-particle drag models are depicted.

From Fig. 5a it can be seen that the bed pressure drop fluctuations predicted by the Schiller-Naumann [10] drag model are as uniform as the measured fluctuations. Furthermore, a dominant peak in the Fourier analysis around 6 - 7Hz can be seen in the simulation and the experiment. However, due to the under-prediction of the bed expansion, the amplitudes of the simulated pressure drop fluctuations are lower than the measured ones. Recapitulating, one can conclude that the spouted bed hydrodynamics are reflected acceptably well by applying the Schiller-Naumann drag [10] model. The drag models of Wen and Yu [14] (Fig. 5b), Syamlal-O`Brien [11] (Fig. 5c) and Gidaspow [3] (Fig. 5d) calculate to high drag forces (see Fig. 1) and over-predict the bed expansion considerably. Also, the bed pressure drop fluctuations do not show the periodicity of the measured fluctuations. Therefore, no dominant peak in the Fourier analysis on these spectra is obtained. It can be concluded that the hydrodynamics of the spouting process cannot be reflected by applying the drag models of Wen and Yu [14], Syamlal-O`Brien [11] and Gidaspow [3]. The drag models by Koch and Hill [6] and van der Hoef et al. [13] simulate fairly uniform and periodic bed pressure drop fluctuations.

However, the frequency of the bursting gas bubbles on the bed surface (around 2-3 Hz, see Fig. 5e and f bottom) and the bed expansion are under-predicted compared to the experiment. The reason for this is that these drag models have been established for particle Reynolds numbers up to 10^2 and 10^3, respectively. However, over 80% of the

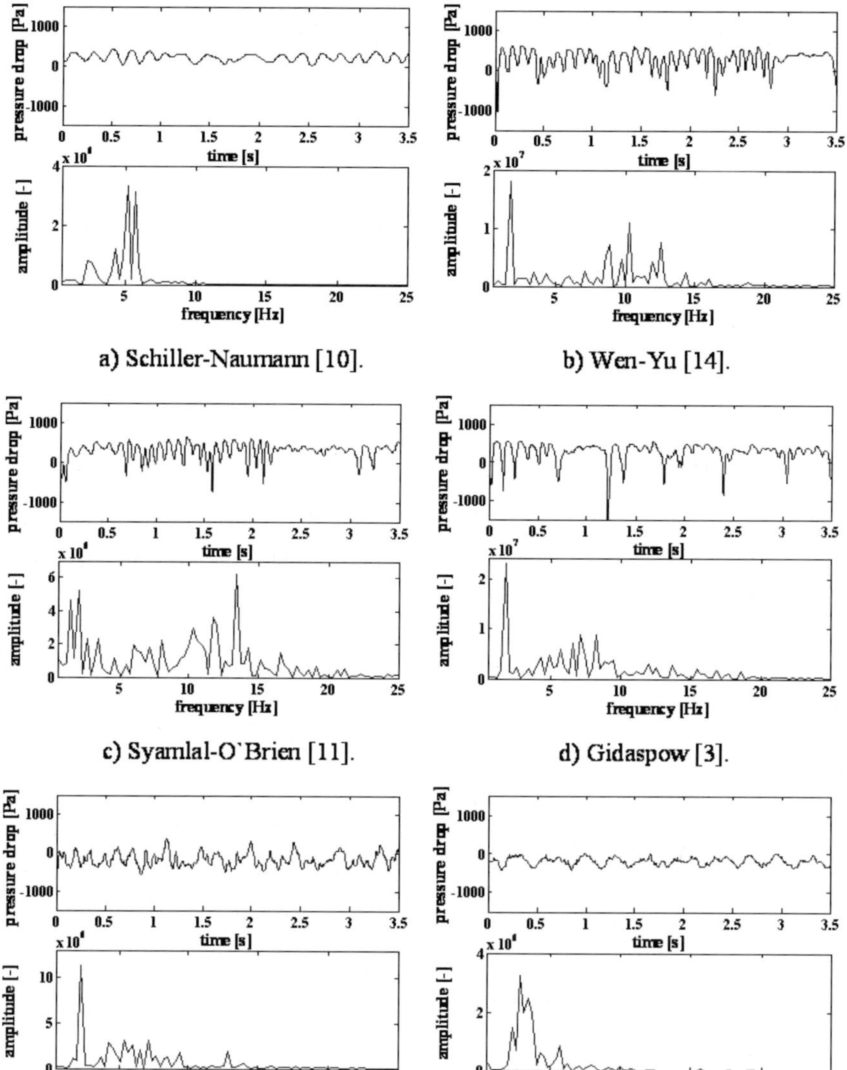

a) Schiller-Naumann [10]. b) Wen-Yu [14].

c) Syamlal-O`Brien [11]. d) Gidaspow [3].

e) Koch-Hill [6]. f) van der Hoef et al. [13].

Fig. 5. Simulated bed pressure drop fluctuations (top) and Fourier-transformation (bottom). γ-Al_2O_3-particles, $d_s = 1.75mm$, $\dot{V}_{gas,in} = 0.043m^3/s$.

particle Reynolds numbers are higher than 10^2 (see Fig. 2) concluding that the confined area of validity is responsible for the unrealistic calculation of the spatial distribution of the granular solid phase.

Just like Goldschmidt et al. [4] already concluded in their investigations of dense gas-solid fluidised beds, one can state that the search for an appropriate drag model is of importance for obtaining realistic flow structures, but other influencing factors should also be focused on (e.g. neglection of particle rotation, long-term particle contacts, frictional viscosity model, collisional contributions…).

5 Conclusions

In this work the hydrodynamic behaviour of a novel spouted bed apparatus is modelled by means of a CFD continuum approach. Different gas-particle drag functions are investigated to obtain realistic flow structures. The gas flow retained constant. Experimental and predicted time-dependent distributions of the granular solid phase in the apparatus and gas phase pressure fluctuations over the entire bed are analysed and contrasted.

Different flow structures are calculated in dependency of the gas-particle drag model. The best resemblance is obtained by the Schiller-Naumann [10] drag model. The drag models of Syamlal-O`Brien [12], Wen and Yu [14] and Gidaspow [3] simulate unrealistic flow structures and over-predict the bed expansion. Contrary to this over-prediction of the bed expansion, the drag models proposed by Koch and Hill [6] and van der Hoef et al. [13] under-predict the bed expansion considerably. Also the measured progressions of the gas phase pressure fluctuations are only reflected qualitatively well by application of the Schiller-Naumann [10] drag model. Chaotic and irregular pressure fluctuations showing no similarities with the experiment are calculated by the drag models by Syamlal-O`Brien [11], Wen and Yu [14] and Gidaspow [3]. The Koch and Hill [6] and the van der Hoef et al. [13] drag models simulate uniform and periodic pressure drop fluctuations, but the frequency of the calculated fluctuations differ considerably compared to the measurement.

Finally, it can be concluded that the CFD continuum modelling containing closures of the kinetic theory of granular flow constitutes a reasonable simulation tool for the investigated spouted bed apparatus to gain qualitative realistic results. However, more physically based closures for, e.g., the calculation of the frictional solids viscosity should be incorporated in the continuum models in future.

References

[1] Chapman, S., Cowling, T.G.: The mathematical theory of non-uniform gases, 3rd edn. Cambridge University Press, Cambridge (1970)

[2] Ergun, S.: Flow through packed columns. Chemical Engineering Progress 48(2), 89 (1952)

[3] Gidaspow, D., Bezburuah, R., Ding, J.: Hydrodynamics of circulating fluidized beds, kinetic theory approach. In: Fluidization 7, Proceedings of the 7th Engineering Foundation Conference on Fluidization, pp. 75–82 (1992)

[4] Goldschmidt, M.J.V., Beetstra, R., Kuipers, J.A.M.: Hydrodynamic modeling of dense gas-fluidized beds: comparison and validation of 3D discrete particle and continuum models. Powder Technology 142(1), 23–47 (2004)

[5] Gryczka, O., Heinrich, S., Deen, N.G., van Sint Annaland, M., Kuipers, J.A.M., Jacob, M., Mörl, L.: Characterization and CFD-modeling of the hydrodynamics of a prismatic spouted bed apparatus. Chemical Engineering Science (in preparation, 2008)

[6] Koch, D.L., Hill, R.J.: Inertial effects in suspension and porous-media flows. Annu. Rev. Fluid Mech., 619–647 (2001)
[7] Laux, H.: Modeling of dilute and dense dispersed fluid–particle flow. PhD thesis, NTNU Trondheim, Trondheim, Norway (1998)
[8] Lun, C.K.K., Savage, S.B., Jeffrey, D.J., Chepurniy, N.: Kinetic theories for granular flow: Inelastic particles in curette flow and slightly inelastic particles in a general flow field. J. Fluid Mech. 140, 223–256 (1984)
[9] Schaeffer, R.W.: Instability in the evolution equations describing incompressible granular flow. J. Diff. Equ. 66, 19–50 (1987)
[10] Schiller, L., Naumann, Z.: A drag coefficient correlation. Z. Ver. Deutsch. Ing. 77, 318 (1935)
[11] Syamlal, M., O'Brien, T.J.: Computer simulation of bubbles in a fluidized bed. AIChE Symp. Series 85, 22–31 (1989)
[12] Syamlal, M., Rogers, W., O' Brien, T.J.: MFIX Documentation: vol. 1, Theory guide. National technical information service. Springfield, VA, DOE/METC-9411004, NTIS/DE9400087 (1993)
[13] van der Hoef, M.A., Beestra, R., Kuipers, J.A.M.: Lattice Boltzmann simulations of low-Reynolds-number flow past mono- and bidisperse arrays of spheres: results for the permeability and drag force. J. Fluid Mech. 528, 233–254 (2005)
[14] Wen, C.-Y., Yu, Y.H.: Mechanics of fluidization. Chem. Eng. Prog. Symp. Series 62, 100–111 (1966)

Numerical Study of the Influence of Diffusion of Magnetic Particles on Equilibrium Shapes of a Free Magnetic Fluid Surface

S. Beresnev[1], V. Polevikov[2], and L. Tobiska[1]

[1] Institut für Analysis und Numerik, Otto-von-Guericke-Universität Magdeburg
[2] Department of Computational Mathematics, Belarusian State University, Minsk

Abstract. A general mathematical model and a computational method is proposed for studying the influence of the particle diffusion on equilibrium shapes of a magnetic liquid. It is applied on the ferrohydrostatic problem of equilibrium shapes of a magnetic liquid in a cylindrical cavity subject to a magnetic field created by a cylindrical permanent magnet located below the cavity. Numerical simulations show that the usual assumption of a uniform concentration field does not apply in the case of high-gradient magnetic fields.

1 Introduction

Because of their ability for ponderomotive interaction with an external magnetic field, magnetic fluids have not only provoked the development of a new direction in fluid mechanics but have become a new technological material which found a wide application in engineering [1, 3, 4, 9]. A magnetic fluid is a stable colloidal suspension of ferromagnetic particles in a carrier liquid (oil, water, bio-compatible liquid). The size of particles is of the order of 10^{-8}m, and they are in the Brownian motion state in the carrier liquid. In modelling magnetic fluids often a uniform distribution of particles has been assumed. However, owing to the fact that the particle possess magnetic properties, not only Brownian motion but also a magneto-phoresis diffusion process takes place in a magnetic fluid [5, 9]. This diffusion process becomes significant when the magnetic fluid is under the influence of a high-gradient magnetic field.

The main objective of this work is the investigation of the influence of diffusion of magnetic particles on equilibrium axisymmetric shapes of a free magnetic-fluid surface. As an example, we will consider the problem of equilibrium shapes of a magnetic fluid in a cylindrical cavity subject to a magnetic field created by a cylindrical permanent magnet located below the cavity. As the fluid magnetization value is directly proportional to the particle concentration in the fluid volume [2, 3, 7, 9], which is determined by the magnetic field structure, the diffusion effect is expected to become appreciable under a strongly non-uniform magnetic field.

2 Mathematical Model

Under the assumptions that the influence of the gravity force on the diffusion of Brownian particles is negligible and that the magnetic particles are of spherical form and of equal size, the magnetic particle mass transfer in a magnetic fluid can be described by the equation [2, 7]

$$\frac{\partial C}{\partial t} + \mathbf{v}\cdot\nabla C = D\nabla\cdot\left(\nabla C - CL(\xi H)\nabla(\xi H)\right), \tag{1}$$

where C is the volume concentration of the particles in the colloid; t the time variable; $\mathbf{v}$ the velocity of a convective motion; D the diffusion coefficient; H the magnetic field intensity; $\xi = \frac{\mu_0 m_m}{kT}$, $\mu_0 = 4\pi\times10^{-7}\mathrm{Hm}^{-1}$ is the magnetic constant (magnetic permeability of vacuum); m_m the magnetic moment of a particle; $k = 1.3806568\times10^{-23}\mathrm{JK}^{-1}$ the Boltzmann constant; T the particle temperature; m the mass of a particle; and

$$L(\xi H) = \coth \xi H - \frac{1}{\xi H}$$

the Langevin function. We assume that the fluid is incompressible and the boundary is impermeable, thus $\operatorname{div}\mathbf{v} = 0$ inside the fluid, $\mathbf{v}\cdot\mathbf{n} = 0$ on fixed walls, $\mathbf{v}\cdot\mathbf{n} = v_\Gamma$ on the free surface, where $\mathbf{n}$ denotes the outer unit normal at the boundary and v_Γ the velocity of the free surface. Equation (1) is supplemented with the condition of impermeability of boundaries by particles

$$\frac{\partial C}{\partial n} - C\,\xi\, L(\xi\, H)\frac{\partial H}{\partial n} = 0\,. \tag{2}$$

Moreover, a uniform concentration at the initial state,

$$C = C_0 = const\,, \quad t = 0 \tag{3}$$

is prescribed. Equation (1) together with condition (2) and (3) represent the mathematical model of the diffusion process of ferromagnetic particles in a magnetic fluid. Notice that the solution of the problem (1)-(3) satisfies the condition of conservation of the mean concentration:

$$\frac{1}{V}\int_V C dV = C_0 \quad \text{for all} \quad t \ge 0\,,$$

where V is the fluid volume (or the spatial domain of definition of the problem).

For $t \to \infty$ we obtain the steady-state concentration problem with $\mathbf{v} = \mathbf{0}$ which can be written in the form

$$\begin{gathered} \nabla\left(\nabla C - C\nabla(\ln\varphi)\right) = 0 \ \text{inside the fluid}, \\ \frac{\partial C}{\partial n} - \frac{\partial(\ln\varphi)}{\partial n} = 0 \ \text{at the boundary}, \\ \int_V C dV = VC_0\,, \ \text{where} \ \ \varphi = \frac{\sinh \xi H}{\xi H}\,. \end{gathered} \tag{4}$$

As show in [7], the problem (4) admits an analytical solution given by

$$C = \varphi \frac{C_0 V}{J_0}, \quad J_0 = \int_V \varphi dV . \tag{5}$$

The magnetic properties of the magnetic fluid are determined by its magnetization M which depends on both the magnetic field intensity H and the particle concentration C. In ferrohydrodynamics [2, 3, 7, 9], Langevin's magnetization law for a non-uniformly concentrated magnetic fluid is defined by the formula

$$M = M(H, C) = \frac{M_s}{C_0} L(\xi H) C \tag{6}$$

where M_s is the magnetic-fluid saturation magnetization; C_0 the mean concentration corresponding to a uniform distribution of the particles.

The equilibrium shapes of a free magnetic-fluid surface are described by the Young-Laplace equation. In the static case, it takes the form

$$\sigma_0 K = \frac{1}{2} \mu_0 \left(\frac{M}{H} H_n \right)^2 + \mu_0 \int_0^H M dH + p_f - p_0 \tag{7}$$

where K is the sum of principal free-surface curvatures which is positive if the surface is convex; σ_0 the surface tension coefficient, p_0 the pressure in a surrounding non-magnetic medium; p_f the thermodynamic pressure in the fluid.

Equation (7) is supplemented with boundary conditions as well as with the non- local (integral) condition of fluid volume conservation. As boundary conditions we have either conditions where the fluid contacts a solid wall specified by the wall geometry and a given wetting angle, or the symmetry condition at the rotation axis.

Let us now apply our model to a specific example. We consider a fixed volume of a magnetic fluid filling a cylindrical cavity. Let R_c be the radius of the cavity and

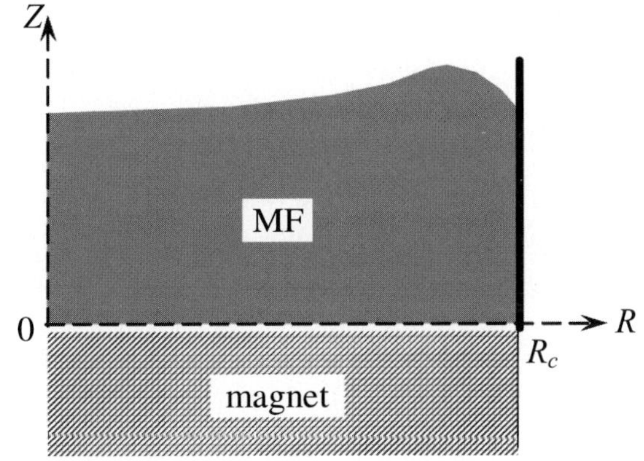

Fig. 1. Illustration of the problem

magnet, and let V denote the fluid volume. The free surface is supposed to by axisymmetric, its shape is completely determined by the equilibrium line. We introduce cylindrical coordinates R, Z and parametrise the equilibrium line with respect to the arc length S that takes the value S=0 at the symmetry axis Z=0 and $S=l$ at the solid wall $R= R_c$. Then, the equilibrium line is given by the parametric functions ($R(S)$, $Z(S)$). Note that the tangent vector $\tau = (R', Z')$ is oriented in the direction of increasing S. The surface curvature is defined by $K = -(RZ')'/(RR')$, where the prime stands for differentiation with respect to S. Natural boundary conditions corresponding to a contact angle of $\pi/2$ for (7) are the following

$$\begin{aligned} R(0) &= 0, \quad R'(0) = 1, \quad Z'(0) = 0, \\ R(l) &= R_c, \quad R'(l) = 1, \quad Z'(l) = 0. \end{aligned} \tag{8}$$

Note that also different values of the contact angle can be handled by our method. The given volume V is determined as the volume of a body of revolution and gives the constraint

$$V = 2\pi \int_0^l ZRR'dS . \tag{9}$$

The magnetic field is created by a permanent magnet having the form of a circular cylinder, see Fig1. In this case an analytical solution is known [8].

$$\begin{aligned} H_z &= \frac{M_c}{2\pi} \sum_{k=1}^{2} (-1)^k \int_{R-R_c}^{R+R_c} \frac{Z_k}{X^2 + Z_k^2} \sqrt{\frac{R_c^2 - (X-R)^2}{R_c^2 + Z_k^2 + 2XR - R^2}} dX, \\ H_r &= \frac{M_c}{2\pi} \sum_{k=1}^{2} (-1)^k \int_{R-R_c}^{R+R_c} \frac{X}{X^2 + Z_k^2} \sqrt{\frac{R_c^2 - (X-R)^2}{R_c^2 + Z_k^2 + 2XR - R^2}} dX, \end{aligned} \tag{10}$$

$$Z_1 = Z - h_c, \qquad Z_2 = Z - h_c .$$

Here, h_c denotes the half-height of the magnet and M_c is its magnetization. These formulas satisfy the Maxwell equation and describe the field in the whole space except in a narrow neighbourhood of the magnet. We will neglect the magnetization of the ferrofluid due to $M<<H$, and apply these formulas in the domain filled with the ferrofluid.

Next we choose the arclength l of the equilibrium line as a characteristic dimension and introduce dimensionless variables

$$s = \frac{S}{l}, \quad z = \frac{Z}{l}, \quad r = \frac{R}{l}, \quad h = \frac{H}{M_c}, \quad \overline{C} = \frac{C}{C_0}.$$

Then, we can reformulate (7)-(9) in the dimensionless form

$$\begin{aligned} & z''=r'F, && r''=-z'F, && 0\le s\le 1, \\ & r(0)=0, && r'(0)=1, && z'(0)=0, \\ & r(1)=\left(\frac{I_0}{U}\right)^{1/3}, && r'(1)=1, && z'(1)=0. \end{aligned} \tag{11}$$

A new condition for z at $s = 1$ can be obtained by integrating (9) by parts $z(1)=\dfrac{1}{\pi r^2(1)}(I_0+I_1)$.

Finally, the dimensionless magnet field can be derived form (10) to be

$$\begin{aligned} h_z &= \frac{r^2(1)}{2\pi}\int_{-1}^{1}[\sqrt{1-t^2}(z^-\xi^- - z^+\xi^+)]\,dt, \\ h_r &= \frac{r^2(1)}{2\pi}\int_{-1}^{1}\left[\sqrt{1-t^2}(r(1)t+r)(\xi^- - \xi^+)\right]dt. \end{aligned} \tag{12}$$

Here the following abbreviations have been used

$$F=\left[-\phi(\overline{C},h,s)+\gamma\right]-\frac{z'}{r},\quad \phi(\overline{C},h,s)=\frac{A_1}{r(1)}\overline{C}+\frac{1}{2r(1)}A_3\left(\overline{C}L(A_2h)\frac{h_n}{h}\right)^2,$$

$$\overline{C}=\varphi(A_2h)\frac{I_0}{J_0},\quad J_0=\int_v \varphi dv,\quad I_0=2\pi\int_0^1 zrr'ds,\quad I_1=2\pi\int_0^1 z'\frac{r^2}{2}ds,$$

$$\xi^\pm=\frac{1}{\left[(r(1)t+r)^2+(z^\pm)^2\right]\sqrt{r^2(1)+(z^\pm)^2+r^2+2r(1)tr}},\qquad z^\pm=z\pm r(1)P,$$

$$A_1=\frac{\mu_0 M_s R_c}{\xi\sigma_0},\quad A_2=\xi M_c,\quad A_3=\frac{\mu_0 M_s^2 R_c}{\sigma_0},\quad U=\frac{V}{R_c^{\,3}},\quad P=\frac{h_c}{R_c}.$$

The solution $z(s)$, $r(s)$ of the dimensionless problem (11), (12) is determined by dimensionless parameters A_1, A_2, A_3, U, P and the undefined constant γ.

3 Computational Algorithm

Following the strategy in [6], we construct a difference scheme of second-order approximation on the uniform grid $\{s_i = ih|\ i = 0,1,\dots,N;\ h=1/N\}$ for the problem (11)-(12). We will denote the solution of the difference scheme by the same letters as the corresponding solution of the differential equation

$$\begin{aligned}
&\Lambda_1(r,z,F)|_i = r_{\bar{s}s,i} + z_{\dot{s},i}F_i = 0, \quad i = 2,\dots,N-1\\
&\Lambda_2(r,z,F)|_i = z_{\bar{s}s,i} - r_{\dot{s},i}F_i = 0; \quad i = 1,\dots,N-1\\
&r_1 = h, \qquad r_{\bar{s},N} = 1;\\
&z_{\bar{s},0} = \frac{h}{2}F_0, \quad z_N = \frac{1}{\pi}\left(\frac{U}{I_0}\right)^{2/3}(I_0 + I_1);
\end{aligned} \tag{13}$$

where

$$\begin{aligned}
&F_i = F(r_i, z_i, r_{\dot{s}}, z_{\dot{s}}), \quad i = 1,\dots,N-1;\\
&F_0 = F(z_0), \quad i = 0;\\
&r_{s,i} = (r_{i+1} - r_i)/h, \qquad r_{\bar{s},i} = (r_i - r_{i-1})/h,\\
&r_{\dot{s},i} = (r_{i+1} - r_{i-1})/(2h), \quad r_{\bar{s}s,i} = (r_s - r_{\bar{s}})/h.
\end{aligned}$$

The integrals I_0 and I_1 are evaluated by the trapezoidal rule. For the calculation of the magnetic field components (12) a Gaussian quadrature rule has been used. The non-linear difference problem (13) has been solved by the two-layer iteration scheme

$$\begin{aligned}
&\frac{1}{\tau}(r^{n+1}_{\bar{s}s,i} - r^{n}_{\bar{s}s,i}) + \Lambda_1(r^n, z^n, F^n)|_i = 0; \quad i = 2,\dots,N-1;\\
&r_1^{n+1} = h, \quad r^{n+1}_{\bar{s},N} = 1;\\
&\frac{1}{\tau}(z^{n+1}_{\bar{s}s,i} - z^{n}_{\bar{s}s,i}) + \Lambda_2(r^n, z^n, F^n)|_i = 0; \quad i = 1,\dots,N-1;\\
&z^{n+1}_{\bar{s},0} = \frac{h}{2}F_0^n, \quad z_N^{n+1} = \frac{1}{\pi}\left(\frac{U}{I_0^n}\right)^{2/3}(I_0^n + I_1^n);
\end{aligned} \tag{14}$$

Here, $n = 0,1,\dots$ is the iteration number; $\tau > 0$ is a relaxation parameter. The two-layer iteration scheme (14) is requires in each iteration step the solution of linear tridiagonal systems which has been done by means the three-point elimination method (Thomas algorithm).

4 Numerical Results

The numerical study has been performed for fixed values $A_1 = 8$, $A_3 = 6.5$, $P = 1$, the two values U=0.07 and U=2 corresponding two different fluid volumes, and for a wide range of the parameter A_2 characterizing the magnitude of the magnetization of the permanent magnet. In order to study the influence of the diffusion process, computation have been carried out by assuming both a uniform distribution of the particle in the fluid and taking the diffusion effect into consideration.

The diffusion effect of particles can clearly be seen in Fig. 2 which represents the axisymmetric equilibrium shapes for a small volume U=0.07 and A_2=4. The magnetic field strength along the free surface shape differs for different fluid volumes and influences its shape as can be seen by comparing Fig. 2 and Fig. 3. Moreover, in Fig. 3,

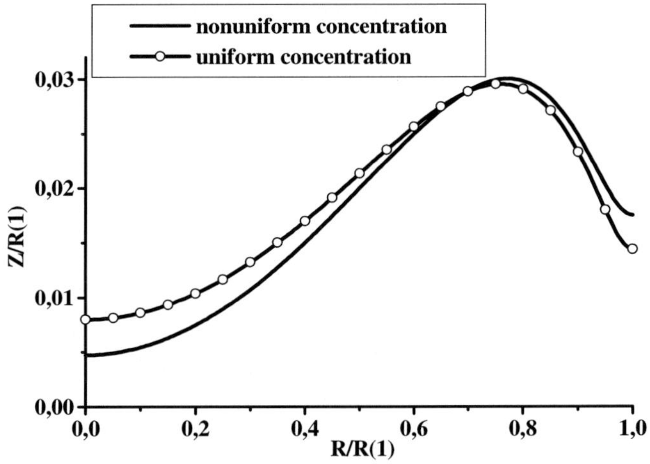

Fig. 2. Free surface shapes for $A_2 = 4$ and U=0.07

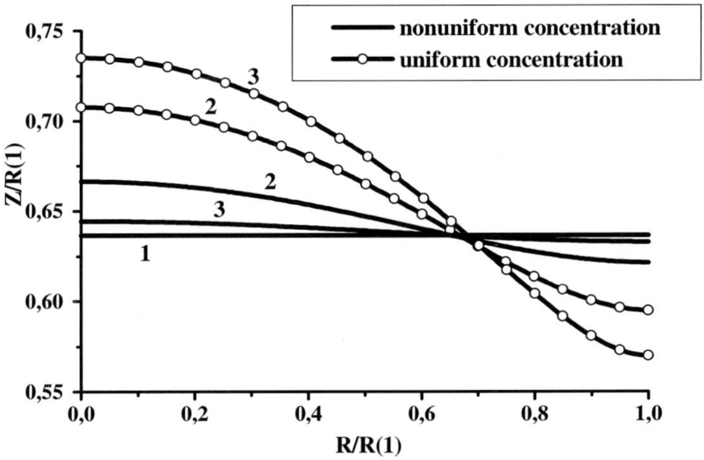

Fig. 3. Free surface shapes U=2: $1 - A_2 = 0$; $2 - A_2 = 2$; and $3 - A_2 = 4$

in which characteristic equilibrium shapes for three values of A_2 are shown, we detect a new property. For higher values of the parameter A_2 a distinction between the free surfaces with uniform and non-uniform particle concentration can be observed. Starting with small values of the parameter A_2 and increasing it, both curves move in the same direction. Then, the curve corresponding to a non-uniform particle concentration reaches a "critical" position, after which it starts to move in the opposite direction. As we can see in Fig. 4 the particle concentration far from the magnet becomes close to zero for increasing values of A_2 and does not influence considerably the free surface shape.

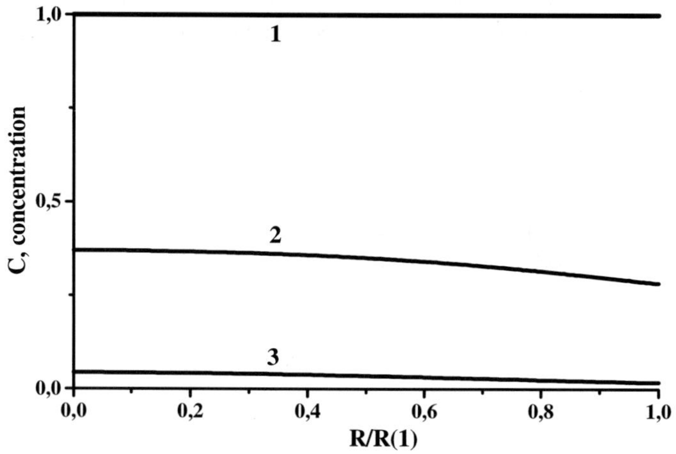

Fig. 4. Distribution of concentration at a free surface for U=2: 1 – A_2 = 0; 2 – A_2 = 2; and 3 – A_2 = 4

Acknowledgments. The authors would like to thank the German Academic Exchange Service (DAAD) and the Graduate School 828 (DFG) for partially supporting the research in this paper.

References

[1] Bashtovoi, V.G., Berkovski, B.M., Vislovich, A.N.: Introduction to thermomechanics of magnetic fluid. Hemisphere Publ., Washington (1988)

[2] Bashtovoi, V.G., Polevikov, V.K., Suprun, A.E., Stroots, A.V., Beresnev, S.A.: Influence of Brownian diffusion on statics of magnetic fluid. Magnetohydrodynamics 43(1), 17–25 (2007)

[3] Berkovski, B.M., Bashtovoi, V.G.: Magnetic Fluids and Applications Handbook. Begell House Publ., New York (1996)

[4] Berkovski, B.M., Medvedev, V.F., Krakov, M.S.: Magnetic Fluids: Engineering and Application. Oxford University Press, Berlin (1997)

[5] Blums, E., Cebers, A., Maiorov, M.M.: Magnetic Fluids. W. de Gryuter, Berlin (1997)

[6] Polevikov, V.K.: Methods of numerical modeling of two-dimensional capillary surfaces. Computational Methods in Applied Mathematics 4(1), 66–93 (2004)

[7] Polevikov, V., Tobiska, L.: On the solution of the steady-state diffusion problem for ferromagnetic particles in a magnetic fluid. Mathematical Modeling and Analysis 13(2), 233–240 (2008)

[8] Pshenichnikov, A.F.: Magnetic field in the vicinity of a single magnet. Magnetohydrodynamics 29, 33–36 (1993)

[9] Rosensweig, R.E.: Ferrohydrodynamics. Cambridge University Press, New York (1985)

A Note on Sectional and Finite Volume Methods for Solving Population Balance Equations

J. Kumar[1], G. Warnecke[1], M. Peglow[2], and E. Tsotsas[2]

[1] Institut für Analysis und Numerik, Otto-von-Guericke-Universität Magdeburg
[2] Institut für Verfahrenstechnik, Otto-von-Guericke-Universität Magdeburg

Abstract. In this work, we survey some numerical methods for solving population balance equations (PBEs). Among several numerical methods available in the literature, sectional and finite volume methods offer distinct advantages like conservation of mass and the accurate prediction of moments. The sectional methods are obtained from the standard form of PBE while the finite volume method is applied by transforming the PBE into a mass conservation law. Here we summarize these discretisation methods, mainly the cell average technique and the finite volume technique for solving aggregation-breakage PBEs. The cell average technique was developed by the authors. Furthermore, we also discuss possible extensions of the methods for solving higher dimensional population balance equations. There are two different approaches to deal with an n-dimensional PBE: computation on a reduced model and on a complete model. In contrast to the complete model we perform computation on a set of n one-dimensional PBEs in the reduced model approach. The computation time reduces drastically in the reduced model approach but it is not possible to capture the complete information of the particle property distribution.

1 Introduction

Population balances are widely known in high shear granulation, crystallization, atmospheric science and many other particles related engineering problems. The general form of population balance equation for combined aggregation and breakage is given as [1]

$$\frac{\partial n(t,x)}{\partial t} = \frac{1}{2}\int_0^x \beta(x-u,u)\,n(t,x-u)\,n(t,u)\,du - \int_0^\infty \beta(x,u)\,n(t,u)\,n(t,x)\,du + \int_x^\infty b(t,x,u)\,S(u)\,n(t,u)\,du - S(x)\,n(t,x). \tag{1}$$

Here n(t, x) denotes the number density distribution and x as well as u represent the particle volume. The first two terms on the right hand side are due to aggregation while the last two terms model the breakage process. The nature of the aggregation process is governed by the coagulation kernel $\beta(x, y)$ representing properties of the physical medium. It is a non-negative and satisfies the symmetry condition $\beta(x, y) = \beta(y, x)$. The breakage function b(t, x, y) is the probability density function for the formation of particles of size x from a particle of size y at time t. The selection function S(x) describes the rate at which particles are selected to break.

Besides the information of particle number density distribution, some integral properties like moments are also of great interest in several particulate systems. The jth moment of the particle size distribution is defined as

$$\mu_j(t) = \int_0^\infty x^j\, n(t,x)\,dx\,. \tag{2}$$

The first two moments are of special interest. The zeroth (j=0) and first (j=1) moments are proportional to the total number and total mass of particles, respectively. Furthermore, for certain particulate systems the second moment is proportional to the light scattered by particles in the Rayleigh limit [2, 3].

Various numerical techniques for solving the population balance equation can be found in the literature. The stochastic methods (Monte-Carlo) [4, 5] are very efficient for solving multi-dimensional population balance equations, since other numerical techniques become computationally very expensive in such cases. Wide varieties of finite element methods, weighted residuals, the method of orthogonal collocation and Galerkin's method are also used for solving population balance equations [6, 7]. In these methods, the solution is approximated by linear combinations of basis functions over a finite number of sub-domains. In the moment method [8, 9] the population balance equation is transformed into a system of ordinary differential equations (ODEs) describing the evolution of the moments of the particle size distributions. In recent times, the sectional methods [2, 23] have become computationally very attractive. A detailed review of sectional methods has been recently given by Vanni [10]. Very recently, a new method, called *cell average technique*, has been developed which gives more accurate results than existing sectional methods, see Kumar et al. [13, 16] and Kostoglou [18].

Since aggregation and breakage terms are mass conservative, one can also rewrite them in a conservative form of mass density g(t, x) = x n(t, x) as

$$\frac{\partial g(t,x)}{\partial t} + \frac{\partial}{\partial x}\left(F_{agg}(t,x) + F_{break}(t,x)\right) = 0\,. \tag{3}$$

The abbreviations *agg* and *break* are used for aggregation and breakage terms, respectively. The flux functions F_{agg} and F_{break} are given by

$$F_{agg}(t,x) = \int_0^x \int_{x-u}^\infty \frac{\beta(u,v)}{v} g(t,u)\,g(t,v)\,dv\,du\,, \tag{4}$$

and

$$F_{break}(t,x) = -\int_x^\infty \int_0^x u \frac{b(u,v)}{v} S(v)\,g(t,v)\,du\,dv. \tag{5}$$

It should be noted that both the forms of population balance equations mentioned above are interchangeable. For details about conservative forms of aggregation readers are referred to [14]. Furthermore, we have the following relationship from the Eqs. (1) and (3) as

$$-\frac{1}{x}\frac{\partial F_{agg}}{\partial x}=\frac{1}{2}\int_0^x \beta(x-u,u)\,n(t,x-u)\,n(t,u)\,du-\int_0^\infty \beta(x,u)\,n(t,u)\,n(t,x)\,du \quad (6)$$

and

$$-\frac{1}{x}\frac{\partial F_{agg}}{\partial x}=\int_x^\infty b(t,x,u)\,S(u)\,n(t,u)\,du-S(t,x)\,n(t,x)\,. \quad (7)$$

Finite volume schemes are frequently used for solving conservation laws, see for example LeVeque [21]. However, Filbet and Laurençot [19] were the first to apply this approach for solving population balance equations using the conservative form (3) for pure aggregation problems. Later the scheme has been applied to solve the combined Eq. (3) by Kumar [14] and Kumar et al. [11]. Furthermore the scheme has been extended to solve two-dimensional aggregation problems by Qamar and Warnecke [20]. Finally it has been observed that the finite volume scheme (FVS) is a good alternative to solve the population balance equations due to its automatic conservation property.

2 Numerical Methods for One-Dimensional PBEs

In this section two different types of numerical methods for solving one-dimensional population balances are summarized. The first method approximates the number density in terms of Dirac point masses and is based on an exact prediction of some selected moments. The second method treats aggregation and breakage as mass conservation laws and makes use of the finite volume type scheme to solve the population balance equation.

2.1 The Cell Average Technique

Here we present the general idea of the cell average technique. The entire size domain is divided into a finite number I of cells. The lower and upper boundaries of the ith cell are denoted by $x_{i-1/2}$ and $x_{i+1/2}$ respectively. All particles belonging to a cell are identified by a representative size in the cell, also called grid point. The representative size of a cell can be chosen at any position between the lower and upper boundaries of the cell. A typical discretised size domain is shown in Fig. 1. The representative of the ith cell is represented by $x_i = (x_{i-1/2} + x_{i+1/2})/2$. The width of the ith cell is denoted by $\Delta x_i = x_{i+1/2} - x_{i-1/2}$. The size of a cell can be fixed arbitrarily depending upon the process of application. In most applications, however, geometric type grids are preferred.

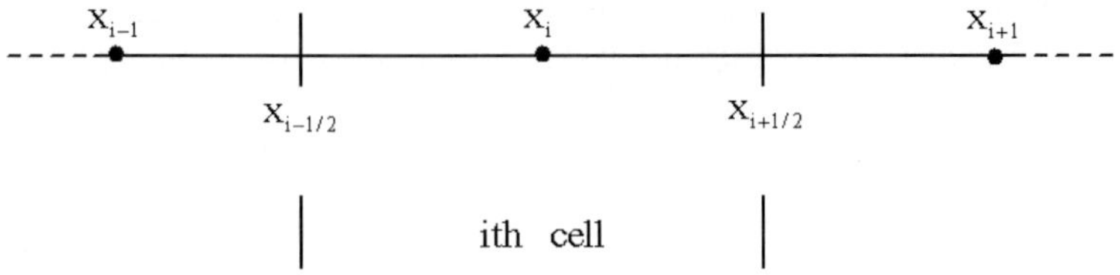

Fig. 1. A discretised size domain

We wish to transform the general continuous population balance equation into a set of I ODEs that can be solved using any standard ODE solver. Denoting the total number in the ith cell by N_i, we seek a set of ODEs of the following form

$$\frac{dN_i}{dt} = \underbrace{B_i^{CA}}_{\text{birth due to partuculate events}} - \underbrace{D_i^{CA}}_{\text{death due to particulate events}}, \quad i = 1, \ldots, I. \tag{8}$$

The particulate events that may change the number concentration of particles include breakage, aggregation, growth, nucleation etc. However, here we consider only aggregation and breakage. The abbreviation CA stands for cell average. Note that this general formulation is not similar to the traditional sectional formulation where birth terms corresponding to each process are summed up to determine the total birth. Here all particulate events will be considered in a similar fashion as we treat individual discrete processes.

The first step is to compute particle birth and death in each cell. Consideration of all possible events that lead to the formation of new particles in a cell provides the birth term. Similarly all possible events that lead to the loss of a particle from a cell give the death rate of particles. The new particles in the cell may either appear at some discrete positions or they may be distributed continuously in accordance with the distribution function. For example, in a binary aggregation process particles appear at discrete points in the cell whereas in the breakage process they are often distributed everywhere according to a continuous breakage function. Due to non-uniform grids, it is then possible that the size of a newborn particle in a cell does not match exactly with the representative size of that cell. The newborn particles, the sizes of which do not match with any of the representative sizes, cause an inconsistency of moments in the formulation. Therefore, the aim is to remedy the inconsistency in an efficient manner.

Let us demonstrate the basic concepts of the cell average technique by the following example. Particle births $B_i^1, B_i^2, \ldots, B_i^{I_i}$ take place at positions $y_i^1, y_i^2, \ldots, y_i^{I_i}$, respectively, due to some particulate processes like aggregation, breakage in the cell i. Here we considered the purely discrete case but analogous steps can be performed for continuous appearance of the particles in the cell. First we compute the total birth of particles in the ith cell as

$$B_i = \sum_{j=1}^{I_i} B_i^j. \tag{9}$$

Since we know the positions of the newborn particles inside the cell, it is easy to calculate the average volume of newborn particles $\overline{v_i}$. It is given by the following formula

$$\overline{v_i} = \frac{\sum_{j=1}^{I_i} y_i^j B_i^j}{B_i}. \tag{10}$$

Now we may assume that B_i particles are sitting at the position $\overline{v_i}$. It should be noted that the averaging process still maintains consistency with the first two moments. If the average volume $\overline{v_i}$ matches with the representative size x_i then the total birth B_i can be assigned to the node x_i. But this is rarely possible and hence the total particle birth B_i has to be reassigned to the neighbouring nodes such that the total number and mass remain conserved. Considering that the average volume $\overline{v_i} > x_i$, the assignment of particles must be performed by considering the following equations

$$\begin{aligned} a_1(\overline{v_i}, x_i) + a_2(\overline{v_i}, x_{i+1}) &= B_i, \\ x_i\, a_1(\overline{v_i}, x_i) + x_{i+1}\, a_2(\overline{v_i}, x_{i+1}) &= B_i\, \overline{v_i}. \end{aligned} \tag{11}$$

Here $a_1(\overline{v_i}, x_i)$ and $a_2(\overline{v_i}, x_{i+1})$ are the fractions of the birth B_i to be assigned at x_i and x_{i+1}, respectively. Solving the above equations we obtain

$$\begin{aligned} a_1(\overline{v_i}, x_i) &= B_i \frac{\overline{v_i} - x_{i+1}}{x_i - x_{i+1}} = B_i\, \lambda_i^+(\overline{v_i}), \\ a_2(\overline{v_i}, x_{i+1}) &= B_i \frac{\overline{v_i} - x_i}{x_{i+1} - x_i} = B_i\, \lambda_{i+1}^-(\overline{v_i}), \end{aligned} \tag{12}$$

where

$$\lambda_i^{\pm}(x) = \frac{x - x_{i\pm1}}{x_i - x_{i\pm1}}. \tag{13}$$

There are 4 possible birth fractions that may add a birth contribution at the node x_i: two from the neighbouring cells and two from the ith cell. Collecting all the birth contributions, the birth term for the cell average technique is given by

$$\begin{aligned} B_i^{CA} = {} & B_{i-1}\, \lambda_i^-(\overline{v_{i-1}})\, H(\overline{v_{i-1}} - x_{i-1}) + B_i\, \lambda_i^-(\overline{v_i})\, H(x_i - \overline{v_i}) \\ & + B_i\, \lambda_i^+(\overline{v_i})\, H(\overline{v_i} - x_i) + B_{i+1}\, \lambda_i^+(\overline{v_{i+1}})\, H(x_{i+1} - \overline{v_{i+1}}). \end{aligned} \tag{14}$$

The *Heaviside step function* is a discontinuous function also known as *unit step function* and is here defined by

$$H(x)=\begin{cases}1, & x>0\\ \frac{1}{2}, & x=0\\ 0, & x<0.\end{cases} \tag{15}$$

Substituting the values of B_i^{CA} and D_i^{CA} into Eq. (8) we obtain a set of ordinary differential equation. It will be then solved by any higher order ODE solver. Note that there is no need to modify the death term since particles are just removed from the grid points and therefore the formulation remains consistent with all moments due to discrete death. As a result the death term in the cell average formulation D_i^{CA} is equal to sum of total death in the ith cell. For detail description of the scheme and numerical results, readers are referred to [13, 14, 16].

2.2 The Finite Volume Technique

As mentioned before, finite volume schemes can be applied directly to conservation laws (3). We first discretise the space into cells $\Lambda_i = [x_{i-1/2}, x_{i+1/2}[$ and define an approximation of the cell averages of the solution on cell Λ_i as

$$g_i \approx \frac{1}{\Delta x_i}\int_{x_{i-1/2}}^{x_{i+1/2}} g(t,x)\,dx. \tag{16}$$

Integrating Eq. (3) over a cell Λ_i, we obtain

$$\frac{dg_i(t)}{dt} = -\frac{1}{\Delta x_i}\left[J_{i+1/2}^{agg} + J_{i+1/2}^{break} - J_{i-1/2}^{agg} - J_{i-1/2}^{break}\right], \tag{17}$$

where the numerical flux corresponding to aggregation and breakage problems are given by

$$\begin{aligned} J_{j+1/2}^{agg} &= \sum_{k=1}^{i} \Delta x_k g_k \left(\sum_{j=\alpha_{i,k}}^{I} \int_{\Lambda_j} \frac{\beta(u,x_k)}{u}du\, g_j + \int_{x_{i+1/2}-x_k}^{x_{\alpha_{i,k}-1/2}} \frac{\beta(u,x_k)}{u}du\, g_{\alpha_{i,k}-1} \right), \\ J_{i+1/2}^{break} &= -\sum_{k=i+1}^{I} g_k \int_{\Lambda_k} \frac{S(\varepsilon)}{\varepsilon}d\varepsilon \int_0^{x_{i+1/2}} ub(u,x_k)\,du. \end{aligned} \tag{18}$$

The integer $\alpha_{i,k}$ corresponds to the index of the cell such that $x_{i+1/2} - x_k \in \Lambda_{\alpha_{i,k}-1}$. Here the functions $J_{i+1/2}^{agg}$ and $J_{i+1/2}^{break}$ are call numerical fluxes for aggregation and breakage, respectively. They are obtained by approximating the continuous fluxes (4) and (5). Other details of the formulation, convergence and stability can be found in Filbet and Laurençot [21] and Kumar [14]. Several numerical comparisons of the cell average and finite volume scheme are recently presented by the authors [11]. It is concluded there that the finite volume scheme predicts more accurate results for particle number density on fine grids, on the other hand quite reasonable results for number density as well as for its moments can be obtained using the cell average scheme even on coarse grids.

3 Numerical Methods for Two-Dimensional PBEs

Now we mention briefly numerical schemes for the solutions of a multi-dimensional PBE. As mentioned before we will investigate numerical methods on two major classes of multi-dimensional PBE: reduced models and complete models. The reduced model is based on a reduction of a complete n-dimensional population balance to a set of n one-dimensional PBEs. Model reduction is done to get an approximation of some average values at low computational cost. On the other hand a complete model provides the entire property distribution at a very high computational cost.

3.1 Reduced Model Approach

A two-dimensional particle property distribution is defined as f(t, v, c), where v and c are two distinct properties, granule volume (size) and tracer volume, respectively. Thus, the total number in a domain D is given by

$$\int_D f(t,v,c)\,dv\,dc. \tag{19}$$

It should be noted that granule volume v contains volume of tracer and volume of particles, that is, $c < v$. The two-dimensional PBE can be obtained by extending the classical one-dimensional PBE [22] to two-dimensional space as

$$\begin{aligned}\frac{df(t,v,c)}{dt} &= \frac{1}{2}\int_0^v \int_{\max(0,c-v+\varepsilon)}^{\min(c,\varepsilon)} \hat{\beta}(v-\varepsilon,\varepsilon,c-\gamma,\gamma)\,f(t,v-\varepsilon,c-\gamma)\,f(t,\varepsilon,\gamma)\,d\gamma\,d\varepsilon \\ &\quad - \int_0^\infty \int_0^\varepsilon \hat{\beta}(v,\varepsilon,c,\gamma)\,f(t,v,c)\,f(t,\varepsilon,\gamma)\,d\gamma\,d\varepsilon.\end{aligned} \tag{20}$$

Let us assume that the aggregation kernel $\hat{\beta}$ depends only on size of the granules, but not on the tracer contents within the granules, i.e. $\hat{\beta} = \beta(v,\varepsilon)$. Then the 2D PBE can be converted into two 1D PBEs corresponding to the conventional number density n(t, v) and mass of tracer within granules m(t, v). The number density n(t, v) can be obtained from f by integrating over all possible tracer masses

$$n(t,v) = \int_0^v f(t,v,c)\,dc. \tag{21}$$

The mass of tracer within the granules of size v is given analogously by

$$m(t,v) = \int_0^v c\, f(t,v,c)\, dc. \tag{22}$$

From (20) the following two one-dimensional PBEs can easily be obtained [14]

$$\frac{\partial n(t,v)}{\partial t} = \frac{1}{2}\int_0^v \beta(v-u,u)\, n(t,v-u)\, n(t,u)\, du - \int_0^\infty \beta(x,u)\, n(t,u)\, n(t,v)\, du\,, \tag{23}$$

and

$$\frac{\partial m(t,v)}{\partial t} = \frac{1}{2}\int_0^v \beta(x-u,u)\, m(t,v-u)\, n(t,u)\, du - \int_0^\infty \beta(v,u)\, n(t,u)\, m(t,v)\, du\,. \tag{24}$$

Equations (23) and (24) are ordinary integro-differential equations, which can be solved numerically using one of the schemes discussed before. Eq. (23) is exactly the same which we dealt with in the previous section. But the numerical discretisation of Eq. (24) is slightly more complicated. Hounslow et al. [23] developed a discretised method for solving PBE (23) and later they extended the discretisation for the solution of Tracer Population Balance Equation (TPBE) (24), see Hounslow et al. [24]. The discretisation of TPBE was applicable only for a special geometric grid of the type $v_{i+1} = 2v_i$. In the next section we first briefly discuss the Discretised Tracer Population Balance Equation (DTPBE) and then propose a new idea for the better accuracy and implementation.

One can use the DTPBE of Hounslow et al. [24] for the computation of various extensive properties (amount of water within the particle, enthalpy of particles etc.) of aggregating systems. The equations predict the total amount of extensive properties exactly but it fails to predict intensive properties which are proportional to the ratio of extensive properties and mass of granules. For the purpose of illustration let us consider an example where a particle system is described by the two properties: volume of particles and amount of water within the particles. Our interest is to calculate the particle moisture content as the ratio of water mass to particle mass – an intensive property. We assume that the density is constant and particle volume is equal to the particle mass. For simplicity let us assume that initially the water mass inside a particle is equal to the particle mass. In other words, the ratio between water mass and particle mass or equivalently particle moisture content is constant over the particle size range and is equal to one. Since agglomeration is the only governing mechanism which changes the particle size, the particle moisture content should be constant throughout the process.

Let us consider the following initial condition for the PSD with volume as the distributed property

$$n(v) = \frac{N_0}{v_0}\exp\left(-\frac{v}{v_0}\right). \tag{25}$$

In the DTPBE [24], the particle size domain is divided into discrete size ranges using a geometric discretisation of the type $v_{i+1} = 2v_i$. Integration of Eq. (25) over an interval $[v_i, v_{i+1}[$ gives the total number of particles within the interval

$$N_i = \frac{N_0}{2}\left[\exp\left(-\frac{v_i}{v_0}\right) - \exp\left(-\frac{v_{i+1}}{v_0}\right)\right]. \qquad (26)$$

The mass of particles in an interval $[v_i, v_{i+1}[$ is approximated as

$$M_{P,i} = N_i\left(\frac{v_i + v_{i+1}}{2}\right). \qquad (27)$$

Here we make an assumption of constant density, so that the mass could be replaced by volume. The initial condition for water mass distribution is chosen in such a way that the total mass of water $M_{w,i}$ within the particles in the interval $[v_i, v_{i+1}[$ is equal to the total mass of particles in this interval $M_{w,i} = M_{P,i}$.

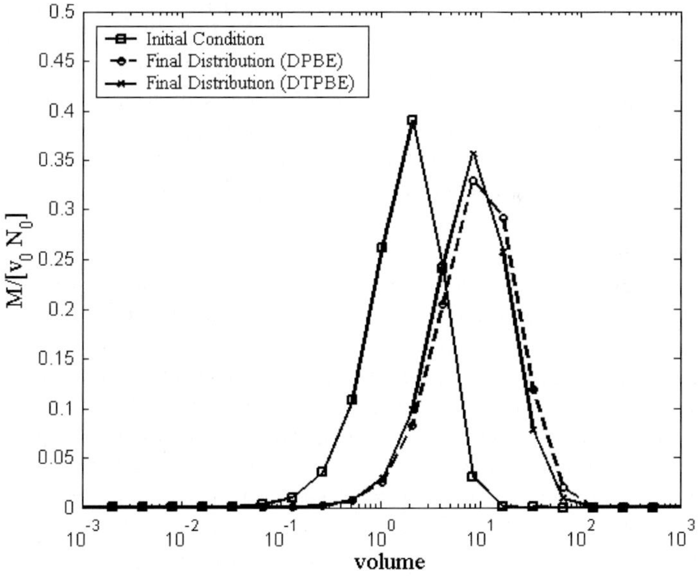

Fig. 2. Initial and final distribution for size independent aggregation, $I_{agg} = 4/5$ (from [15])

The computation is made for a size independent kernel. We calculated the particle mass and the water content within the particles as a function of particle volume using Hounslow's discrete population balance equation [23] and Hounslow's DTPBE [24], respectively. The numerical results at the extent of aggregation $I_{agg} = 0.8$ together with initial condition have been plotted in Fig. 2. Clearly the DTPBE fails to predict the water distribution within the particles correctly since both water distribution and mass distribution must be the same during the process. Furthermore, Figure 3 includes initial and final ratio of water mass and particle mass within the intervals. It can be seen from the figure that the final ratio is not constant. However, it should be pointed out that both discretised formulations conserve mass.

We presented in [15] a new formulation of the DTPBE of Hounslow et al. [24]. The new version retains all the advantages of the original version. It has been tested for a simple aggregation problem in a batch system. Contrary to the original version, the new version predicts a constant ratio of tracer mass and granule mass during aggregation in a batch system. It has been shown that both versions predict the same results for the total tracer mass, while the original version is more accurate for tracer-weighted mean particle volume. Moreover, the DTPBE has been extended geometric grids of the type $v_{i+1} = 2^{1/q}v_i$ and validated by many problems where analytical solutions are available.

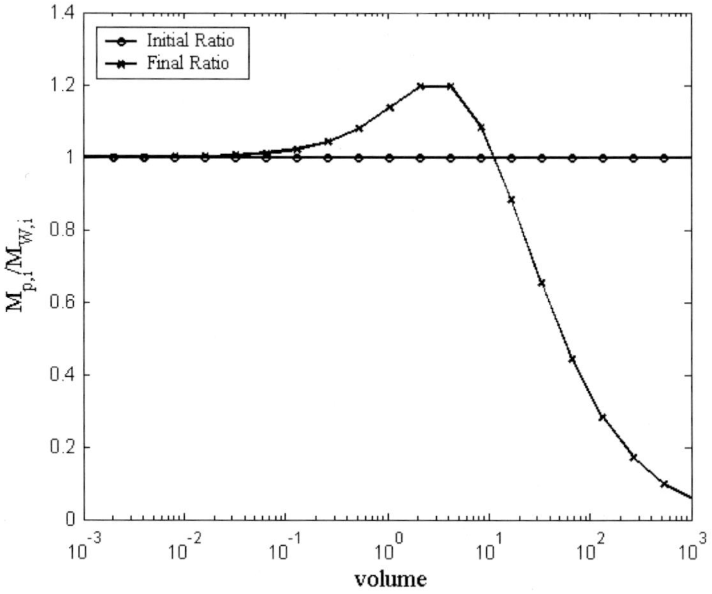

Fig. 3. Initial and final ratio of particle mass and water mass distribution for the results of Fig. 2 (from [15])

Additionally, a new discretisation for tracer population balance equations based on the cell average technique is developed [17]. It is compared to the modified discretised tracer population balance equation of Peglow et al. [15]. The new formulation provides excellent prediction of the tracer mass distribution in all test cases. Furthermore, the new formulation is more efficient from a computational point of view. It takes less computational effort and is able to give a very good prediction on a coarser grid. Again, it is independent of the type of grid chosen for computation, i.e. the scheme can be implemented using any type of grid. For finer grids, both formulations tend to produce the same results. The performance of the new formulation is illustrated by the comparison with various analytically tractable problems. Moreover, the new formulation preserves all the advantages of the modified discretised tracer population balance equation and provides a significant improvement in predicting tracer mass distribution and tracer-weighted mean particle volume during an aggregation process.

3.2 Complete Model Approach

The numerical solution of the above two dimensional PBE (20) is difficult due to the double integral and the non-linear behaviour of the equation. There are only few numerical techniques available in the literature to compute the complete property distribution but all of them either have problems regarding the preservation of properties of the distribution or they are computationally very expensive. The accurate computation of total mass and number is essential in many applications. In order to overcome the computational load, an attempt has been made in the previous section. The reduced model is, of course, computationally less expensive but it is not possible to capture the complete two-dimensional behaviour of the population with the model. In the reduced model, it has been assumed that particles of the same size contain the same amount of the second property. Therefore the objective of this work is to calculate the complete two-dimensional distribution of the population at a relatively low computational cost.

The cell average discretisation has been recently extended for solving a two-dimensional population balance equation in [12]. Similar to the one-dimensional case, the scheme is based on an exact prediction of certain moments of the population. The formulation is quite simple to implement, computationally less expensive than previous approaches and highly accurate. Numerical diffusion is a common problem with many numerical methods while applied on coarse grids. The presented technique nearly eliminates numerical diffusion and predicts three moments of the population at high accuracy. The technique can be implemented on any type of grid. The accuracy of the scheme has been analysed by comparing analytical and numerical solutions of some test problems. The numerical results are in excellent agreement with the analytical results and show the ability to predict higher moments very precisely. Additionally, an extension of the proposed technique to higher dimensional problems is also discussed in [12].

4 Conclusions

This work presents a recently developed numerical scheme, the cell average technique, for solving a general population balance equation (1D & 2D) which assigns particles within the cells more precisely. The technique follows a two step strategy: one is to calculate the average size of the newborn particles in a cell and the other to assign them to neighbouring nodes such that the properties of interest are exactly preserved. The new technique preserves all the advantages of conventional discretised methods and provides a significant improvement in predicting the particle size distribution and higher moments. The technique allows the convenience of using non-homogeneous, geometric- or equal-size cells.

A finite volume approach for solving aggregation-breakage PBEs is discussed. This technique treats aggregation and breakage as mass conservation laws and makes use of the finite volume scheme to solve the population balance equations. It is concluded that the number density distribution calculated by the finite volume scheme is more accurate than that obtained by the cell average technique. On the other hand, in contrast to the cell average technique, it has been observed that the finite volume scheme produces poor results for the zeroth moment.

Furthermore, the discretised tracer population balance equations of Hounslow et al. [24] for aggregation problems are discussed and modified. It is shown that the original version is not entirely consistent with the associated discretised population balance equation. These inconsistencies are remedied in a new formulation [15] that retains the advantages of the original discretised tracer population balance equation, such as conservation of total tracer mass, prediction of tracer-weighted mean particle volume, and so on.

Moreover, a new discretisation (see [17]) for the tracer population balance equations has been developed. It is compared to the modified discretised tracer population balance equation of Peglow et al. [15]. The new formulation provides an excellent prediction of the tracer mass distribution in all test cases. Furthermore, the new formulation is more efficient from a computational point of view. It takes less computational effort and is able to give a very good prediction on a coarser grid. Again, it is independent of the type of grid chosen for computation, i.e. the scheme can be implemented using any type of grid.

Acknowledgments. This work was supported by the Graduiertenkolleg-828, "Micro-Macro-Interactions in Structured Media and Particles Systems", Otto-von-Guericke-Universität Magdeburg. The authors gratefully acknowledge for funding through this PhD program.

References

[1] Ramkrishna, D.: Population balances. Theory and applications to particulate systems in Engineering, 1st edn. Academic Press, New York (2000)
[2] Kumar, S., Ramkrishna, D.: On the solution of population balance equations by discretization-I. A fixed pivot technique. Chem. Eng. Sci. 51, 1311–1332 (1996)
[3] Seifert, A., Beheng, K.D.: A double-moment parameterization for simulating auto conversion, accretion and self collection. Atmos. Res. 59, 265–281 (2001)
[4] Kruis, F.E., Maisels, A., Fissan, H.: Direct simulation Monte Carlo method for particle coagulation and aggregation. AIChE J. 46, 1735–1742 (2000)
[5] Lee, K., Matsoukas, T.: Simultaneous coagulation and breakage using constant-N Monte Carlo. Powder Technol. 110, 82–89 (2000)
[6] Mahoney, A.W., Ramkrishna, D.: Efficient solution of population balance equations with discontinuities by finite elements. Chem. Eng. Sci. 57, 1107–1119 (2002)
[7] Nicmanis, M., Hounslow, M.J.: A finite element analysis of the steady state population balance equation for particulate systems: Aggregation and growth. Comput. Chem. Eng. 20, 261–266 (1996)

[8] Madras, G., McCoy, B.J.: Reversible crystal growth-dissolution and aggregation breakage: Numerical and moment solutions for population balance equations. Powder Technol. 143, 297–307 (2004)
[9] Marchisio, D.L., Fox, R.O.: Solution of population balance equations using the direct quadrature method of moments. J. Aerosol Sci. 36, 43–73 (2005)
[10] Vanni, M.: Discretization procedure for the breakage equation. AIChE J. 45, 916–919 (1999)
[11] Kumar, J., Warnecke, G., Peglow, M., Heinrich, S.: Comparison of numerical methods for solving population balance equations incorporating aggregation and breakage. Powder Technol. (accepted for publication, 2008)
[12] Kumar, J., Warnecke, G., Peglow, M., Heinrich, S.: The cell average technique for solving multi-dimensional aggregation population balance equation. Comput. Chem. Eng. 32, 1810–1830 (2008)
[13] Kumar, J., Warnecke, G., Peglow, M., Heinrich, S.: An efficient numerical technique for solving population balance equation involving aggregation, breakage, growth and nucleation. Powder Technol. 179, 205–228 (2008)
[14] Kumar, J.: Numerical approximations of population balance equations in particulate systems. In: Dissertation, Otto-von-Guericke University Magdeburg, Germany, pages. 241. Docupoint-Verlag Magdeburg (2006), `http://diglib.uni-magdeburg.de/Dissertationen/2006/jitkumar.pdf` ISBN 3-939665-13-4
[15] Peglow, M., Kumar, J., Warnecke, G., Heinrich, S., Tsotsas, E., Mörl, L., Hounslow, M.: An improved discretized tracer mass distribution of Hounslow et al. AIChE J. 52, 1326–1332 (2006)
[16] Kumar, J., Peglow, M., Warnecke, G., Heinrich, S., Mörl, L.: Improved accuracy and convergence of discretized population balances: The cell average technique. Chem. Eng. Sci. 61, 3327–3342 (2006)
[17] Kumar, J., Peglow, M., Warnecke, G., Heinrich, S., Mörl, L.: A discretized model for tracer population balance equation: Improved accuracy and convergence. Comput. and Chem. Eng. 30, 1278–1292 (2006)
[18] Kostoglou, M.: Extended cell average technique for the solution of coagulation equation. J. Colloid Interface Sci. 306, 72–81 (2007)
[19] Filbet, F., Laurençot, P.: Numerical simulation of the Smoluchowski coagulation equation. SIAM J. Sci. Comp. 25, 2004–2028 (2004)
[20] Qamar, S., Warnecke, G.: Solving population balance equations for two-component aggregation by a finite volume scheme. Chem. Eng. Sci. 62, 679–693 (2007)
[21] LeVeque, R.J.: Finite Volume Methods for Hyperbolic Problems, 1st edn. Cambridge University Press, Cambridge (2002)
[22] Hulburt, H.M., Katz, S.: Some problems in particle technology. A statistical mechanical formulation. Chem. Eng. Sci. 19, 555–578 (1964)
[23] Hounslow, M.J., Ryall, R.L., Marshall, V.R.: A discretized population balance for nucleation, growth and aggregation. AIChE J. 38, 1821–1832 (1988)
[24] Hounslow, M.J., Pearson, J.M.K., Instone, T.: Tracer studies of high shear granulation: II. Population balance modeling. AIChE J. 47, 1984–1999 (2001)

Population Balance Modelling for Agglomeration and Disintegration of Nanoparticles

Y.P. Gokhale[1], J. Kumar[2], W. Hintz[1], G. Warnecke[2], and J. Tomas[1]

[1] Institut für Verfahrenstechnik, Otto-von-Guericke-Universität Magdeburg
[2] Institut für Analysis und Numerik, Otto-von-Guericke-Universität Magdeburg

Abstract. To control the particle size and morphology of nanoparticles is of crucial importance from a fundamental and an industrial point of view. Nanoparticle precipitation in the batch reactor is investigated experimentally as well as by simulations based on population balance equations combined with the model using titanium dioxide as the material under investigation. The superposition of the population balance models for agglomeration and disintegration with the different kernels leads to a system of partial differential equations, which can be numerically solved by various methods. This includes a comparison of the derived particle size distributions, moments and their accuracy depending on the initial particle size distribution. Furthermore, the capability of the precipitation model is evaluated, achieving good agreement of the particle sizes between experimental and simulation results. Finally, the computational effort of both methods in comparison to the prior mentioned parameters is evaluated in terms of practical applications.

1 Introduction

There is a tremendous number of high-tech applications of nanostructure metal oxide material devices such as dye-sensitised solar cells [1], displays and smart windows [2], chemical [3], gas [4], and biosensors [5], lithium batteries [6], and supercapacitors [7]. Hence, it is crucial, both, from the fundamental and the industrial point of view, to control the particle size and morphology of nanoparticles. Indeed, further development, improvement, and optimisation of such devices will be reached, if the following two factors exist. One, better understanding of the unique physical properties and the complex electronic structure; second, a better match of the materials design and the required device applications for the sake of miniaturization and performance [8]. One of the fundamental issues that need to be addressed in modelling macroscopic mechanical behaviour of nanostructured materials based on molecular structure is the large difference in length scales. On the opposite ends of the length scale, the spectrum of computational chemistry and solid mechanics consists of highly developed and reliable modelling methods. Computational chemistry models predict molecular properties based on known quantum interactions, while computational solid mechanics models predict the macroscopic mechanical behaviour of materials idealized as continuous media based on known bulk material properties. However, a corresponding model does not exist in the intermediate length scale range. If a hierarchical approach is used to model the macroscopic behaviour of nanostructured materials, then a methodology must be developed to link the molecular structure and macroscopic properties. Many properties of solid particles are not only a function of the materials bulk properties but also depend on the particle size distribution (PSD). These property changes arise from the increasing influence of surface properties in comparison to volumetric bulk properties as the particle size decreases. Especially nanoscaled particles show altered properties and have therefore widespread applications like pigments, pharmaceuticals, cosmetics, ceramics, catalysts and filling materials. Since the desired product properties might vary with particle size as well as with the degree of aggregation or the aggregate structure, controlling of the PSD and the aggregate structure is a key criterion for product quality. New and improved products can then be designed by adjusting and optimising the PSD and the particle structure. Precipitation is a promising method for the economic production of commercial quantities of nanoparticles as it is fast and operable at an ambient temperature. However, process control due to the rapidity of the involved processes and especially the control of aggregation through stabilization represent a challenge. Nanoparticle precipitation with the scope on the influence of mixing and super-saturation, the thermodynamic driving force of solid formation, has been investigated experimentally as well as theoretically by the authors [0 -13].

2 Experimental Condition

Titanium dioxide nanoparticles have been prepared by sol-gel precipitation processing. The reaction starts with titanium tetra isopropoxide (TTIP) as organic precursor on a laboratory scale [14]. Two simultaneous reactions, namely hydrolysis and polycondensation, take place during reaction of TTIP with water in presence of nitric acid. This process has been characterised by a rapid precipitation of large aggregates on a millisecond time scale, followed by a slow redispersion (peptisation) induced by the presence of nitric acid and shear stress applying a turbulent hydrodynamic regime inside the stirred tank reactor.

Intermediate stages for manufacturing titanium dioxide nanoparticles have been written as shown in Figure 1. Production of nanosized titanium dioxide has been carried out on a laboratory scale using a 250ml baffled, stirred batch reactor with confirmed standard configuration. The reaction suspension has been stirred continuously (6-blade stirrer). The centre of the impeller has been positioned at 1/3 height of the tank, the rotational speed has been measured. A thermostat has been used to keep a constant temperature of 50°C inside of the batch reactor.

For generating titanium dioxide nanoparticles via a sol-gel process, the procedure is as follows. A specified amount of 0.1 M HNO_3 (141ml, p.a.) is placed into the batch reactor. The organic precursor titanium tetra isopropoxide (~9.8g TTIP, 98%, Alfa Aesar) is added to the heated solution under constant stirring. Precipitation is observed to be occurring immediately due to the presence of dilute nitric acid in the reaction mixture. Temperature is held constant for the rest of the redispersion reaction, accordingly optimal reaction conditions for the titanium dioxide nanoparticles synthesis. The variation of the stirrer speed is made to investigate the influence of turbulent hydrodynamic conditions on particle size distribution and particle structure

Hydrolysis

$$Ti(OC_3H_7)_4 + 4\,H_2O \xrightarrow[\text{pH 1,3 (0,1 M HNO}_3)]{\text{aqueous suspension, 50 °C}} Ti(OH)_4 + 4\,C_3H_7OH$$

Titanium tetra isopropoxide (TTIP) — Titanium hydroxide — Isopropanol

Polycondensation

$$Ti(OH)_4 \xrightarrow[\text{pH 1,3 (0,1 M HNO}_3)]{\text{aqueous suspension, 50 °C}} TiO_2 + 2\,H_2O$$

Titanium hydroxide — Titanium dioxide

Redispersion

$$TiO_2\ (\text{Gel}) \xrightarrow[\text{pH 1,3 (0,1 M HNO}_3)]{\text{aqueous suspension, 50 °C}} \text{nano-}TiO_2\ (\text{Sol})$$

Titanium dioxide — Titanium dioxide

Fig. 1. Intermediate reactions for preparation of nanosized titanium dioxide

during reaction. The rotational speed of stirrer (shear rates from $\dot{\gamma}=370s^{-1}$ to $\dot{\gamma}=2515s^{-1}$) is chosen to be large in order to maximize redispersion of agglomerates and to keep particles away from sedimentation settlement.

3 Characterisation of TiO_2 Nanoparticles

Particle sizes smaller than 1μm have been measured via dynamic light scattering (Model Zetasizer, Malvern Instruments, UK) using a He-Ne laser as light source (λ = 633nm). All size measurements has been performed at a scattering angle of 90° and at 25°C. For particle size distributions in the micrometer size range, a laser diffraction method is used (Model Mastersizer 2000, Malvern Instruments, UK, He-Ne laser as red light source, λ = 633nm, solid state laser as blue light source, λ = 466nm)[14]. Samples withdrawn are mild stirred and dispersed at the same concentration and pH conditions of nitric acid inside in a small volume sample dispersion unit HYDRO SM of Mastersizer 2000 before being measured. For evaluating particle stability, zeta-potential measurements based on electrophoretical mobility have been carried out with a Malvern Zetasizer instrument.

a. Particle size distributions during redispersion

At the initial time of redispersion, the obtained agglomerate size distributions of titania are shown in Figure 2. Particle sizes in the micron range have been observed at the beginning of experiment due to agglomeration micro process. Here the fluid velocity gradients of the applied shear rate bring the particles close enough to collide and adhere. The agglomeration continues to form large, porous, and open structures [15]. The redispersion process becomes more significant as the agglomerates become larger, and will limit the agglomerates growth, smaller particles will be created by disintegration.

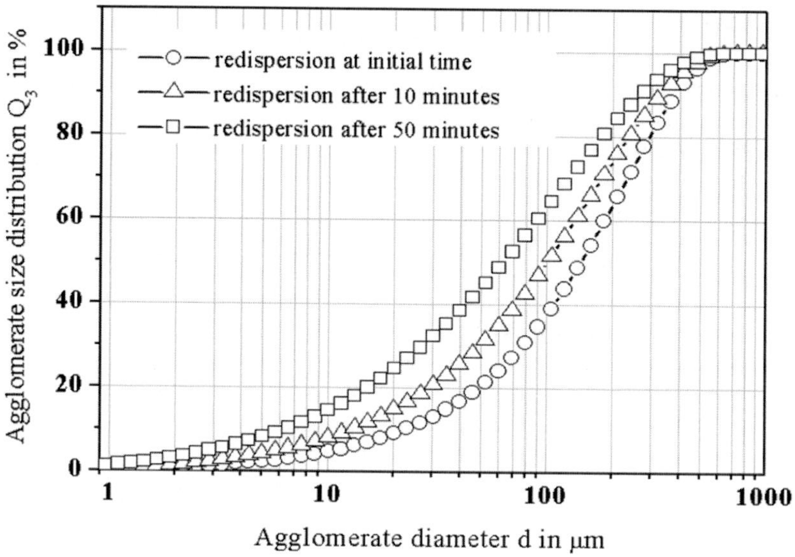

Fig. 2. Cumulative agglomerate size distributions Q_3 (d) at initial time, 10 and 50 minutes of redispersion reaction ($\dot{\gamma}$=2663s^{-1})

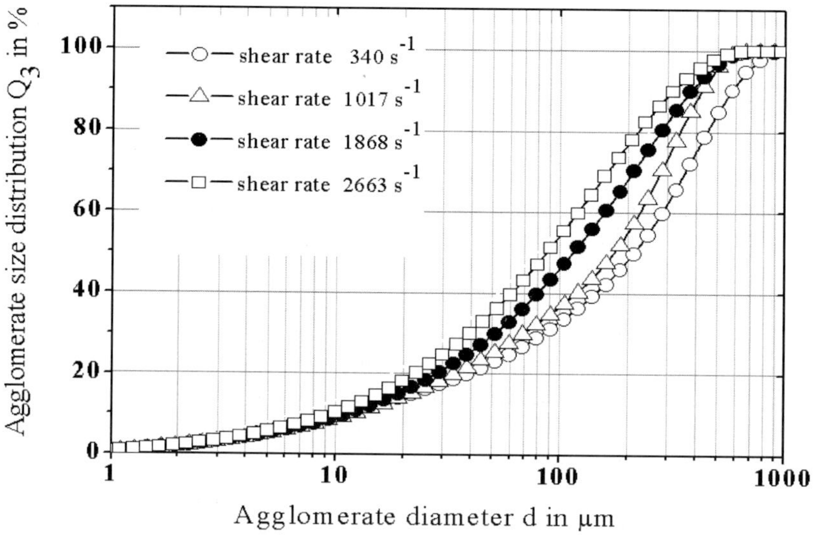

Fig. 3. Cumulative agglomerate size distributions Q_3 (d) for different shear rates $\dot{\gamma}$ at the initial state (20 min) of redispersion reaction

b. Effect of shear rate on particle size distribution

To find the influence of the shear rate $\dot{\gamma}$ on particle size distribution and particle structure, different shear rates are applied to stress mechanically the agglomerates and to enhance the redispersion (shear rate from $\dot{\gamma}$=340 to $\dot{\gamma}$=2663s^{-1} equals stirrer tip speed from 0.6ms^{-1} to 2.2ms^{-1}).

Figure 3 shows that a dependency between particle sizes characterized by median diameter $d_{50,3}$ to the applied shear rates $\dot{\gamma}$. The graph shows an obvious tendency in which higher shear rates create lower mean particle diameters. This shows the sequence of applied shear rates in relation to its effects in creating smaller particle size is as follows 2663s^{-1} > 1868s^{-1} > 1017s^{-1} > 340s^{-1} at the initial state (20min) of redispersion.

4 Population Balance Equation

Population balance equation are the most frequently used modelling tool to describe and control a wide range of particulate processes like precipitation, crystallization, granulation, flocculation, and polymerisation. An extensive review of the application of population balances to particulate systems in engineering is given by Ramkrishna [18].

$$\begin{aligned}\frac{\partial n(t,x)}{\partial t} &= \frac{1}{2}\int_0^x \beta(t,x-y,y)n(t,x-y)n(t,y)dy \\ &-n(t,x)\int_0^\infty \beta(t,x,y)n(t,y)dy \\ &+\int_x^\infty b(t,x,y)S(t,y)n(t,y)dy - S(t,x)n(t,x)\end{aligned} \tag{1}$$

In the above Eq. (1), the first term on the right hand side represents the birth of particles of size x as a result of the agglomeration of particles of sizes (x- y) and y. The factor 1/2 prevents the double counting of collisions of the particles.

The second term describes the merging of particles of size x with any other particles. The second term is called the death term due to aggregation. The faction β is known to be agglomeration kernel. The disintegration function b (t, x, y) is the probability density function for the formation of particles of size x from particle of size y. The selection function S (t, x) describes the rate at which particles of size x are selected to disintegrate. In terms of disintegration, the last two terms are called the birth and death terms, respectively [19].

a. Agglomeration micro process

Agglomeration is a micro process where two or more particles combine together to form a larger particle. The total number of particles reduces during an aggregation process while the mass remains conserved. This process is most common in powder processing industries.

Table 1. Different agglomeration kernels

Name	kernel $\beta(x, y)$	source
turbulent shear	$\sqrt{\frac{8\pi\varepsilon\rho_F}{10\eta}}(x+y)^3$	Sommer et al. [20]
Brownian motion	$\frac{2kT}{3\mu}\left(\frac{(x+y)^2}{xy}\right)$	Smoluchowski [21]
Thompson kernel	$\frac{(x^3-y^3)^2}{x^3+y^3}$	Thompson [22]

The aggregation kernel is a measure of the frequency with which a particle of size x aggregates with one size y, as above Eq. (1) In this agglomeration process, it is assumed that all particles collisions are binary in nature, which means particle concentrations are sufficiently low. The aggregation kernel is a product of the two factors. Thus kernel may be written as

$$\beta(t, x, y) = \beta_0(t) \cdot f(x, y), \tag{2}$$

Where $\beta_0(t)$ depends on the chemical reaction conditions as the fluid velocity as well as shear rate. It is independent of size. The second factor $f(x, y)$ is some function of particle size. Various kernels are commonly used to model the aggregation during precipitation. This is shown in Table 1.

These kernels involve a wide range of different assumptions about the mechanical properties of the particles and the system characteristics. Besides Brownian motion and turbulent shear as reasons for particle–particle collisions, the effect of particle–particle interactions on the number of collisions is included by a stability ratio W. This ratio describes the relative number of effective collisions without and with interactions in addition to the omnipresent van der Waals forces. Additional interactions, such as electrostatic repulsion forces, are attributed to the adsorption of ions on the particles and the resulting formation of an electrical double layer. Depending on the collision mechanism, there are different models to calculate the stability ratio W [23]. For Brownian motion, W is frequently calculated based on a transport approach by Fuchs on particle diffusion under the influence of particle-particle interaction, whereas for shear induced collisions force-based approaches are to be preferred.

b. Disintegration micro process

In a disintegration process, particles are stressed and may break into two or many fragments. Disintegration has a significant effect on the number of particles. The total number of particles in disintegration process increases while the total mass remains constant [20, 24]. The selection function (disintegration probability) denotes the fraction of particles which are disintegrating after a certain stressing event. Theory concerning disintegration process is not as developed as that of agglomeration process. The power law disintegration distribution expression is the form usually used when more accurate predictions are required. The exponent p describes the number of particles per disintegration event and c is for the shape of the daughter distribution. Other parameters like S_0, μ, β_0 are constants. The selection function (disintegration probability) is [20]

$$S(x) = S_0 \cdot x^{\mu}. \tag{3}$$

The power law disintegration distribution is given as [20]

$$b(x, y) = \frac{\left(py^c (x-y)^{c+(c+1)(p-2)}\right)c + (c+1)(p-1)!}{x^{pc+(p-1)} c! \cdot (c + (c+1)(p-2)!)}. \tag{4}$$

c. The moment form of the population balance

The moments of the particle size distribution (PSD) can be obtained from writing the population balance in terms of moments. The jth moment of the number density function, $n(t, x)$, with respect to its internal coordinate x is defined as

$$m_j(t) = \int_0^{\infty} x^j n(t, x) dx. \tag{5}$$

If the internal coordinates x are taken as length, then the zeroth moment is equal to the total number of particles, and first, second, and third moments are proportional to the length, area and volume of particulate process, respectively. On the other hand if x denotes volume of the particles, then the zeroth and first moments are proportional to the total number and total mass of particles respectively. The second moment is in this case proportional to the light scattered by particles in the Rayleigh limit [25, 26]. The moment forms of the population balance can be very powerful.

5 Results and Discussion

The precipitation processes of titanium dioxide nanoparticles have been analysed using batch reactor to determine the agglomeration rates with different forms of agglomeration kernels listed in Table 1. In this model numerical results are obtained using cell average technique (CAT) [27, 28].

The different agglomeration kernels and selection functions as well as power law fits with super-saturation are shown in Figure 4, 5, 6 respectively.

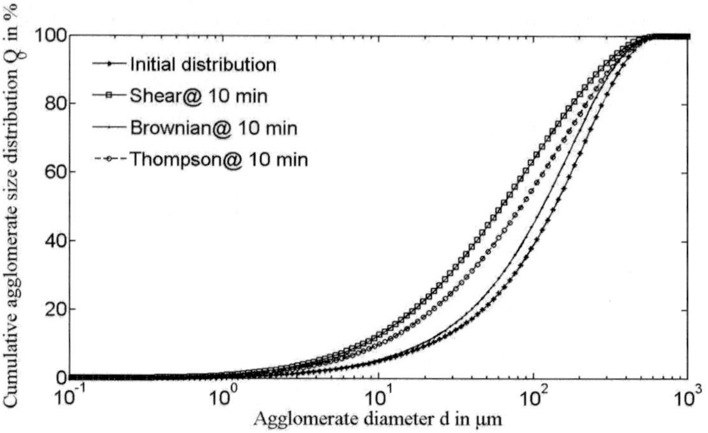

Fig. 4. Agglomerate size distribution Q_0 for different kernels.

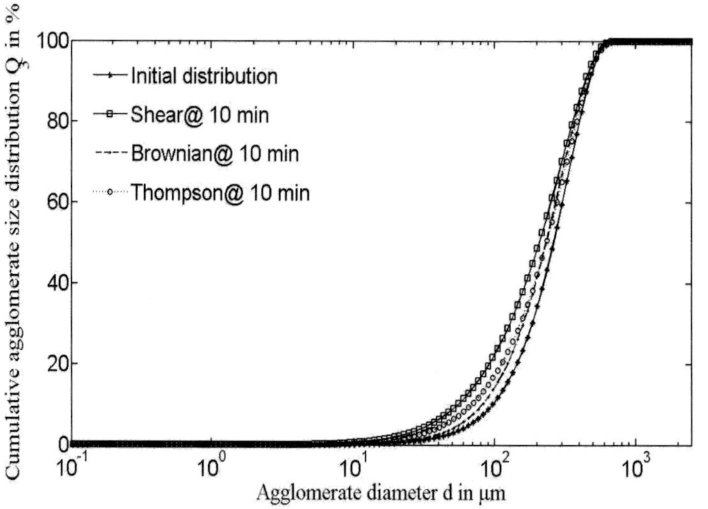

Fig. 5. Agglomerate size distribution Q_3 for different kernels

Fig. 4 shows the comparison of different kernels on particle size distribution during the sol-gel precipitation process. In Fig. 5 the agglomeration is investigated considering both shear induced and Brownian aggregation.

In Fig. 7 results from numerical simulations are in good agreement with experimental observation of the particle size distribution. The shear kernel accurately describes the experimental results and the variation of the zeroth moment.

In Fig. 6 the particle size distribution of the zeroth moment is plotted on different kernels. The prediction of the particle size distribution is done well by the cell average technique. At the end of the sol-gel reaction, a large number of particles is generated and mass remains conserved.

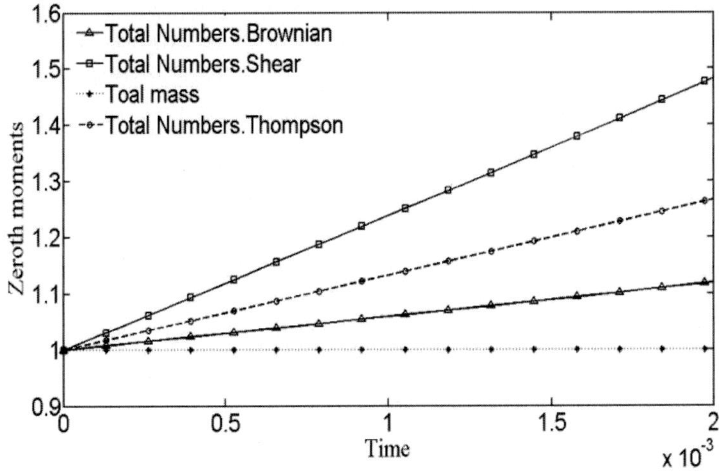

Fig. 6. Moments for different kernels

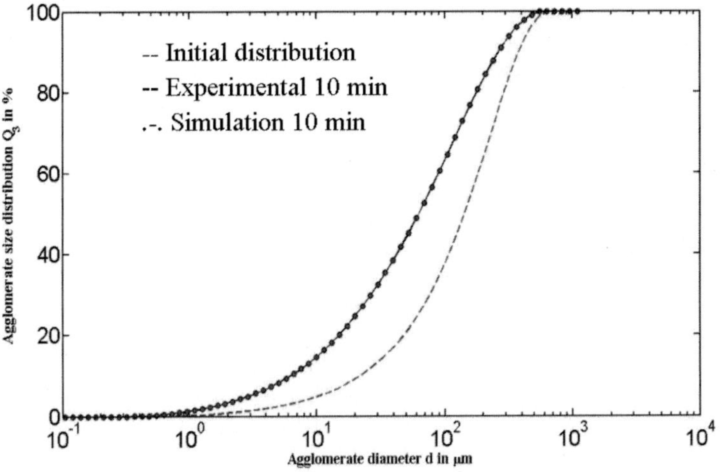

Fig. 7. Comparison of experimental results and shear kernel of particle size distribution

6 Conclusions

The reversible agglomeration and disintegration processes based on repulsive particle-particle interactions are investigated for precipitated nanoparticles using titanium dioxide as model substance. From experimental methods, agglomeration rates are determined by measuring the evolution of particle size distribution with time. A modelling framework is developed by using population balance equations to investigate different agglomeration kernels. As important parameters in the modelling of nanoparticle precipitation, the physicochemical features in the calculation of the shear rate, the pH value has been included in the mathematical form of solution of the kernel. Numerical results from a population balance model that accounts for agglomeration and disintegration are in reasonable agreement with experimental observations. From the population balance model it is possible to distinguish which of the kernel best describes the experimental data based on comparison of the particle size distributions and their moments. It is found that the shear kernel stands as the best fit to the experimental data. The computational features for this method are such that, this model can be used easily on a personal computer.

References

[1] Gratzel, M.: Light-induced redox reactions in nanocrystalline systems. Chem. Rev. 95, 49 (1995)
[2] Granqvist, G.: Handbook of Inorganic Electrochromic Materials. Elsevier Science, Amsterdam (1995)
[3] Janata, J., Josowicz, M., Vanysek, P., DeVaney, D.M.: Chemical Sensors. Anal. Chem. 70(12), 179R (1998)
[4] Kupriyanov, L.Y. (ed.): Semiconductor Sensors in Physico-Chemical Studies: Handbook of Sensors and Actuators 4. Elsevier, Amsterdam (1996)
[5] Kress-Rogers, E. (ed.): Handbook of Biosensors and Electronic Noses. CRC, Boca Raton (1997)
[6] Manthiram, A., Kim, J.: Nanosized manganese oxide as cathode material for lithium batteries: Influence of carbon mixing and grinding on cyclability. Journal of power sources 146, 294–299 (2005)
[7] Sarangapani, S., Tilak, B.V., Chen, C.P.: Materials for Electrochemical Capacitors. Journal of Electrochem. Soc. 143(11), 3791 (1996)
[8] Vayssieres, L., Hagfeldt, A., Lindquist, S.E.: Purpose-built metal oxide nanomaterials. The emergence of a new generation of smart materials. Pure. Appl. Chem. 72(1), 47–52 (2000)
[9] Schwarzer, H.C., Peukert, W.: Prediction of aggregation kinetics based on surface properties of nanoparticles. Chemical Engineering Science 60, 11–25 (2005)
[10] Schuetz, S., Piesche, M.: A model of the coagulation process with solid particles and flocs in turbulent flow. Chemical Engineering Science 57, 4357–4368 (2002)
[11] Schwarzer, H.C., Peukert, W.: Experimental investigation into the influence of mixing on nanoparticle precipitation. Chemical Engineering Technology 25(6), 657–661 (2002)
[12] Schwarzer, H.C., Peukert, W.: Tailoring particle size through nanoparticle precipitation. Chemical Engineering Communication 191(4), 580–606 (2004)
[13] Gregory, M., Odegard, S., Thomas, G., Lee, M., Nicholsonc, K., Wise, E.: Composites Science and Technology 62, 1869–1880 (2002)
[14] Hintz, W., Nikolov, T., Jordanova, V., Tomas, J.: Preparation of titanium dioxide nanoparticles and their characterization. Nanoscience & nanotechnology - nanostructured materials, application and innovation transfer (Sofia) 3, 73–76 (2003)
[15] Nikolov, T., Hintz, W., Jordanova, V., Tomas, J.: Synthesis and characterisation of titanium dioxide nanoparticles. Journal of the University of Chemical Technology and Metallurgy (Sofia) 38(3), 725–734 (2003)
[16] McCoy, J.B.: A population balance framework for nucleation, growth, and aggregation. Chemical Engineering Science 57, 2279–2285 (2002)
[17] Peukert, W., Schwarzer, H.C., Stenger, F.: Control of aggregation in production and handling of nanoparticles. Chemical Engineering and Processing 44, 245–252 (2005)
[18] Ramkrishna, D.: Population balances. Theory and applications to particulate systems in engineering, 1st edn. Academic Press, New York (2000)

[19] Kumar, J.: Numerical approximations of population balance equations in particulate systems, Ph.D. Thesis. Otto-von-Guericke-University Magdeburg, Germany (2006)
[20] Sommer, M., Stenger, F., Peukert, W., Wagner, N.J.: Agglomeration and breakage of nanoparticles in stirred media mills a comparison of different methods and models. Chemical Engineering Science 61, 135–148 (2006)
[21] Smoluchowski, M.V.: Versuch einer mathematischen theorie der koagulationskinetic kolloider losunggen. Z. Phys. Chem. 92, 129 (1917)
[22] Thompson, P.D.: Proceedings International Conference Cloud Physics, Toronto, p. 115 (1968)
[23] Fuchs, N.: Über die Stabilität und Aufladung der Aerosole. Z. Phys. 89, 736 (1934)
[24] Melis, S., Verduyn, M., Storti, G., Morbidelli, M., Baldyga, J.: Effect of fluid motion on the aggregation of small particles subject to interaction forces. AIChE J. 45, 1383 (1999)
[25] Kumar, S., Ramkrishna, D.: On the solution of population balance equations by discretization – I. A fixed pivot technique. Chemical Engineering Science 51, 1311–1332 (1996)
[26] Seifert, A., Beheng, K.D.: A double-moment parameterization for simulating autoconversion, accreation and selfcollection. Atmospheric Research 59(60), 265–281 (2001)
[27] Kumar, J., Peglow, M., Warnecke, G., Heinrich, S., Mörl, L.: Improved accuracy and convergence of discretized population balance for aggregation: the cell average technique. Chemical Engineering Science 61, 282–292 (2006)
[28] Kumar, J., Peglow, M., Warnecke, G., Heinrich, S.: An efficient numerical technique for solving population balance equation involving aggregation, breakage, growth and nucleation. Powder Technology 182, 81–104 (2008)

Nomenclature

b	Disintegration function	m^{-3}
c	Shape of daughter distribution	--
k	Boltzmann constant $1.380\ 6504 \times 10^{-23}$	J·K-1
p	Number of particles per disintegration event	--
$q_r\,(d)$	Particle size frequency distribution	m^{-1}
$Q_r\,(d)$	Cumulative particle size distribution	%
S_n	Selection function	s^{-1}
t	Time	s
T	Absolute temperature	K
x, y	Number of particles	m^{-3}

Greek Symbols

β	One-dimensional agglomeration kernel	$m^3 s^{-1}$
ε	Turbulent energy dissipation rate	m^2/s^3
ρ_F	Density of the fluid	kg/m^3
η	Kinematic viscosity of the fluid	m^2/s
$\dot{\gamma}$	Shear rate	s^{-1}
μ	Viscosity of the fluid	kg/m s
x, y	Number of particles	m^{-3}

Author Index

Printing: Krips bv, Meppel, The Netherlands
Binding: Stürtz, Würzburg, Germany